Integrating Clinical and Translational Research Networks—Building Team Medicine - Series 2

Integrating Clinical and Translational Research Networks—Building Team Medicine - Series 2

Editors

Ravi Salgia
Prakash Kulkarni

MDPI

Basel • Beijing • Wuhan • Barcelona • Belgrade • Novi Sad • Cluj • Manchester

Editors
Ravi Salgia
City of Hope National Medical Center
Duarte, CA, USA

Prakash Kulkarni
City of Hope National Medical Center
Duarte, CA, USA

Editorial Office
MDPI
St. Alban-Anlage 66
4052 Basel, Switzerland

This is a reprint of articles from the Special Issue published online in the open access journal *Journal of Clinical Medicine* (ISSN 2077-0383) (available at: https://www.mdpi.com/journal/jcm/special_issues/Team_Medicine_2).

For citation purposes, cite each article independently as indicated on the article page online and as indicated below:

Lastname, A.A.; Lastname, B.B. Article Title. *Journal Name* **Year**, *Volume Number*, Page Range.

ISBN 978-3-0365-8472-0 (Hbk)
ISBN 978-3-0365-8473-7 (PDF)
doi.org/10.3390/books978-3-0365-8473-7

Contents

About the Editors

Ravi Salgia

Ravi Salgia, MD, PhD, is the Arthur and Rosalie Kaplan Chair of Medical Oncology & Therapeutics Research at City of Hope National Medical Center, in Duarte, California. Previously, Dr. Salgia was Professor and Director of Translational Research at the University of Chicago. His research interests focus on novel therapeutics against lung cancer. Dr. Salgia has been honored with numerous awards, and prior to his tenure at the University of Chicago, Dr. Salgia was faculty at the Dana-Farber Cancer Institute and Harvard Medical School. Dr. Salgia earned his undergraduate summa cum laude in mathematics, biology, and chemistry, and then his MD and PhD degrees from Loyola University in Chicago, IL, where he also completed fellowships in neurochemistry and physiology. He continued his postgraduate training with an internship and residency in internal medicine at The Johns Hopkins University School of Medicine in Baltimore, MD, followed by a fellowship in medical oncology at the Dana-Farber Cancer Institute in Boston, MA, during which time he also served as a clinical fellow at Harvard Medical School in Boston.

Prakash Kulkarni

Prakash Kulkarni, PhD, is a Research Professor and Director of Translational Research, Department of Medical Oncology, and Department of Systems Biology at the City of Hope National Medical Centre. He completed postdoctoral training at New York University School of Medicine. He began his independent academic career as an Assistant Professor of urology and oncology at Johns Hopkins University, where he was named the Irene and Bernard L. Schwartz Scholar of the Patrick C Walsh Prostate Research Fund. He then moved to become an Associate Research Professor at the W. M. Keck Laboratory for Structural Biology, University of Maryland Institute for Bioscience and Biotechnology Research. Prior to Johns Hopkins, Dr. Kulkarni held Staff Scientist positions in the Division of Chemistry & Chemical Engineering and Division of Biology & Biological Engineering at the California Institute of Technology, and in the Department of Genetics at the Yale University School of Medicine. He is Associate Editor-in-Chief of *Biomolecules*. His research interests are interdisciplinary and are focused on understanding how conformational dynamics of intrinsically disordered proteins contributes to phenotypic switching, especially in the evolution of multicellularity, disease pathology and non-genetic heterogeneity in cancer. He is a Fellow of the Royal Society of Biology, UK.

Journal of
Clinical Medicine

MDPI

Editorial

On the Virtues of "Team Medicine"—A City of Hope Perspective

Prakash Kulkarni [1,2] and Ravi Salgia [1,*]

1 Department of Medical Oncology, City of Hope National Medical Center, 1500 Duarte Rd., Duarte, CA 91010, USA; pkulkarni@coh.org
2 Department of Systems Biology, City of Hope National Medical Center, 1500 Duarte Rd., Duarte, CA 91010, USA
* Correspondence: rsalgia@coh.org

Our first Special Issue of the *Journal of Clinical Medicine*, entitled 'Integrating Clinical and Translational Research Networks—Building Team Medicine,' highlighted the collective experience of the City of Hope and was a tremendous success. Buoyed by the enthusiastic response from our peers and colleagues, we embarked on bringing out the second Special Issue. The basic theme has remained the same, namely, integrating academic medical centers with their clinical network in various geographic locations to ensure that all patients, regardless of their physical proximity to major medical institutions, can benefit from recent clinical advances, and implementing exceptional care.

Like its predecessor, the collection of papers in this volume underscores the importance of integrating basic research scientists together with clinicians to enable a systems-level perspective. The two articles on addressing drug resistance by Kulkarni et al. [1] and Ramisetty et al. [2] provide examples of how this approach has uncovered new therapeutic strategies; they describe how the team medicine spirit helped address cisplatin resistance in NSCLC and identify a previously approved compound to alleviate cisplatin resistance in NSCLC, respectively.

Similarly, the article by Heater et al. [3] describes a new model utilizing online platforms to expand the reach of clinical expertise in the treatment of advanced soft-tissue sarcoma. Likewise, the paper by Sattler et al. [4] discusses the genetic/non-genetic duality in EGFR inhibitor drug resistance and underscores how team medicine approaches, wherein clinical developments work hand in hand with drug development research, drive potential opportunities for combination therapy.

For an expert clinical/research network with complementary expertise and the capability of multidisciplinary care, it is obvious that appropriate infrastructure would be necessary to empower this network to deliver personalized precision care to their patients. Thus, as cancer care becomes exponentially more complex with new diagnostic and therapeutic choices, providing decision support remains challenging. In an article offering a unique perspective, Bosserman et al. [5] describe how City of Hope has developed a pyramidal decision support framework to address these challenges, which have been exacerbated by the COVID pandemic, health plan restrictions, and growing geographic site diversity. They demonstrate how optimizing efficient and targeted decision support, backed by multidisciplinary cancer expertise, can improve individual patient treatment plans and thus achieve improved care and survival wherever patients are treated.

The paper by Mambetsariev et al. [6] discusses how cross-collaboration and integration between individual academic sites, national cancer networks, and community practices can enhance true personalized medicine. The implementation of these ideas, powered by recent advances in artificial intelligence and machine learning, could, in the future, allow for personalized high-throughput drug screenings that would yield faster drug discoveries and approved therapeutics. These are just a few examples highlighting the content of this

Citation: Kulkarni, P.; Salgia, R. On the Virtues of "Team Medicine"—A City of Hope Perspective. *J. Clin. Med.* **2023**, *12*, 4897. https://doi.org/10.3390/jcm12154897

Received: 18 July 2023
Accepted: 24 July 2023
Published: 26 July 2023

thematic issue. Together, the 17 papers provide an overview of the state of the art of team medicine and its virtues in clinical practice.

With >125 faculty members in the Department of Medical Oncology, and 31 clinical network centers in Southern California, Arizona, Georgia, and Illinois, the City of Hope is an encompassing enterprise in which we have inculcated the team medicine philosophy in order to integrate basic and translational research, along with clinical medicine in academic centers and their clinical networks. Thus, we trust our colleagues across the US and around the world will find the approach described in the articles included herein—as well as those included in the predecessor volume [7]—useful for guiding their own approaches to treating cancer patients.

Author Contributions: Conceptualization, P.K. and R.S.; writing—original draft preparation, P.K. and R.S.; writing—review and editing, P.K. and R.S. All authors have read and agreed to the published version of the manuscript.

Acknowledgments: We take this opportunity to thank the leadership at City of Hope for their support and encouragement; the authors for readily agreeing to share their unique experience, vision, and ideas via their papers; and the patients and their families for their participation and enduring spirit. We also appreciate and thank all of our healthcare professionals as well as our own families for their strong support of the team medicine vision.

Conflicts of Interest: The authors declare no conflict of interest.

References

1. Kulkarni, P.; Mohanty, A.; Bhattacharya, S.; Singhal, S.; Guo, L.; Ramisetty, S.; Mirzapoiazova, T.; Mambetsariev, B.; Mittan, S.; Malhotra, J.; et al. Addressing Drug Resistance in Cancer: A Team Medicine Approach. *J. Clin. Med.* **2022**, *11*, 5701. [CrossRef] [PubMed]
2. Ramisetty, S.; Kulkarni, P.; Bhattacharya, S.; Nam, A.; Singhal, S.S.; Guo, L.; Mirzapoiazova, T.; Mambetsariev, B.; Mittan, S.; Malhotra, J.; et al. A Systems Biology Approach for Addressing Cisplatin Resistance in Non-Small Cell Lung Cancer. *J. Clin. Med.* **2023**, *12*, 599. [CrossRef] [PubMed]
3. Heater, N.K.; Okuno, S.; Robinson, S.; Attia, S.; Seetharam, M.; Siontis, B.L.; Yoon, J.; Chawla, S.; Milhem, M.M.; Monga, V.; et al. The Midwest Sarcoma Trials Partnership: Bridging Academic and Community Networks in a Collaborative Approach to Sarcoma. *J. Clin. Med.* **2023**, *12*, 2561. [CrossRef] [PubMed]
4. Sattler, M.; Mambetsariev, I.; Fricke, J.; Tan, T.; Liu, S.; Vaidehi, N.; Pisick, E.; Mirzapoiazova, T.; Rock, A.G.; Merla, A.; et al. A Closer Look at EGFR Inhibitor Resistance in Non-Small Cell Lung Cancer through the Lens of Precision Medicine. *J. Clin. Med.* **2023**, *12*, 1936. [CrossRef] [PubMed]
5. Bosserman, L.D.; Mambetsariev, I.; Ladbury, C.; Barzi, A.; Johnson, D.; Morse, D.; Deaville, D.; Smith, W.; Rajurkar, S.; Merla, A.; et al. Pyramidal Decision Support Framework Leverages Subspecialty Expertise across Enterprise to Achieve Superior Cancer Outcomes and Personalized, Precision Care Plans. *J. Clin. Med.* **2022**, *11*, 6738. [CrossRef] [PubMed]
6. Mambetsariev, I.; Fricke, J.; Gruber, S.B.; Tan, T.; Babikian, R.; Kim, P.; Vishnubhotla, P.; Chen, J.; Kulkarni, P.; Salgia, R. Clinical Network Systems Biology: Traversing the Cancer Multiverse. *J. Clin. Med.* **2023**, *12*, 4535. [CrossRef] [PubMed]
7. Salgia, R.; Kulkarni, P. Integrating Clinical and Translational Research Networks—Building Team Medicine. *J. Clin. Med.* **2020**, *9*, 2975. [CrossRef] [PubMed]

Journal of
Clinical Medicine

MDPI

Article

Leveraging the Academic Artificial Intelligence Silecosystem to Advance the Community Oncology Enterprise

Kevin J. McDonnell

Center for Precision Medicine, Department of Medical Oncology & Therapeutics Research, City of Hope Comprehensive Cancer Center, Duarte, CA 91010, USA; kemcdonnell@coh.org

Abstract: Over the last 75 years, artificial intelligence has evolved from a theoretical concept and novel paradigm describing the role that computers might play in our society to a tool with which we daily engage. In this review, we describe AI in terms of its constituent elements, the synthesis of which we refer to as the AI Silecosystem. Herein, we provide an historical perspective of the evolution of the AI Silecosystem, conceptualized and summarized as a Kuhnian paradigm. This manuscript focuses on the role that the AI Silecosystem plays in oncology and its emerging importance in the care of the community oncology patient. We observe that this important role arises out of a unique alliance between the academic oncology enterprise and community oncology practices. We provide evidence of this alliance by illustrating the practical establishment of the AI Silecosystem at the City of Hope Comprehensive Cancer Center and its team utilization by community oncology providers.

Keywords: artificial intelligence; City of Hope; oncology; community practice

1. Introduction

Artificial intelligence (AI) plays an ever-increasing role in our daily lives most immediate to us in our use of entertainment, consumer and communication products [1,2]. Less immediately obvious to the oncology patient, AI has become an important tool to assist the clinical management of and guide therapy for cancer [3–5]. Within the academic oncology sphere, AI already has a significant impact. For example, AI has substantial, established roles in precision oncology [6–8], clinical oncology decision-making [9–11], digital cancer pathology [12–16] and radiology [17–19]. For community oncology practice, the role of AI remains limited but continues to emerge [20–22]. In this review, we seek to further expand knowledge of the role that AI plays in the community practice of oncology. We organize this manuscript into two parts. In Part I, we review the history, current state and emerging innovations relating to the computer hardware, data and software components that make AI possible. For conceptual simplicity and coherence, we refer to the synthesis of these components as the AI Silecosystem. We trace the emergence of the AI Silecosystem, its current state and future directions within the context of a Kuhnian scientific paradigm. In Part II, we provide a case example of the establishment and application of the AI Silecosystem in community oncology practice. We review the historical role and current integral position that academic medical institutions occupy in facilitating utilization of the AI Silecosystem by the community oncologist. We describe and place special emphasis on our experience at the City of Hope COH) Comprehensive Cancer Center to advance community oncology team utilization of the AI Silecosystem.

2. The AI Silecosystem as Kuhnian Paradigm

By AI Silecosystem we mean the synthesis of data, hardware and software that undergird the operation, make available the use, and fuel the growth of AI (Figure 1). To conceptually appreciate the history, progress and future trajectory of the AI Silecosystem, we may conceive and provide description of the AI Silecosystem as a Kuhnian paradigm [23]. As a

Citation: McDonnell, K.J. Leveraging the Academic Artificial Intelligence Silecosystem to Advance the Community Oncology Enterprise. *J. Clin. Med.* **2023**, *12*, 4830. https://doi.org/10.3390/jcm12144830

Academic Editors: Ravi Salgia and Prakash Kulkarni

Received: 7 June 2023
Revised: 5 July 2023
Accepted: 7 July 2023
Published: 21 July 2023

Kuhnian paradigm, the AI Silecosystem has disrupted and shifted the original paradigm of computer as finite *computational* machine to the novel paradigm of computer as versatile, multipotent *thinking* machine. This paradigm shift characteristically matures through three discrete, iterative stages: inception, intermission and invigoration.

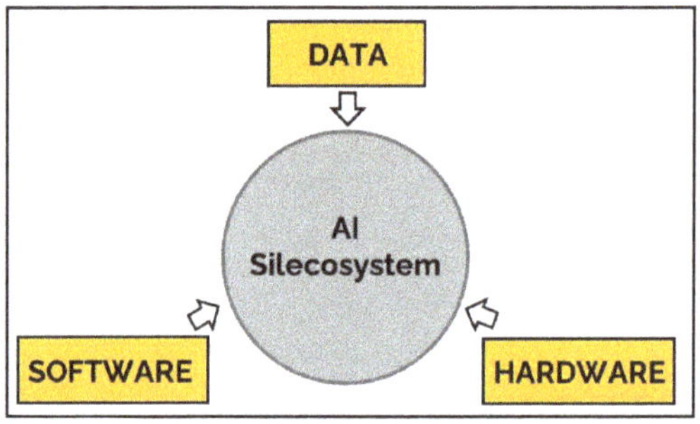

Figure 1. The AI Silecosystem comprises hardware, data and software components. The integrated components of the AI Silecosystem facilitated the development, utilization and evolution AI.

3. Origins of the AI Silecosystem: A Chronicle of an Emergent Paradigm

3.1. Inception: Articulation Anticipates Actualization

McCulloch and Pitts defined the incipient notion of computer as a thinking machine, suggesting that engineers might design computers to functionally mimic the operation of the human nervous system. In this theoretic nervous system model, an individual neuronal logic element achieves its ultimate activation state through cumulative summation of weighted inputs generated from a syndicate of contiguous neuronal logic elements [24]. This proposal represented an important architectural anlage preceding physical construction of Rosenblatt's early neural network, the Perceptron [25,26]. Rosenblatt's Mark 1 Perceptron neural network machine demonstrated the ability to perform basic visual pattern recognition. These early insights and accomplishments gave rise to an inchoate AI Silecosystem that Alan Turing further accelerated with his proposition that machines might "think" through serial adjudication of true and false logic states [27] (Figure 2). Formal AI development acquired significant academic interest and gained further momentum in 1956 when the early pioneers, McCarthy, Minsky and Shannon, convened a summer research convention at Dartmouth College where they sought critical evaluation of the assertion that "every aspect of learning or any other feature of intelligence can in principle be so precisely described that a machine can be made to simulate it" [28]. Historians credit McCarthy as one of the originators of the term "artificial intelligence". Consistent with previous Kuhnian paradigms, articulation of the AI Silecosystem paradigm anticipated its practical implementation.

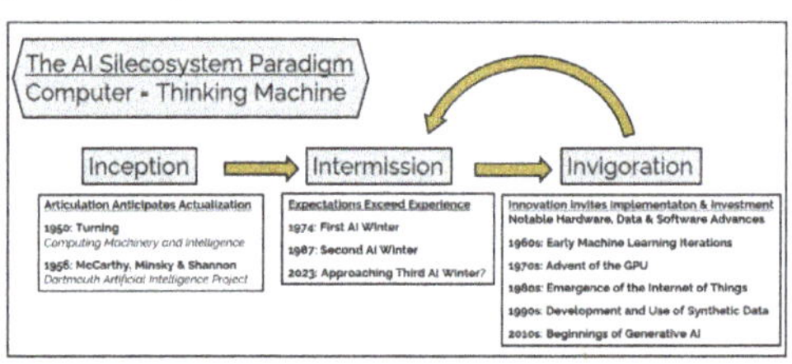

Figure 2. The AI Silecosystem paradigm conceives of the computer as thinking machine. As proposed by Turning [27] and McCarthy et al. [28], the computer may function as bone fide thinking machine, rather than mere computational machine. In accordance with a Kuhnian paradigm, the AI Silecosystem undergoes a series of stages: Inception, Intermission and Invigoration. Characteristically, the paradigm experiences a series of iterative Intermission and Invigoration cycles as new expectations develop and innovations occur.

3.2. Intermission: Expectations Exceed Experience

Initial efforts to create and implement the AI Silecosystem experienced setbacks. Between 1970 and 1990, a series of pivotal, adverse events led to intermittent intermissions in AI Silecosystem utilization, research and development. The inability of the AI Silecosystem to deliver its promise to perform complex, traditionally human-only tasks such as language translation, speech recognition and advanced image analysis efficiently and accurately muted expectations for AI-based approaches. These shortcomings prompted sponsors to withdraw financial support from several prominent AI initiatives. During this two-decade period, the Defense Advanced Research Projects Agency (DARPA) reduced funding for Carnegie Mellon's AI speech recognition program, and the United States National Research Council ended its financing of AI language translation efforts [29]. Following the Lighthill report, the United Kingdom halted further public AI development [30], and Japan curtailed AI investment after its Fifth Generation project failed to meet its articulated goals [31]. These setbacks instigated widespread public disillusionment with AI and precipitated a series of intermissions in further AI discovery and advancement, that is, the "AI Winters". Intermissions, such as the AI Winters, Kuhn would recognize as expected phases in the lifecycle of a paradigm shift. Full acceptance of a paradigm often must await creation of the technology and evaluation tools to permit complete use, valid assessment and thorough validation of the novel paradigm. Kuhn notes, for example, that many years passed after Newton and Einstein first introduced their mechanics and relativity paradigms until the availability of experimental verification protocols allowed scientists to fully understand, confirm and accept their revolutionary ideas [23]. Ultimately, innovation and insight facilitate endorsement and adoption of emerging paradigms, and, specifically, in the case of the AI Silecosystem, led to thawing of the AI Winters.

3.3. Invigoration: Innovation Invites Implementation and Investment

Innovation of and transformational progress within three core elements of the AI Silecosystem, i.e., computer hardware, data acquisition and processing and software algorithms, hastened thawing of the AI Winters. The following sections survey these key, instrumental innovations and advances.

3.3.1. Advances in Computer Hardware: The Engines That Power the AI Silecosystem

If we view the AI Silecosystem as a computational vehicle, its hardware elements function as the engines powering AI algorithmic processing. The invention of the silicon chip [32], introduction of multicore constructs [33] and development of ultrahigh capacity data storage systems [34], among other hardware innovations, enabled efficient, inexpensive performance of computationally complex, data-dense AI algorithms. The following more recent advances promise to further boost adoption and expansion of the AI Silecosystem.

Quantum Computing

Quantum computing uses the quantum bit (qubit) as its fundamental unit of information in contrast to conventional digital computing which employs the binary bit. Two different value states define the classic binary bit, and these value states exhibit mutual exclusivity (either 1 or 0). The qubit, however, may retain both values states simultaneously (1 and 0) in a quantum condition known as superposition. Superposition enables more rapid completion of complex, intensive computational tasks by quantum computation; digital computation cannot complete these tasks within a meaningful time frame. The computational superiority of the quantum computer, termed "quantum supremacy", was first demonstrated by Google in 2019 using a programable superconducting processor [35]. Quantum supremacy has the potential to amplify the power and practical utility of the AI Silecosystem. For example, computational scientists have developed and now apply AI algorithms to solve complicated combinatoric problems such as those encountered in molecular oncology drug design [36] and cancer diagnostics [37]. Processing of such AI algorithms on traditional computer platforms, however, might require exorbitant, cost- and time-prohibitive computational resources; implementation of quantum computation may allow tractable, economic solutions for combinatoric and other equally complex oncologic questions. Oncologists have successfully used quantum computing, together with AI applications, in the prediction of breast cancer [38], the application of radiotherapy [39] and cancer histologic assessment [40].

Artificial-Intelligence-Boosted Internet of Things (AIoT)

The internet of things (IoT) describes a system of local and remote physical instruments with communication, data processing, computational, memory storage and sensor capabilities interconnected via the internet and/or a local network [41,42]. The IoT aims to leverage the full potential of modern digital resources to optimize and assist with the activities and pursuits of daily living. Domestic examples of the IoT include smart speakers, home security systems and integrated, residential thermostat devices. The IoT has the potential for broad societal utilization. Specifically, within the sphere of health care, the IoT, i.e., the internet of medical things (IoMT), has enabled new, vital medical services, for instance, distance clinical assessment and monitoring [43,44] and remote health emergency notification [45]. In addition, investigators have proposed using the IoMT to enhance breast cancer detection [46], patient-centric healthcare [47,48], and the performance of health-care-related deep learning models [49].

With the advent of AI, the next iteration of the IoT emerged: artificial-intelligence-boosted IoT (AIoT) [50]. The AIoT underpins a range of familiar IoT applications such as autonomous driving vehicles [51], industrial robots [52] and surveillance drones [53]. The AIoT has provided impetus for several AI-based initiatives, for example, the development of anticipatory manufacturing machine maintenance, automated optimization of commercial operational efficiency and machine-learning-based urban safety monitoring and traffic control. Hospitals have begun using the AIoT to maintain efficient daily facility functioning and provide centralized patient monitoring. At COH, researchers have harnessed the AIoT to ensure safe, timely and effective post-surgery recovery for the patient after return to their home [54].

Distributive Edge Computing

Shared, centralized high-performance computer centers (HPCCs) have made available to a multitude of scientists the computer resources required to perform highly complex, computationally intense analyses. A HPCC may be located at a significant physical distance from the data source; moreover, as a shared resource, HPCC analytic jobs enter a work queue and process them in a serial fashion. The geographical and operational architecture of the HPCC results in "in due time" job completion. A complementary data analytic approach, edge computing, redistributes data processing, computations and memory storage from HPCC hubs to smaller, local computer nodes contiguous with the data source [55]. Edge computer nodes excel at "now time" processing of smaller discrete data parcels. For certain applications, most notably IoT platforms, edge computing offers distinct advantages over centralized HPCC processing: improved efficiency, low latency and increased agility; further, for large institutions, with often immensely large HPCC computational demands, edge computing helps alleviate computational backlog and obviate compromise of network bandwidth. Currently, edge computing plays an indispensable role in healthcare, processing data originating from local clinics as well as patient wearable monitoring devices. [56,57]. Researchers have begun to leverage the AI Silecosystem to catalyze new discoveries in and applications of edge computing. Recent efforts seek to bring the power, versatility and efficacy of AI to the edge in order to enhance local analytic capabilities [58–60]; specific initiatives seek to apply AI to edge immune-oncology and precision oncology computational efforts [61,62].

Cloud Computing

Cloud computing refers to as-needed, subscription use of off-site computer services, typically utilizing an internet connected network. Cloud computing allows organizations to rapidly adapt to and accommodate their changing computational needs. Cloud computing mitigates the often-substantial transitional financial and time lag costs associated with start-up or rapidly expanding computer needs. As the owners of the cloud computer services manage and maintain their product, subscribers avoid administrative and custodian cost burdens. Further, in the event of abrupt computational deceleration or change in operational goals, cloud computing eliminates organizational depreciation costs associated with dormant or obsolete equipment and software. Even stably established and well-resourced HPCCs may utilize cloud computing services to buffer acute fluxes in computer needs. Cloud computing currently plays a pivotal role in supporting the healthcare industry, including provision of the off-site storage of patient electronic medical records, the warehousing of large genomic data sets, the enablement of robust telehealth capabilities and the hosting of patient access portals [63]. Cloud computing utilizes the AI Silecosystem to automate complex healthcare data management protocols and enhance workflows associated with the processing and analysis of patient data [64]. Cloud AI platforms make more immediately available to oncologists and their patients the tremendous power of AI protocols [65]. AI-augmented cloud computing helps to advance tumor board operations, cancer therapeutics, patient management, diagnostics and oncology services [66].

Neuromorphic Computing

Neuromorphic computing adapts the physical architecture and functionality of the human central nervous system to enhance computer design and operation [67–70]. The artificial neuron constitutes the fundamental functional unit of neuromorphic computing. The construction and implementation of the artificial neuron and neuromorphic computers rely on interdisciplinary collaboration among neurobiologists, electrical engineers, computer scientists and computational specialists. Neuromorphic computing provided the basis for the invention and utilization of neuromorphic sensors such as artificial retinas and cochleae. Neuromorphic computing research inspired specialized subdisciplines, for example, neuromemrestive initiatives that utilize electromagnetic memristors to create CNS-computer interfaces [71]. Neuromorphic computing plays an ever-increasingly important role in healthcare applications such as patient safety monitoring [72], neurorehabilitation [73] and interactive health care robotics [74]. Recently, computer researchers have incorporated neuromorphic computing approaches into AI platforms to boost their effectiveness and efficiency [75–77]. Cancer scientists and oncologists have implemented AI-based neuromorphic computing to enrich their research [78–80] and improve clinical patient care [78,81].

Analog Neural Networks

As with neuromorphic computing, analog neural networks seek to mimic, more closely, the biochemical and neurophysiological functioning of the biological nervous system. Because biologic neuronal inputs comprise parallel converged signals originating from a multitude of neighboring neurons, the inputs do not occur within discrete time episodes, nor do the strength of signals have categorical quantitative values. Therefore, a nervous system model with analog continuous, rather than digital, input values more closely approximates actual nervous system functioning. Analog neural networks require less energy and less computational time compared with digital networks [82–85]. Analog neural networks now play central roles in the operation of numerous healthcare and medical software applications, e.g., those related to medical imaging [86], mimicking of the olfactory function [87] and modeling of mastoid bone pathologic events [88]. Investigators observe that analog neural networks may be used to support AI-based platforms such as vector machine learning [89], advanced edge computing [90] and natural language processing [91]. Cancer computational specialists have adapted analog neural networks to strengthen AI-informed oncology research, including the development of efficient cancer classification workflows [92,93], cancer histological analytic approaches [94] and oncology drug design pathways [95].

Monolithic-3D AI Systems

Electrical engineers originally designed the integrated circuit (IC) as a two-dimensional, flat semiconductor device containing a vast array of electronic elements such as transistors, capacitors and resistors. The IC has the capability to perform a wide range of data processing and computational operations. Relative to a collection of discrete circuit elements, ICs carry out operations more rapidly and use less energy. Recent advancements in IC design have led to the development of a three-dimensional (3D) IC configuration in which engineers vertically layer two-dimensional IC units [96]. This innovative design allowed construction of monolithic 3D ICs that contain within a single chip the necessary electronic components to carry out increasingly complex, advanced computational tasks [97]. Monolithic 3D ICs demonstrate improved efficiency of operation and allow for construction of ever more compact electronic instrumentation. The introduction of monolithic 3D ICs rapidly accelerated practical implementation of often very complicated AI machine learning and deep neural network algorithms in IoT devices such as personal, wearable medical devices and point-of-service health equipment [98].

The Graphics Processing Unit

The central processing unit (CPU) provides global program execution instructions for the computer; typically, the CPU performs its operational tasks in a serial fashion, one following another. CPUs normally contain a modest number of individual processing units (most often fewer than one hundred). Electrical engineers designed the CPU to complete dedicated large-scale computer operational tasks. In comparison, the graphics processing unit (GPU) has more limited operation execution responsibilities related to specific tasks [99]. The GPU can execute functions in a parallel fashion, handling multiple tasks simultaneously; facilitating parallel execution, the GPU may contain thousands of processing units. Although originally designed to perform video and graphics functions, computer scientists realized that vis-à-vis the CPU, the GPU performs AI-related tasks (e.g., machine learning and neural network operations) more proficiently. Oncologists have utilized GPU-based devices to augment their ability to implement radiation therapy [100] and interpret neuro-oncology MRI images [101]

Analog, Non-Volatile Memory Devices

Analog memory devices can store continuous data values. Volatile memory requires a continuous power source to retain data; non-volatile memory devices retain and stably store data after power discontinuation. The profound interest in implementing AI-based approaches, such as neuromorphic computing, that require durable and continuously valued data sets, has intensified the need for analog, non-volatile memory devices. Recently, engineers have innovated memory storage with the introduction of analog, nonvolatile ferroelectric field-effect [102,103], resistive random access memory [104–106], magnetic random access memory [107,108] and phase change memory technologies [109–111]. Analog, non-volatile memory has been instrumental in the continuing maturation of AI-based neural networks [84,112,113], image analytic platforms [114] and bio-sensor devices [115,116].

3.3.2. Advances in Data

Data fuels the engine of the AI Silecosystem vehicle [117]; historically, several data-related innovations contributed to thawing of the AI Winters. Increasing the size of a data set characteristically elevates performance of an AI algorithm [118,119]. The advent of systematized large-scale data acquisition, concomitant with convergent informational and technical advances such as data compression [120], solid state memory [121] and random access memory [122], contributed to improved AI algorithmic functionality and abetted the awakenings of the AI Silecosystem from its early hibernations. In the following section, we examine additional data innovations that have driven forward the evolution and growth of the AI Silecosystem.

Synthetic Data

Synthetic data refer to information originating from an intentionally engineered process, in contrast to authentic data generated spontaneously from actual, real-world events. The desire for optimized AI algorithmic operability and larger data sets drove the development of synthetic data fabrication protocols.

Synthetic data production typically requires application of stringent statistical analytic procedures, precise data sampling approaches and rigorous testing methods to ensure accuracy and validity [123,124]. Synthetic data offer several key advantages over real-world data. For very large data sets, synthetic data avoid the often-tremendous financial costs associated with real-world data collection. Moreover, synthetic data, as they do not originate from actual patients, do not pose privacy risks and, additionally, eliminate the potential financial liability associated with a data breech. In addition, because of anonymity, synthetic data collections may allow their unrestricted use as open-source data repositories. The collection of real-world data may expose investigators to physical hazard. Data arising from natural disaster areas, associated with dangerous chemical or biologic agents, or originating from an unsafe physical environment (e.g., an active military combat zone or crime-challenged neighborhood) may all threaten the safety of data collection personnel. The surrogate production of synthetic data obviates such threats.

Within the AI Silecosystem, synthetic data have acquired increased prominence as recognition of their utility has grown. Synthetic data have driven forward innovations within the healthcare space. Synthetic data undergird many current initiatives in medical education [125,126], clinical training [127,128], epidemiology research [129,130] and disease prevention [131,132]. Cancer researchers now use synthetic data resources to bolster their work including precision medicine [133] and palliative care [134].

Facilitating Culturally Representative AI Data Sets

Experts identify cultural inequity and lack of diversity as ongoing and significant challenges in our society specifically impacting healthcare and medical outcomes [135–137]. As AI gains increasing currency as a tool to direct healthcare decision-making, and recognizing that patient data set composition influences AI algorithmic outcomes, consideration of the racial and ethnic composition of patient data sets has become important in order to ensure equity of healthcare outcomes, specifically within the sphere of cancer care [138]. Nevertheless, despite legal requirements for representative inclusion of racial and ethnic minorities in health research, disparities persist; data sets used in AI-based algorithms continue to employ non-representative patient populations, undermining the validity of algorithmic decision-making [139,140]. Novel initiatives aim to improve and maintain broad population representation within health care data sets and across AI platforms. These initiatives include the implementation of intentionally diverse data sets [141], the enactment of more effective legislative guidelines to promote equity and diversity [142] and initiation of proactive community programs to promote health research participation [143].

Optimizing Data Deposition and Engineering

In order to optimize functioning of the Silecosystem and performance of downstream applications, computer engineers and scientists require tractable access to high-quality, large-volume data [144,145]. For example, machine learning algorithms for drug discovery [146], diagnostic prediction [147] and oncology medical imaging [148] demonstrate significant improvement with enhancement of data quantity and quality. The construction of national federated data repositories seeks to establish direct, streamlined public access to large data warehouses [149–153]. Data engineering aims to modify and format data to facilitate AI model building and the completion of analytic tasks [154,155]. Recent data engineering efforts have sought to automate data quality improvement protocols such as eliminating bias in and assessing the integrity of large data sets [156–158].

Together, the careful generation of synthetic data, increased attention to equitable data representation and the facilitation of high-quality data access have promoted the saliency and amplified the currency of the AI Silecosystem. In the section that follows, we chronicle the role of software algorithms in mitigating past AI winters and their continuing role to solidify collective adoption of the AI Silecosystem.

3.3.3. Advances in Software Algorithms: Piloting the AI Ecosystem

If hardware functions as engine, and data serve as fuel, then the software algorithm operates as pilot to direct the AI Silecosystem. As a pilot, the software algorithm directs the operational flow, direction and output of the AI Silecosystem. The AI computer scientist may choose among a variety of software algorithms; most frequently, the scientist utilizes machine learning or neural network algorithms [159,160].

Machine learning algorithms employ either supervised or unsupervised protocols [161]. With supervised protocols, input data have assigned labels that link with an output result; using this label, the algorithm then "learns" the rule that governs the relationship between the input and output data. With unsupervised protocols, the data lacks labels, and the algorithm must devise its own associative rules to understand patterns in the data. Among a range of practical applications, supervised machine learning has been used to predict customer behavior [162,163], differentiate cells of different histologies [164,165] and recognize faces [166,167]. With unsupervised machine learning, the algorithm seeks to cluster entities based upon some discoverable property of the entities, for example, grouping anonymous individuals within a large crowd based upon biometric or acquired physical variables [168,169].

Neural network algorithms, subsets of machine learning, generally supervised, work by mimicking the workings of the nervous system; within a neural network, an artificial neuron receives multiple inputs from neighboring neurons and then generates a resultant output based upon combined input [170]. In turn, the neuron transmits its output signal to other neighboring neurons, culminating, ultimately, in a final, consolidated output value from the system. The neural network algorithm "learns" the necessary rules that govern the correct association between input and output values. For example, computer scientists have adapted neural networking to interpret handwriting; this task entails making the correct association between a handwritten word and the ground truth, intended word [171–173].

Building upon the revolutionary impact of machine learning, other software inventions and algorithmic discoveries helped to rejuvenate AI and continue to transform the Silecosystem. A brief synopsis of major innovations follows.

Generative AI

Generative AI, an evolutionary offshoot of machine learning, uses rules derived from established instances of creative content to generate novel content such as original, advanced-level written documents [174], music compositions [175] and video game platforms [176], among others. Recently available generative AI applications, Microsoft's ChatGPT [177] and Google's Bard [178], have piqued the public's attention as both tools demonstrate the ability to very quickly generate works that approach the imaginative and technical abilities of human creators [179,180]. ChatGPT and Bard have authored working computer code [181–183], achieved passing scores on professional qualifying and academic exams [184,185] and written jokes [186]. In the health care field, generative AI enables chatbot services [187], carries out natural language processing of medical records [188] and completes medical education tasks [189]. These generative AI applications currently play important roles in cancer drug discovery [190], review of cancer patient medical records [191] and digital pathology [192].

Virtual and Augmented Reality

Virtual reality relies upon AI-empowered three-dimensional viewing devices together with positional tracking to construct and allow participation in a simulated, pseudo-physical existence [193]. Augmented reality combines input originating from physical reality with information generated by a computer device to enrich the conscious experience [194,195]. Providers have utilized both virtual and augmented realities in health care, for example, to improve medical practice and basic science research, advance educational curricula [196–200], refine surgical skills [201,202], guarantee the safety and effectiveness of medical procedures [203,204] and alleviate cancer pain and suffering [205–207]. Future virtual and augmented reality efforts aim to optimize routine, everyday tasks as well as medical professional-related procedures [208–210].

Explainable Machine Learning

Machine learning algorithms achieve their solutions through progression of relationally dependent steps. The underlying logic governing these relations, however, may be abstruse and not readily decipherable by a computer scientist [211]. Disambiguating the machine learning logic yields significant benefits. For just as explaining the mechanism of a biologic process or chemical reaction may reveal secondary insights and lead to additional discovery, so also may explaining the logic of a machine learning solution lead to derivative AI computational breakthroughs [212]. Furthermore, end users of transparent, explainable machine learning algorithms have increased confidence in the predictions of and conclusion made by the algorithm [213,214]. AI computer scientists use a variety of explanatory methods to reveal and illuminate the underlying governing logic of a machine learning behavior [215–218]. For example, gradient methods quantify the effect that a change in a machine input parameter has on the algorithm output at each step of the algorithm [219,220]. Deconvolution protocols provide logical information about the logical relationship between a specific output feature and input variable [221,222]. Local interpretable, model-agnostic explanations work by randomly inactivating model inputs and then observing and collectively analyzing output results [223–225]. These and other explainable methods promise to enhance the intuitive utility of and confidence in machine learning as well as other AI-based methods. For example, oncologists have employed explainable machine learning to boost their ability to perform breast cancer morphological and molecular breast cancer profiling [226] as well as estimate cancer hospital length of stay [227].

Generative Adversarial Networks

Generative adversarial networks (GANs) represent a category of generative machine learning algorithms in which two neural networks, a generator and discriminator, "compete" to achieve a maximized generative outcome, for example, production of an artificial image indistinguishable from an actual image [228,229]. Ground truth data sets train the generator to produce artificial data and also train the discriminator to distinguish between actual and artificial data [230,231]. The GAN algorithm achieves its generative objective when the generator produces artificial data, a majority of which the discriminator fails to distinguish from authentic data [230]. GANs have applications across a variety of disciplines including natural language processing [232–234], cybersecurity [235,236], manufacturing [237–239] and military defense [240,241]. Prominently, science and medicine have adapted GANs to design and analyze biological networks [242], perform medical imaging [243,244], inform precision oncology [245] and prescribe radiation medicine protocols [246–248].

Neuro-Vector-Symbolic Architecture

Illustrative of the rapid transformation of the AI Silecosystem, computer scientists recently introduced a novel AI computer operational structure, neuro-vector-symbolic architecture (NSVA) [249]. NSVA combines two existing, highly impactful AI strategies, deep neural networks (DNNs) and vector symbolic architectures (VSAs). DNNs excel at discerning objects in images, but lack the ability to differentiate among similarly shaped objects with differentiating secondary characteristics [250,251]. VSAs have the capacity to distinguish among entities having a multitude of secondary characteristics; however, they faulter with image perception [252,253]. Thus, neither DNNs nor VSAs can independently solve image-based abstract reasoning problems adequately. NSVAs incorporate the strengths of both SVAs and DNNs without their inherent weaknesses to create an innovative AI architecture capable of solving complex, perceptual problems [254]. Applied architectural synergism, such as the NSVA, provides a model for evolving the AI Silecosystem to accommodate the burgeoning computational complexity brought about by the accelerated societal adoption and use of AI. Cancer specialists have adapted these novel architectures to aid image analysis [255] and tumor classification [256].

The Democratization of Resources/Open-Source AI Software

Open-source software refers to computer software universally available to individuals for unrestricted use, modification and distribution [257]. Open-source software, beyond facile, economic availability, accelerates computer discovery, engenders trust in the software and organically self-improves due to iterative public editing and optimization [258]. The AI community has access to a broad menu of open-source software applications. Two frequently used AI open-source programs, TensorFlow [259] and PyTorch [260], provide platforms for the development of machine learning programs. Computer scientists frequently utilize TensorFlow to develop and train deep neural networks [261,262]. PyTorch has a variety of uses including the construction of natural language processing applications [263,264] and image processing [265,266]. Open-source AI software promotes the free exchange of ideas among users, sustains the democratization and pace of AI Silecosystem maturation, and serves as a catalyst for continuing research, invention and insight. Currently, AI computer scientists employ open-source software solutions to facilitate brain cancer research [267], perform cancer digital pathology [268] and analyze cancer genomic data [269].

In Table 1 below, we provide a summary of the significant historical and ongoing hardware, data and software innovations with regard to their impact on seven key metrics of the AI Silecosystem: AI algorithmic speed, efficiency, utility, agility, accuracy, security and accessibility.

Table 1. Significant Past and Ongoing Advances in Hardware, Data and Software Driving Evolution of the AI Ecosystem and Their Value Impact.

		AI Silecosystem Metric						
Component	**Innovation**	**AI Algorithmic Speed**	**Efficiency/Cost**	**Utility**	**Agility**	**Accuracy/Validity/Reliability**	**Security/Safety**	**Accessibility**
Hardware	Quantum Computing	Greater	Greater	Greater	Lesser	Lesser	Lesser	Lesser
	AI Internet of Things	Greater	Greater	Greater	Greater	Lesser	Lesser	Greater
	Distributive Edge Computing	Greater	Greater	Greater	Greater	Lesser	Lesser	Greater
	Cloud Computing	Greater	Greater	Greater	Greater	Lesser	Lesser	Greater
	Neuromorphic Computing	Greater	Greater	Greater	Greater	Lesser	Lesser	Lesser
	Analog Neural Networks	Greater	Greater	Greater	Greater	Lesser	Lesser	Lesser
	Monolithic 3D AI Systems	Greater	Greater	Greater	Greater	Lesser	Lesser	Greater
	Graphics Processing Unit	Greater	Greater	Greater	Greater	Lesser	Lesser	Greater
	Analog Non-Volatile Memory	Lesser	Lesser	Greater	Lesser	Greater	Lesser	Lesser
Data	Synthetic Data	Greater	Greater	Greater	Greater	Greater	Greater	Greater
	Culturally Representative Data Sets	Lesser	Lesser	Greater	Lesser	Greater	Lesser	Greater
	Data Optimization	Greater	Greater	Greater	Lesser	Greater	Lesser	Lesser
Software	Generative AI	Greater	Greater	Greater	Greater	Lesser	Lesser	Greater
	Virtual and Augmented Reality	Lesser	Lesser	Greater	Lesser	Lesser	Lesser	Greater
	Explainable Machine Learning	Lesser	Lesser	Greater	Lesser	Greater	Lesser	Greater
	Generative Adversarial Networks	Greater	Greater	Greater	Lesser	Lesser	Lesser	Lesser
	Neuro Vector-Symbolic Architecture	Greater	Greater	Greater	Greater	Greater	Lesser	Lesser
	Open-Source AI Software	Greater	Greater	Greater	Greater	Greater	Lesser	Greater

Greater Impact (red) — Lesser Impact (blue)

4. Tribulations of the AI Silecosystem: Impending AI Winter or Early Twilight of a Paradigm in Demise?

Interest in, adoption of and innovation associated with the AI Silecosystem have surged in no small measure due to the recent advances in the field of generative AI. With this surge, however, has come an amplification of concerns over the real and emerging risks and dangers of the AI Silecosystem [270]. Some experts see a more powerful AI Silecosystem as an existential threat to humanity [271]; the Center for AI safety recently advised that "mitigating the risk of extinction from AI should be a global priority alongside other societal-scale risks such as pandemics and nuclear war" [272]. Consequently, some societal leaders and countries have sought to pause or curtail continued AI development and/or use [273,274].

Regarding the use of AI within the healthcare and oncology sphere, leaders have voiced three broad concerns: loss of autonomy, malpractice and loss of compassion.

Scholars envision, on the horizon, ostensibly in the very near future, an AI singularity event wherein the intellectual capabilities of AI surpass that of humans, potentially with AI demonstrating unpredictable and uncontrollable behavior [275,276]. In this scenario, humans may unintentionally cede autonomy over their healthcare decision-making to an AI algorithm based upon actual superior medical insight [277–279], misperceived medical authority [280] or psychological manipulation [281].

Computer scientists and AI end users have expressed concerns over factual errors generated by AI algorithms [183,282–284]. AI-informed healthcare may pose real physical danger for the patient as AI algorithms may be prone to misdiagnosis [285] and incomplete or inaccurate treatment recommendations [286–288]. Healthcare specialists now recommend careful assessment of AI algorithms used for medical decision-making and expert review of AI-generated recommendations to avoid medical mistreatment [289,290].

Many patients do not trust AI [291–293]. Patients feel slighted by AI algorithms as the algorithms may, seemingly without apparent logic, deny patients health care coverage and needed services [294,295]. Patients perceive AI decisions as obdurate, unnuanced and arbitrary [296,297]. AI lacks compassion. The AI Silecosystem may be intelligent, but to many it is not wise.

These challenges, if not timely addressed, may precipitate the next AI intermission. Alternately, and potentially of greater consequence, the recent ascendancy of generative AI may presage an incipient twilight of the paradigm of "computer as thinking machine" along with the dawning of a succeeding, replacement paradigm, "computer as rational, sentient being".

In Part I, we reviewed the primary hardware, data and software components of AI that enable its operation and advancement, encapsulated in the idea of the AI Silecosystem. As well, we chronicled the historical phases of progress and recession of the AI Silecosystem, conceptualized as the Kuhnian paradigm. In Part II that follows, we provide an example of practical utilization of the AI Silecosystem and illustrate its value to advance community oncology practice at the COH Comprehensive Cancer Center. We begin with a short discussion of the academic origins of the AI Silecosystem, and then proceed to detail its application at COH to advance community oncology practice.

5. The Academic Origins and Catalysis of the AI Silecosystem

The AI Silecosystem can trace its origins back to a number of key societal institutions that include commercial enterprises [298–304], the military [304–307] and, arguably, most prominently, academic centers [308–310]. Given their focus on research and education as well as their often substantial financial resources, academic centers became the natural home, incubator and accelerator of the AI Silecosystem. Because of their interdisciplinary and collaborative natures, academic departments often cross-pollinate ideas among departments and anticipate, react to and advance emerging paradigms such as the AI Silecosystem. Examples of notable AI advances originating from academic centers include invention of the Perceptron at the Cornell Aeronautical Laboratory in 1943 [311], conceptualization of the idea of AI at the 1956 Dartmouth Summer Research Project on Artificial Intelligence [312], construction of the first life-like robot at Waseda University in 1970 [313], demonstration of the first autonomous driving vehicle, the Stanford Cart, in 1979 [314] and creation of ImageNet, an annotated image repository, at Princeton University [315].

The emergence of the AI Silecosystem from academic centers accelerated adoption by academic healthcare and further advanced AI discoveries within the healthcare field. AI has established a widespread presence within medicine [316,317]. For instance, radiologists have harnessed AI to assist with interpretation of medical images [16,318,319], cardiologists use AI to diagnose and monitor patients with heart disease [320–322], gastroenterologists leverage AI to enhance the effectiveness of their interventions [323–325] and pulmonologists apply AI algorithms to optimize their diagnoses [326–328]. The AI Silecosystem has demonstrated tremendous value in oncology. Academic AI-based protocols have impacted oncologic approaches to the early diagnosis of cancer [329,330], targeted precision therapeutic recommendations [331] and palliative interventions [332,333]. After early applications in academic oncology, subsequent initiatives aimed to extend the AI Silecosystem paradigm to community oncology practice. Next, we chronicle these various initiatives.

6. Harnessing of the Academic Oncology AI Silecosystem to Advance Community Oncology Practice: The City of Hope Experience

Although the AI Silecosystem has firm footing within academic oncology, its place within community oncology practice continues to mature. The City of Hope Cancer Center (COH) comprises a central, academic campus together with over 30 community satellite oncology practices. The central academic campus hosts COH's AI Silecosystem. In the following section, we describe the hardware, data, and software algorithm resources of the COH Silecosystem, the availability of these resources to the community oncology practices and the efforts to advance AI-empowered oncology care within the COH oncology enterprise (Figure 3).

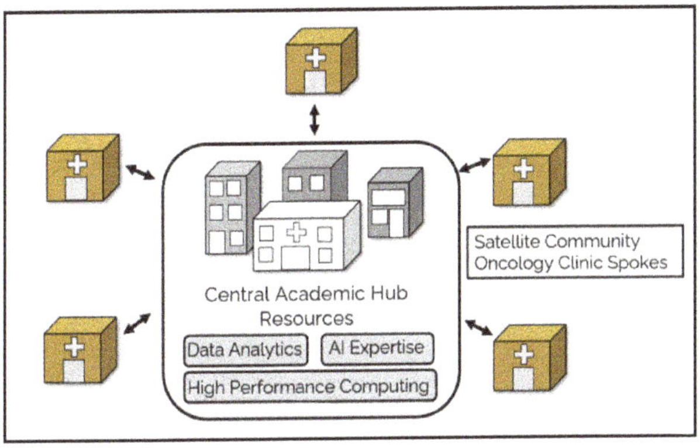

Figure 3. Satellite COH community oncology clinics may access the institutional AI Silecosystem through hub-and-spoke service operations. Community oncology practices may utilize data analytic, AI expert and HPCC resources via centralized network services provided to the COH community.

6.1. Hardware Resources: High-Performance Computer Cluster

To support AI computations, COH maintains a high-performance computer center (HPCC) comprising 7300 CPU cores, 80 TB of memory and 176 GPUs. All COH physicians, faculty, staff and students, including community oncology members, have privileges to access the HPCC remotely through desktop terminal applications. Round-the-clock IT experts provide technical support to assist with access to and utilization of the HPCC.

6.2. Data Resources

The COH Data Center manages and ensures reliable availability of several petabytes of deidentified clinical and genomic data for AI-related projects. To facilitate AI research and clinical projects, the Data Center relies on an institution-wide data repository, POSEIDON (Precision Oncology Software Environment Interoperable Data Ontologies Network), to house patient clinical and genomic data [334]. AI-assisted natural language processing organizes POSEIDON data according to a Common Data Model to optimize and accelerate downstream data input into AI operational workflows. To date, POSEIDON has assembled nearly one quarter million unique real world patient data sets. COH information and health care scientists have instituted and optimized operational protocols to structure efficiently patient-generated data for AI-based applications [335].

6.3. Software Resources

COH maintains a suite of bioinformatics and AI application modules on the HPCC. Clients may utilize HPCC resources and pursue AI investigations independently or collaboratively with COH expert consultants. COH established its Department of Applied Artificial Intelligence and Data Science (AAI/DS) to educate the COH community, facilitate institutional AI-based research and to provide clinical decision support to aid with AI modeling. AAI/DS hosts two forums each month. One forum, a journal club, reviews published manuscripts covering current areas of AI research including image analysis, machine learning and natural language processing. The second forum focuses on machine-learning-related institutional research initiatives, software applications and computational tools.

AAI/DS efforts have resulted in the creation of multiple machine-learning-based models to predict real world clinical events. Following bone marrow transplantation (BMT), the development of severe sepsis has an associated mortality rate exceeding 50%. One AAI/DS project utilized an ensemble approach combining multiple random forest binary classifications models to develop a tool to estimate the risk of patients developing life-threatening sepsis after BMT [336]. COH clinicians have employed this model to improve clinical care, avert sepsis-associated organ damage and ameliorate mortality events after BMT.

Serious complications such as cardiac events, pneumonia, hemorrhage and death many times follow cytoreductive cancer surgeries. Another AAI/DS initiative employed an explainable machine learning strategy to develop a model that predicts complications following cytoreductive surgery [337]. Surgeons at COH currently employ this model to identify patients at risk for post-operative complications and to implement preventive measures to mitigate these risks. For oncologists, time estimation until end of life in terminally ill patients poses a challenge; frequently, oncologists overestimate time until end of life. Such misestimation may negatively impact patient and family emotional and financial planning as well as confound medical management. Working with COH palliative care specialists, AAI/DS used a gradient-boosted trees binary classifier to create a model estimating time to end of life [338]. This model reliably outperformed oncologists for predicting 90-day mortality in terminally ill patients.

Alongside AAI/DS, associate COH departments and institutions further underpin the AI Silecosystem. The COH Center for Informatics, comprising the Divisions of Biostatistics, Clinical Research Information Support, Research Informatics and Mathematical Oncology, provides key computational support to the COH AI Silecosystem. The Center assists with the statistical design of research projects, restructures health and research data to be compatible with computer processing and aids with the visualization and analysis of data. AI projects supported by the Center for Informatics include the use of machine learning approaches to optimize, organize and structure electronic health care records for downstream artificial-intelligence-related projects [339], development of a machine learning platform to visualize and extract computationally employable information from biomedical and clinical data records [340] and utilizing machine learning approaches to advance the study and clinical implementation of immune-oncology [341].

The Translational Genomics Research Institute (TGen), a COH-affiliated center, leverages translational genomics to innovate diagnostic methods, molecular prognostic tools and targeted therapies for cancer through independent and collaborative projects [342]. Implementation of AI and machine learning algorithms have accelerated TGen-driven insights, fortifying the COH AI Silecosystem. One recent TGen-initiated scientific endeavor applied machine learning to develop a novel early cancer detection method, targeted digital sequencing (TARDIS) [343].

The cumulative energies of the AAI/DS, Center for Informatics, TGen, as well as the efforts of independent COH investigators have helped create a rich resource of AI expertise and maintain a robust portfolio of AI research. Examples of other initiatives at COHthat illustrate the depth and breadth of the AI Silecosystem include the use of AI autosegmentation for patients pending bone marrow transplant irradiation [344–346], AI-assisted oncologic drug design [347], expert critical review of clinical AI models [348], AI-based platforms for the evaluation and treatment of lung [349] and breast cancers [350], machine learning enabled pre-surgery physical status scoring [351] and AI-assisted irradiation dose estimation [352].

6.4. *COH AI Silecosystem Engagement with the Community Oncology Network*

Community Oncology patients and physicians at COH interface with and gain advantage from the AI Silecosystem on multiple levels. Every day, COH patients benefit directly from AI-informed institutional clinical care protocols such as the AI-informed diagnostic radiology, radiation oncology, medical oncology and palliative care initiatives described above. Moreover, community oncology patients may qualify for AI-based national clinical trials sponsored by COH. One such trial, currently available at COH, uses machine learning to inform the treatment of high-risk prostate cancer (NCT04513717) [353]. Community oncology patients also collaterally benefit from inclusion of their health care and genomic data in the electronic health record as their data help shape and make more accurate the AI models from which their AI-informed healthcare derives [354].

The COH AI Silecosystem likewise aids community oncologists. The AI Silecosystem provides access to expert AI specialists capable of providing to the community oncologist insights into the clinical serviceability and utilization of AI-based healthcare applications. Additionally, COH community oncologists may avail themselves of the many educational opportunities such as AI-related journal clubs, seminars and lectures. Further, COH community oncologists may employ the AI-Silecosystem data repository and institutional AI-associated hardware and clinical platforms for their own patient care [355]. Moreover, the COH AI Silecosystem helps expand AI-based clinical trial and research opportunities for community oncology providers.

7. Conclusions

The AI Silecosystem operates, innovates and advances as a synthesis of its component hardware, data and software elements. The AI Silecosystem has transformed in accordance with a Kuhnian paradigmatic progression with periods of rapid advancements punctuated by episodes of retreat. Recent signals of possible impending AI recession or even demise notwithstanding, the AI Silecosystem currently enjoys increasing societal currency and practical adoption. The academic oncology healthcare enterprise has significantly leveraged the AI Silecosystem to rapidly advantage cancer care, in particular the clinical management of the community oncology patient. The COH academic-community oncology team alliance demonstrates the practical feasibility and the tangible dividend of such leverage. In the near term, we may reasonably anticipate continued enthusiasm for the AI Silecosystem and its further utilization within community oncology practice.

Funding: This research received no external funding.

Institutional Review Board Statement: Not applicable.

Informed Consent Statement: Not applicable.

Data Availability Statement: Not applicable.

Conflicts of Interest: The author declares no conflict of interest.

References

1. Poola, I. How artificial intelligence in impacting real life everyday. *Int. J. Adv. Res. Dev.* **2017**, *2*, 96–100.
2. Lee, R.S.T. *Artificial Intelligence in Daily Life*; Springer: Singapore, 2020.
3. Bhinder, B.; Gilvary, C.; Madhukar, N.S.; Elemento, O. Artificial Intelligence in Cancer Research and Precision Medicine. *Cancer Discov.* **2021**, *11*, 900–915. [CrossRef] [PubMed]
4. Goldenberg, S.L.; Nir, G.; Salcudean, S.E. A new era: Artificial intelligence and machine learning in prostate cancer. *Nat. Rev. Urol.* **2019**, *16*, 391–403. [CrossRef] [PubMed]
5. Cardoso, M.J.; Houssami, N.; Pozzi, G.; Séroussi, B. Artificial intelligence (AI) in breast cancer care–Leveraging multidisciplinary skills to improve care. *Breast* **2021**, *56*, 110–113. [CrossRef]
6. Bhalla, S.; Laganà, A. Artificial intelligence for precision oncology. In *Computational Methods for Precision Oncology*; Springer: Berlin/Heidelberg, Germany, 2022; pp. 249–268.
7. Dlamini, Z.; Francies, F.Z.; Hull, R.; Marima, R. Artificial intelligence (AI) and big data in cancer and precision oncology. *Comput. Struct. Biotechnol. J.* **2020**, *18*, 2300–2311. [CrossRef]
8. Rompianesi, G.; Pegoraro, F.; Ceresa, C.D.; Montalti, R.; Troisi, R.I. Artificial intelligence for precision oncology: Beyond patient stratification. *NPJ Precis. Oncol.* **2019**, *3*, 6.
9. Rompianesi, G.; Pegoraro, F.; Ceresa, C.D.; Montalti, R.; Troisi, R.I. Artificial intelligence in the diagnosis and management of colorectal cancer liver metastases. *World J. Gastroenterol.* **2022**, *28*, 108. [CrossRef]

10. Christie, J.R.; Lang, P.; Zelko, L.M.; Palma, D.A.; Abdelrazek, M.; Mattonen, S.A. Artificial intelligence in lung cancer: Bridging the gap between computational power and clinical decision-making. *Can. Assoc. Radiol. J.* **2021**, *72*, 86–97. [CrossRef]
11. Derbal, Y. Can artificial intelligence improve cancer treatments? *Health Inform. J.* **2022**, *28*, 14604582221102314. [CrossRef]
12. Ibrahim, A.; Gamble, P.; Jaroensri, R.; Abdelsamea, M.M.; Mermel, C.H.; Chen, P.-H.C.; Rakha, E.A. Artificial intelligence in digital breast pathology: Techniques and applications. *Breast* **2020**, *49*, 267–273. [CrossRef]
13. Jiang, Y.; Yang, M.; Wang, S.; Li, X.; Sun, Y. Emerging role of deep learning-based artificial intelligence in tumor pathology. *Cancer Commun.* **2020**, *40*, 154–166. [CrossRef] [PubMed]
14. Viswanathan, V.S.; Toro, P.; Corredor, G.; Mukhopadhyay, S.; Madabhushi, A. The state of the art for artificial intelligence in lung digital pathology. *J. Pathol.* **2022**, *257*, 413–429. [CrossRef] [PubMed]
15. Försch, S.; Klauschen, F.; Hufnagl, P.; Roth, W. Artificial intelligence in pathology. *Dtsch. Ärzteblatt Int.* **2021**, *118*, 199. [CrossRef]
16. Hosny, A.; Parmar, C.; Quackenbush, J.; Schwartz, L.H.; Aerts, H.J. Artificial intelligence in radiology. *Nat. Rev. Cancer* **2018**, *18*, 500–510. [CrossRef] [PubMed]
17. Tran, W.T.; Sadeghi-Naini, A.; Lu, F.-I.; Gandhi, S.; Meti, N.; Brackstone, M.; Rakovitch, E.; Curpen, B. Computational radiology in breast cancer screening and diagnosis using artificial intelligence. *Can. Assoc. Radiol. J.* **2021**, *72*, 98–108. [CrossRef]
18. Chassagnon, G.; Vakalopoulou, M.; Paragios, N.; Revel, M.-P. Artificial intelligence applications for thoracic imaging. *Eur. J. Radiol.* **2020**, *123*, 108774. [CrossRef]
19. Tagliafico, A.S.; Piana, M.; Schenone, D.; Lai, R.; Massone, A.M.; Houssami, N. Overview of radiomics in breast cancer diagnosis and prognostication. *Breast* **2020**, *49*, 74–80. [CrossRef]
20. Frownfelter, J.; Blau, S.; Page, R.D.; Showalter, J.; Miller, K.; Kish, J.; Valley, A.W.; Nabhan, C. Artificial intelligence (AI) to improve patient outcomes in community oncology practices. *J. Clin. Oncol.* **2019**, *37*, e18098. [CrossRef]
21. Kappel, C.; Rushton-Marovac, M.; Leong, D.; Dent, S. Pursuing Connectivity in Cardio-Oncology Care—The Future of Telemedicine and Artificial Intelligence in Providing Equity and Access to Rural Communities. *Front. Cardiovasc. Med.* **2022**, *9*, 927769. [CrossRef]
22. Ye, P.; Butler, B.; Vo, D.; He, B.; Turnwald, B.; Hoverman, J.R.; Indurlal, P.; Garey, J.S.; Hoang, S.N. The initial outcome of deploying a mortality prediction tool at community oncology practices. *J. Clin. Oncol.* **2022**, *40*, 1521. [CrossRef]
23. Kuhn, T.S. *The Structure of Scientific Revolutions*; University of Chicago Press: Chicago, IL, USA, 1962; Volume XV, p. 172.
24. McCulloch, W.S.; Pitts, W. A logical calculus of the ideas immanent in nervous activity. *Bull. Math. Biol.* **1990**, *52*, 99–115. [CrossRef]
25. Rosenblatt, F. *The Perceptron, a Perceiving and Recognizing Automaton Project Para*; Cornell Aeronautical Laboratory: Buffalo, NY, USA, 1957.
26. Rosenblatt, F. The perceptron: A probabilistic model for information storage and organization in the brain. *Psychol. Rev.* **1958**, *65*, 386–408. [CrossRef]
27. Turing, A.M. Computing machinery and intelligence. *Mind* **1950**, *59*, 433–460. [CrossRef]
28. McCarthy, J.; Minsky, M.; Rochester, N.; Shannon, C.E. A Proposal for the Dartmouth Summer Research Project on Artificial Intelligence, August 31, 1955. *AI Mag.* **2006**, *27*, 12–14.
29. Pierce, J.R.; Carroll, J.B. *Language and Machines: Computers in Translation and Linguistics*; National Academies Press: Washington, DC, USA, 1966.
30. Science Research Council. *Artificial Intelligence; a Paper Symposium*; Science Research Council: London, UK, 1973; p. iv. 45p.
31. ICOT. *Shin-Sedai-Konpyūta-Gijutsu-Kaihatsu-Kikō, FGCS'92. Fifth Generation Computer Systems*; IOS Press: Amsterdam, The Netherlands, 1992; Volume 1.
32. Mack, C.A. Fifty years of Moore's law. *IEEE Trans. Semicond. Manuf.* **2011**, *24*, 202–207. [CrossRef]
33. Gepner, P.; Kowalik, M.K. Multi-core processors: New way to achieve high system performance. In Proceedings of the International Symposium on Parallel Computing in Electrical Engineering (PARELEC'06), Bialystok, Poland, 13–17 September 2006.
34. Goda, K.; Kitsuregawa, M. The history of storage systems. *Proc. IEEE* **2012**, *100*, 1433–1440. [CrossRef]
35. Arute, F.; Arya, K.; Babbush, R.; Bacon, D.; Bardin, J.C.; Barends, R.; Biswas, R.; Boixo, S.; Brandao, F.G.; Buell, D.A. Quantum supremacy using a programmable superconducting processor. *Nature* **2019**, *574*, 505–510. [CrossRef]
36. Thomford, N.E.; Senthebane, D.A.; Rowe, A.; Munro, D.; Seele, P.; Maroyi, A.; Dzobo, K. Natural products for drug discovery in the 21st century: Innovations for novel drug discovery. *Int. J. Mol. Sci.* **2018**, *19*, 1578. [CrossRef]
37. Jain, S.; Ziauddin, J.; Leonchyk, P.; Yenkanchi, S.; Geraci, J. Quantum and classical machine learning for the classification of non-small-cell lung cancer patients. *SN Appl. Sci.* **2020**, *2*, 1088. [CrossRef]
38. Davids, J.; Lidströmer, N.; Ashrafian, H. Artificial Intelligence in Medicine Using Quantum Computing in the Future of Healthcare. In *Artificial Intelligence in Medicine*; Springer: Berlin/Heidelberg, Germany, 2022; pp. 423–446.
39. Niraula, D.; Jamaluddin, J.; Matuszak, M.M.; Haken, R.K.T.; Naqa, I.E. Quantum deep reinforcement learning for clinical decision support in oncology: Application to adaptive radiotherapy. *Sci. Rep.* **2021**, *11*, 23545. [CrossRef]
40. Majumdar, R.; Baral, B.; Bhalgamiya, B.; Roy, T.D. Histopathological Cancer Detection Using Hybrid Quantum Computing. *arXiv* **2023**, arXiv:2302.04633.
41. Madakam, S.; Lake, V.; Lake, V.; Lake, V. Internet of Things (IoT): A literature review. *J. Comput. Commun.* **2015**, *3*, 164. [CrossRef]

42. Čolaković, A.; Hadžialić, M. Internet of Things (IoT): A review of enabling technologies, challenges, and open research issues. *Comput. Netw.* **2018**, *144*, 17–39. [CrossRef]
43. Valsalan, P.; Baomar, T.A.B.; Baabood, A.H.O. IoT based health monitoring system. *J. Crit. Rev.* **2020**, *7*, 739–743.
44. Yuehong, Y.; Zeng, Y.; Chen, X.; Fan, Y. The internet of things in healthcare: An overview. *J. Ind. Inf. Integr.* **2016**, *1*, 3–13.
45. Saloni, S.; Hegde, A. WiFi-aware as a connectivity solution for IoT pairing IoT with WiFi aware technology: Enabling new proximity based services. In Proceedings of the 2016 International Conference on Internet of Things and Applications (IOTA), Pune, India, 22–24 January 2016.
46. Aldhyani, T.H.; Khan, M.A.; Almaiah, M.A.; Alnazzawi, N.; Hwaitat, A.K.A.; Elhag, A.; Shehab, R.T.; Alshebami, A.S. A Secure internet of medical things Framework for Breast Cancer Detection in Sustainable Smart Cities. *Electronics* **2023**, *12*, 858. [CrossRef]
47. Jabarulla, M.Y.; Lee, H.-N. A blockchain and artificial intelligence-based, patient-centric healthcare system for combating the COVID-19 pandemic: Opportunities and applications. *Healthcare* **2021**, *9*, 1019. [CrossRef]
48. Srinivasu, P.N.; Ijaz, M.F.; Shafi, J.; Woźniak, M.; Sujatha, R. 6G driven fast computational networking framework for healthcare applications. *IEEE Access* **2022**, *10*, 94235–94248. [CrossRef]
49. Prayitno; Shyu, C.R.; Putra, K.T.; Chen, H.C.; Tsai, Y.Y.; Hossain, K.T.; Jiang, W.; Shae, Z.Y. A systematic review of federated learning in the healthcare area: From the perspective of data properties and applications. *Appl. Sci.* **2021**, *11*, 11191. [CrossRef]
50. Sung, T.-W.; Tsai, P.-W.; Gaber, T.; Lee, C.-Y. Artificial Intelligence of Things (AIoT) technologies and applications. *Wirel. Commun. Mob. Comput.* **2021**, *2021*, 9781271. [CrossRef]
51. Krasniqi, X.; Hajrizi, E. Use of IoT technology to drive the automotive industry from connected to full autonomous vehicles. *IFAC-Pap.* **2016**, *49*, 269–274. [CrossRef]
52. Jia, W.; Wang, S.; Xie, Y.; Chen, Z.; Gong, K. Disruptive technology identification of intelligent logistics robots in AIoT industry: Based on attributes and functions analysis. *Syst. Res. Behav. Sci.* **2022**, *39*, 557–568. [CrossRef]
53. Wazid, M.; Das, A.K.; Park, Y. Blockchain-Envisioned Secure Authentication Approach in AIoT: Applications, Challenges, and Future Research. *Wirel. Commun. Mob. Comput.* **2021**, *2021*, 3866006. [CrossRef]
54. Perez, F.; Nolde, M.; Crane, T.E.; Kebria, M.; Chan, K.; Dellinger, T.; Sun, V. Integrative review of remote patient monitoring in gynecologic and urologic surgical oncology. *J. Surg. Oncol.* **2023**, *127*, 1054–1061. [CrossRef]
55. Chen, J.; Ran, X. Deep learning with edge computing: A review. *Proc. IEEE* **2019**, *107*, 1655–1674. [CrossRef]
56. Uddin, M.Z. A wearable sensor-based activity prediction system to facilitate edge computing in smart healthcare system. *J. Parallel Distrib. Comput.* **2019**, *123*, 46–53. [CrossRef]
57. Verma, P.; Fatima, S. Smart Healthcare Applications and Real-Tme Analytics through Edge Computing. In *Internet of Things Use Cases for the Healthcare Industry*; Springer: Cham, Switzerland, 2020; pp. 241–270.
58. Cao, B.; Zhang, L.; Li, Y.; Feng, D.; Cao, W. Intelligent offloading in multi-access edge computing: A state-of-the-art review and framework. *IEEE Commun. Mag.* **2019**, *57*, 56–62. [CrossRef]
59. Zhou, Z.; Chen, X.; Li, E.; Zeng, L.; Luo, K.; Zhang, J. Edge intelligence: Paving the last mile of artificial intelligence with edge computing. *Proc. IEEE* **2019**, *107*, 1738–1762. [CrossRef]
60. Deng, S.; Zhao, H.; Fang, W.; Yin, J.; Dustdar, S.; Zomaya, A.Y. Edge intelligence: The confluence of edge computing and artificial intelligence. *IEEE Internet Things J.* **2020**, *7*, 7457–7469. [CrossRef]
61. Chowdhury, A.; Kassem, H.; Padoy, N.; Umeton, R.; Karargyris, A. A Review of Medical Federated Learning: Applications in Oncology and Cancer Research. In *Brainlesion: Glioma, Multiple Sclerosis, Stroke and Traumatic Brain Injuries, proceedings of the 7th International Workshop, BrainLes 2021, Held in Conjunction with MICCAI 2021, Virtual Event, 27 September 2021*; Springer: Cham, Switzerland, 2022; Part I.
62. Rodríguez, C. AIoT for Achieving Sustainable Development Goals. In Proceedings of the 4th International Conference on Recent Trends in Advanced Computing, VIT, Chennai, India, 11–12 November 2021.
63. Rahimi, M.; Navimipour, N.J.; Hosseinzadeh, M.; Moattar, M.H.; Darwesh, A. A systematic review on cloud computing. *J. Supercomput.* **2014**, *68*, 1321–1346.
64. Dang, L.M.; Piran, M.J.; Han, D.; Min, K.; Moon, H. Cloud healthcare services: A comprehensive and systematic literature review. *Trans. Emerg. Telecommun. Technol.* **2022**, *33*, e4473.
65. Raza, K.; Qazi, S.; Sahu, A.; Verma, S. Computational Intelligence in Oncology: Past, Present, and Future. In *Computational Intelligence in Oncology: Applications in Diagnosis, Prognosis and Therapeutics of Cancers*; Springer: Berlin/Heidelberg, Germany, 2022; pp. 3–18.
66. Liu, X.; Luo, X.; Jiang, C.; Zhao, H. Difficulties and challenges in the development of precision medicine. *Clin. Genet.* **2019**, *95*, 569–574. [CrossRef] [PubMed]
67. Schuman, C.D.; Kulkarni, S.R.; Parsa, M.; Mitchell, J.P.; Date, P.; Kay, B. Opportunities for neuromorphic computing algorithms and applications. *Nat. Comput. Sci.* **2022**, *2*, 10–19. [CrossRef]
68. Mead, C. Neuromorphic electronic systems. *Proc. IEEE* **1990**, *78*, 1629–1636. [CrossRef]
69. Indiveri, G.; Linares-Barranco, B.; Hamilton, T.J.; Schaik, A.V.; Etienne-Cummings, R.; Delbruck, T.; Liu, S.-C.; Dudek, P.; Häfliger, P.; Renaud, S. Neuromorphic silicon neuron circuits. *Front. Neurosci.* **2011**, *5*, 73. [CrossRef]
70. Thakur, C.S.; Molin, J.L.; Cauwenberghs, G.; Indiveri, G.; Kumar, K.; Qiao, N.; Schemmel, J.; Wang, R.; Chicca, E.; Olson Hasler, J. Large-scale neuromorphic spiking array processors: A quest to mimic the brain. *Front. Neurosci.* **2018**, *12*, 891. [CrossRef]

71. Bulárka, S.; Gontean, A. Brain-computer interface review. In Proceedings of the 2016 12th IEEE International Symposium on Electronics and Telecommunications (ISETC), Timisoara, Romania, 27–28 October 2016.
72. Yu, Z.; Zahid, A.; Ansari, S.; Abbas, H.; Abdulghani, A.M.; Heidari, H.; Imran, M.A.; Abbasi, Q.H. Hardware-based hopfield neuromorphic computing for fall detection. *Sensors* **2020**, *20*, 7226. [CrossRef] [PubMed]
73. Ceolini, E.; Frenkel, C.; Shrestha, S.B.; Taverni, G.; Khacef, L.; Payvand, M.; Donati, E. Hand-gesture recognition based on EMG and event-based camera sensor fusion: A benchmark in neuromorphic computing. *Front. Neurosci.* **2020**, *14*, 637. [CrossRef]
74. Aitsam, M.; Davies, S.; Di Nuovo, A. Neuromorphic Computing for Interactive Robotics: A Systematic Review. *IEEE Access* **2022**, *10*, 122261–122279. [CrossRef]
75. Yu, Z.; Abdulghani, A.M.; Zahid, A.; Heidari, H.; Imran, M.A.; Abbasi, Q.H. An overview of neuromorphic computing for artificial intelligence enabled hardware-based hopfield neural network. *IEEE Access* **2020**, *8*, 67085–67099. [CrossRef]
76. Sun, B.; Guo, T.; Zhou, G.; Ranjan, S.; Jiao, Y.; Wei, L.; Zhou, Y.N.; Wu, Y.A. Synaptic devices based neuromorphic computing applications in artificial intelligence. *Mater. Today Phys.* **2021**, *18*, 100393. [CrossRef]
77. Roy, K.; Jaiswal, A.; Panda, P. Towards spike-based machine intelligence with neuromorphic computing. *Nature* **2019**, *575*, 607–617. [CrossRef] [PubMed]
78. Pierangeli, D.; Palmieri, V.; Marcucci, G.; Moriconi, C.; Perini, G.; De Spirito, M.; Papi, M.; Conti, C. Optical neural network by disordered tumor spheroids. In Proceedings of the 2019 Conference on Lasers and Electro-Optics Europe & European Quantum Electronics Conference (CLEO/Europe-EQEC), Munich, Germany, 23–27 June 2019.
79. Rong, G.; Mendez, A.; Assi, E.B.; Zhao, B.; Sawan, M. Artificial intelligence in healthcare: Review and prediction case studies. *Engineering* **2020**, *6*, 291–301. [CrossRef]
80. Pierangeli, D.; Palmieri, V.; Marcucci, G.; Moriconi, C.; Perini, G.; De Spirito, M.; Papi, M.; Conti, C. Optical Neural Network for Cancer Morphodynamics Sensing. In *Nonlinear Optics*; Optica Publishing Group: Washington, DC, USA, 2019.
81. Topol, E.J. High-performance medicine: The convergence of human and artificial intelligence. *Nat. Med.* **2019**, *25*, 44–56. [CrossRef]
82. DasGupta, B.; Schnitger, G. Analog versus discrete neural networks. *Neural Comput.* **1996**, *8*, 805–818. [CrossRef]
83. Kakkar, V. Comparative study on analog and digital neural networks. *Int. J. Comput. Sci. Netw. Secur.* **2009**, *9*, 14–21.
84. Xiao, T.P.; Bennett, C.H.; Feinberg, B.; Agarwal, S.; Marinella, M.J. Analog architectures for neural network acceleration based on non-volatile memory. *Appl. Phys. Rev.* **2020**, *7*, 031301. [CrossRef]
85. Cramer, B.; Billaudelle, S.; Kanya, S.; Leibfried, A.; Grübl, A.; Karasenko, V.; Pehle, C.; Schreiber, K.; Stradmann, Y.; Weis, J.; et al. Surrogate gradients for analog neuromorphic computing. *Proc. Natl. Acad. Sci. USA* **2022**, *119*, e2109194119. [CrossRef]
86. Chandrasekaran, S.T.; Jayaraj, A.; Karnam, V.E.G.; Banerjee, I.; Sanyal, A. Fully integrated analog machine learning classifier using custom activation function for low resolution image classification. *IEEE Trans. Circuits Syst. I Regul. Pap.* **2021**, *68*, 1023–1033. [CrossRef]
87. Pan, C.-H.; Hsieh, H.-Y.; Tang, K.-T. An analog multilayer perceptron neural network for a portable electronic nose. *Sensors* **2012**, *13*, 193–207. [CrossRef]
88. Odame, K.; Nyamukuru, M.; Shahghasemi, M.; Bi, S.; Kotz, D. Analog Gated Recurrent Unit Neural Network for Detecting Chewing Events. *IEEE Trans. Biomed. Circuits Syst.* **2022**, *16*, 1106–1115. [CrossRef]
89. Perfetti, R.; Ricci, E. Analog neural network for support vector machine learning. *IEEE Trans. Neural Netw.* **2006**, *17*, 1085–1091. [CrossRef]
90. Krestinskaya, O.; AJames, P.; Chua, L.O. Neuromemristive circuits for edge computing: A review. *IEEE Trans. Neural Netw. Learn. Syst.* **2019**, *31*, 4–23. [CrossRef]
91. Moon, S.; Shin, K.; Jeon, D. Enhancing reliability of analog neural network processors. *IEEE Trans. Very Large Scale Integr. (VLSI) Syst.* **2019**, *27*, 1455–1459. [CrossRef]
92. Geske, G.; Stupmann, F.; Wego, A. High speed color recognition with an analog neural network chi. In Proceedings of the IEEE International Conference on Industrial Technology, Maribor, Slovenia, 10–12 December 2003.
93. Kieffer, C.; Genot, A.J.; Rondelez, Y.; Gines, G. Molecular Computation for Molecular Classification. *Adv. Biol.* **2023**, *7*, 2200203. [CrossRef]
94. Pattichis, C.; Schnorrenberg, F.; Schizas, C.; Pattichis, M.; Kyriacou, K. A Modular Neural Network System for the Analysis of Nuclei in Histopathological Sections. In *Computational Intelligence Processing in Medical Diagnosis*; Physica: Heidelberg, Germany, 2002; pp. 291–322.
95. Morro, A.; Canals, V.; Oliver, A.; Alomar, M.L.; Galan-Prado, F.; Ballester, P.J.; Rossello, J.L. A stochastic spiking neural network for virtual screening. *IEEE Trans. Neural Netw. Learn. Syst.* **2017**, *29*, 1371–1375. [CrossRef]
96. Jiang, J.; Parto, K.; Cao, W.; Banerjee, K. Ultimate monolithic-3D integration with 2D materials: Rationale, prospects, and challenges. *IEEE J. Electron Devices Soc.* **2019**, *7*, 878–887. [CrossRef]
97. Wong, S.; El-Gamal, A.; Griffin, P.; Nishi, Y.; Pease, F.; Plummer, J. Monolithic 3D integrated circuits. In Proceedings of the 2007 International Symposium on VLSI Technology, Systems and Applications (VLSI-TSA), Hsinchu, Taiwan, 23–25 April 2007.
98. Torres-Mapa, M.L.; Singh, M.; Simon, O.; Mapa, J.L.; Machida, M.; Günther, A.; Roth, B.; Heinemann, D.; Terakawa, M.; Heisterkamp, A. Fabrication of a monolithic lab-on-a-chip platform with integrated hydrogel waveguides for chemical sensing. *Sensors* **2019**, *19*, 4333. [CrossRef] [PubMed]
99. Dematté, L.; Prandi, D. GPU computing for systems biology. *Brief. Bioinform.* **2010**, *11*, 323–333. [CrossRef] [PubMed]

100. Zaki, G.; Plishker, W.; Li, W.; Lee, J.; Quon, H.; Wong, J.; Shekhar, R. The utility of cloud computing in analyzing GPU-accelerated deformable image registration of CT and CBCT images in head and neck cancer radiation therapy. *IEEE J. Transl. Eng. Health Med.* **2016**, *4*, 4300311. [CrossRef]
101. Chakrabarty, S.; Abidi, S.A.; Mousa, M.; Mokkarala, M.; Hren, I.; Yadav, D.; Kelsey, M.; LaMontagne, P.; Wood, J.; Adams, M. Integrative Imaging Informatics for Cancer Research: Workflow Automation for Neuro-Oncology (I3CR-WANO). *JCO Clin. Cancer Inform.* **2023**, *7*, e2200177. [CrossRef] [PubMed]
102. Khan, A.I.; Keshavarzi, A.; Datta, S. The future of ferroelectric field-effect transistor technology. *Nat. Electron.* **2020**, *3*, 588–597. [CrossRef]
103. Ajayan, J.; Mohankumar, P.; Nirmal, D.; Joseph, L.L.; Bhattacharya, S.; Sreejith, S.; Kollem, S.; Rebelli, S.; Tayal, S.; Mounika, B. Ferroelectric Field Effect Transistors (FeFETs): Advancements, Challenges and Exciting Prospects for Next Generation Non-Volatile Memory (NVM) Applications. *Mater. Today Commun.* **2023**, *35*, 105591. [CrossRef]
104. Pan, F.; Gao, S.; Chen, C.; Song, C.; Zeng, F. Recent progress in resistive random access memories: Materials, switching mechanisms, and performance. *Mater. Sci. Eng. R Rep.* **2014**, *83*, 1–59. [CrossRef]
105. Gupta, V.; Kapur, S.; Saurabh, S.; Grover, A. Resistive random access memory: A review of device challenges. *IETE Tech. Rev.* **2020**, *37*, 377–390. [CrossRef]
106. Wu, H.; Wang, X.H.; Gao, B.; Deng, N.; Lu, Z.; Haukness, B.; Bronner, G.; Qian, H. Resistive random access memory for future information processing system. *Proc. IEEE* **2017**, *105*, 1770–1789. [CrossRef]
107. Girard, P.; Cheng, Y.; Virazel, A.; Zhao, W.; Bishnoi, R.; Tahoori, M.B. A survey of test and reliability solutions for magnetic random access memories. *Proc. IEEE* **2020**, *109*, 149–169. [CrossRef]
108. Sethu, K.K.V.; Ghosh, S.; Couet, S.; Swerts, J.; Sorée, B.; De Boeck, J.; Kar, G.S.; Garello, K. Optimization of Tungsten β-phase window for spin-orbit-torque magnetic random-access memory. *Phys. Rev. Appl.* **2021**, *16*, 064009. [CrossRef]
109. Chen, A. A review of emerging non-volatile memory (NVM) technologies and applications. *Solid-State Electron.* **2016**, *125*, 25–38. [CrossRef]
110. Si, M.; Cheng, H.-Y.; Ando, T.; Hu, G.; Ye, P.D. Overview and outlook of emerging non-volatile memories. *MRS Bull.* **2021**, *46*, 946–958. [CrossRef]
111. Noé, P.; Vallée, C.; Hippert, F.; Fillot, F.; Raty, J.-Y. Phase-change materials for non-volatile memory devices: From technological challenges to materials science issues. *Semicond. Sci. Technol.* **2018**, *33*, 013002. [CrossRef]
112. Ambrogio, S.; Narayanan, P.; Tsai, H.; Mackin, C.; Spoon, K.; Chen, A.; Fasoli, A.; Friz, A.; Burr, G.W. Accelerating deep neural networks with analog memory devices. In Proceedings of the 2020 2nd IEEE International Conference on Artificial Intelligence Circuits and Systems (AICAS), Genova, Italy, 31 August 2020–2 September 2020.
113. Abunahla, H.; Halawani, Y.; Alazzam, A.; Mohammad, B. NeuroMem: Analog graphene-based resistive memory for artificial neural networks. *Sci. Rep.* **2020**, *10*, 9473. [CrossRef] [PubMed]
114. Zheng, X.; Zarcone, R.V.; Levy, A.; Khwa, W.-S.; Raina, P.; Olshausen, B.A.; Wong, H.P. High-density analog image storage in an analog-valued non-volatile memory array. *Neuromorphic Comput. Eng.* **2022**, *2*, 044018. [CrossRef]
115. Byun, S.-J.; Kim, D.-G.; Park, K.-D.; Choi, Y.-J.; Kumar, P.; Ali, I.; Kim, D.-G.; Yoo, J.-M.; Huh, H.-K.; Jung, Y.-J.; et al. A Low-Power Analog Processor-in-Memory-Based Convolutional Neural Network for Biosensor Applications. *Sensors* **2022**, *22*, 4555. [CrossRef]
116. Tzouvadaki, I.; Gkoupidenis, P.; Vassanelli, S.; Wang, S.; Prodromakis, T. Interfacing Biology and Electronics with Memristive Materials. *Adv. Mater.* **2023**, e2210035, *early view*. [CrossRef]
117. Greener, J.G.; Kandathil, S.M.; Moffat, L.; Jones, D.T. A guide to machine learning for biologists. *Nat. Rev. Mol. Cell Biol.* **2022**, *23*, 40–55. [CrossRef]
118. Choi, R.Y.; Coyner, A.S.; Kalpathy-Cramer, J.; Chiang, M.F.; Campbell, J.P. Introduction to Machine Learning, Neural Networks, and Deep Learning. *Transl. Vis. Sci. Technol.* **2020**, *9*, 14.
119. Zou, J.; Huss, M.; Abid, A.; Mohammadi, P.; Torkamani, A.; Telenti, A. A primer on deep learning in genomics. *Nat. Genet.* **2019**, *51*, 12–18. [CrossRef]
120. Sayood, K. *Introduction to Data Compression*; Morgan Kaufmann: Burlington, MA, USA, 2017.
121. Dirik, C.; Jacob, B. The performance of PC solid-state disks (SSDs) as a function of bandwidth, concurrency, device architecture, and system organization. *ACM SIGARCH Comput. Archit. News* **2009**, *37*, 279–289. [CrossRef]
122. Meena, J.S.; Sze, S.M.; Chand, U.; Tseng, T.-Y. Overview of emerging nonvolatile memory technologies. *Nanoscale Res. Lett.* **2014**, *9*, 526. [CrossRef] [PubMed]
123. Bolón-Canedo, V.; Sánchez-Maroño, N.; Alonso-Betanzos, A. A review of feature selection methods on synthetic data. *Knowl. Inf. Syst.* **2013**, *34*, 483–519. [CrossRef]
124. Raghunathan, T.E. Synthetic data. *Annu. Rev. Stat. Its Appl.* **2021**, *8*, 129–140. [CrossRef]
125. Arora, A.; Arora, A. Generative adversarial networks and synthetic patient data: Current challenges and future perspectives. *Future Healthc. J.* **2022**, *9*, 190. [CrossRef] [PubMed]
126. Rajotte, J.-F.; Bergen, R.; Buckeridge, D.L.; El Emam, K.; Ng, R.; Strome, E. Synthetic data as an enabler for machine learning applications in medicine. *Iscience* **2022**, *25*, 105331. [CrossRef] [PubMed]
127. Delaney, A.M.; Brophy, E.; Ward, T.E. Synthesis of realistic ECG using generative adversarial networks. *arXiv* **2019**, arXiv:1909.09150.

128. Ramesh, V.; Vatanparvar, K.; Nemati, E.; Nathan, V.; Rahman, M.M.; Kuang, J. Coughgan: Generating synthetic coughs that improve respiratory disease classification. In Proceedings of the 2020 42nd Annual International Conference of the IEEE Engineering in Medicine & Biology Society (EMBC), Montreal, QC, Canada, 20–24 July 2020.
129. Braddon, A.E.; Robinson, S.; Alati, R.; Betts, K.S. Exploring the utility of synthetic data to extract more value from sensitive health data assets: A focused example in perinatal epidemiology. *Paediatr. Perinat. Epidemiol.* **2022**, *37*, 292–300. [CrossRef]
130. Thomas, J.A.; Foraker, R.E.; Zamstein, N.; Morrow, J.D.; Payne, P.R.; Wilcox, A.B. Demonstrating an approach for evaluating synthetic geospatial and temporal epidemiologic data utility: Results from analyzing >1.8 million SARS-CoV-2 tests in the United States National COVID Cohort Collaborative (N3C). *J. Am. Med. Inform. Assoc.* **2022**, *29*, 1350–1365. [CrossRef]
131. Hernandez, M.; Epelde, G.; Beristain, A.; Álvarez, R.; Molina, C.; Larrea, X.; Alberdi, A.; Timoleon, M.; Bamidis, P.; Konstantinidis, E. Incorporation of synthetic data generation techniques within a controlled data processing workflow in the health and wellbeing domain. *Electronics* **2022**, *11*, 812. [CrossRef]
132. Gonzales, A.; Guruswamy, G.; Smith, S.R. Synthetic data in health care: A narrative review. *PLoS Digit. Health* **2023**, *2*, e0000082. [CrossRef]
133. D'Amico, S.; Dall'Olio, D.; Sala, C.; Dall'Olio, L.; Sauta, E.; Zampini, M.; Asti, G.; Lanino, L.; Maggioni, G.; Campagna, A. Synthetic data generation by artificial intelligence to accelerate research and precision medicine in hematology. *JCO Clin. Cancer Inform.* **2023**, *7*, e2300021. [CrossRef]
134. Hahn, W.; Schütte, K.; Schultz, K.; Wolkenhauer, O.; Sedlmayr, M.; Schuler, U.; Eichler, M.; Bej, S.; Wolfien, M. Contribution of Synthetic Data Generation towards an Improved Patient Stratification in Palliative Care. *J. Pers. Med.* **2022**, *12*, 1278. [CrossRef] [PubMed]
135. Elias, A.; Paradies, Y. The costs of institutional racism and its ethical implications for healthcare. *J. Bioethical Inq.* **2021**, *18*, 45–58. [CrossRef] [PubMed]
136. Taylor, J. Racism, Inequality, And Health Care for African Americans. *The Century Foundation*, 19 December 2019.
137. Matalon, D.R.; Zepeda-Mendoza, C.J.; Aarabi, M.; Brown, K.; Fullerton, S.M.; Kaur, S.; Quintero-Rivera, F.; Vatta, M.; Social, E.A.; Issues, C.L.; et al. Clinical, technical, and environmental biases influencing equitable access to clinical genetics/genomics testing: A points to consider statement of the American College of Medical Genetics and Genomics (ACMG). *Genet. Med.* **2023**, *25*, 100812. [CrossRef]
138. Dankwa-Mullan, I.; Weeraratne, D. Artificial intelligence and machine learning technologies in cancer care: Addressing disparities, bias, and data diversity. *Cancer Discov.* **2022**, *12*, 1423–1427. [CrossRef] [PubMed]
139. Henry, B.V.; Chen, H.; Edwards, M.A.; Faber, L.; Freischlag, J.A. A new look at an old problem: Improving diversity, equity, and inclusion in scientific research. *Am. Surg.* **2021**, *87*, 1722–1726. [CrossRef]
140. Eubanks, V. *Automating Inequality: How High-Tech Tools Profile, Police, and Punish the Poor*; St. Martin's Press: New York, NY, USA, 2018.
141. Holstein, K.; Vaughan, J.W.; Daumé, H.; Dudik, M.; Wallach, H. *Improving Fairness in Machine Learning Systems: What Do Industry Practitioners Need?* Association for Computing Machinery: New York, NY, USA, 2019; pp. 1–16.
142. Cachat-Rosset, G.; Klarsfeld, A. Diversity, Equity, and Inclusion in Artificial Intelligence: An Evaluation of Guidelines. *Appl. Artif. Intell.* **2023**, *37*, 2176618. [CrossRef]
143. Washington, V.; Franklin, J.B.; Huang, E.S.; Mega, J.L.; Abernethy, A.P. Diversity, equity, and inclusion in clinical research: A path toward precision health for everyone. *Clin. Pharmacol. Ther.* **2023**, *113*, 575–584. [CrossRef]
144. Al-Jarrah, O.Y.; Yoo, P.D.; Muhaidat, S.; Karagiannidis, G.K.; Taha, K. Efficient machine learning for big data: A review. *Big Data Res.* **2015**, *2*, 87–93. [CrossRef]
145. Anh, T.T.; Luong, N.C.; Niyato, D.; Kim, D.I.; Wang, L.-C. Efficient training management for mobile crowd-machine learning: A deep reinforcement learning approach. *IEEE Wirel. Commun. Lett.* **2019**, *8*, 1345–1348. [CrossRef]
146. Vamathevan, J.; Clark, D.; Czodrowski, P.; Dunham, I.; Ferran, E.; Lee, G.; Li, B.; Madabhushi, A.; Shah, P.; Spitzer, M. Applications of machine learning in drug discovery and development. Nature reviews. *Drug Discov.* **2019**, *18*, 463–477. [CrossRef]
147. Ng, K.; Steinhubl, S.R.; DeFilippi, C.; Dey, S.; Stewart, W.F. Early detection of heart failure using electronic health records: Practical implications for time before diagnosis, data diversity, data quantity, and data density. *Circ. Cardiovasc. Qual. Outcomes* **2016**, *9*, 649–658. [CrossRef] [PubMed]
148. Tseng, H.-H.; Wei, L.; Cui, S.; Luo, Y.; Haken, R.K.T.; El Naqa, I. Machine learning and imaging informatics in oncology. *Oncol.* **2020**, *98*, 344–362. [CrossRef] [PubMed]
149. Corrie, B.D.; Marthandan, N.; Zimonja, B.; Jaglale, J.; Zhou, Y.; Barr, E.; Knoetze, N.; Breden, F.M.; Christley, S.; Scott, J.K. iReceptor: A platform for querying and analyzing antibody/B-cell and T-cell receptor repertoire data across federated repositories. *Immunol. Rev.* **2018**, *284*, 24–41. [CrossRef] [PubMed]
150. Shakhovska, N.; Bolubash, Y.J.; Veres, O. Big data federated repository model. In Proceedings of the Experience of Designing and Application of CAD Systems in Microelectronics, Lviv, Ukraine, 24–27 February 2015.
151. Barnes, C.; Bajracharya, B.; Cannalte, M.; Gowani, Z.; Haley, W.; Kass-Hout, T.; Hernandez, K.; Ingram, M.; Juvvala, H.P.; Kuffel, G. The Biomedical Research Hub: A federated platform for patient research data. *J. Am. Med. Inform. Assoc.* **2022**, *29*, 619–625. [CrossRef] [PubMed]
152. Lin, D.; Crabtree, J.; Dillo, I.; Downs, R.R.; Edmunds, R.; Giaretta, D.; De Giusti, M.; L'Hours, H.; Hugo, W.; Jenkyns, R. The TRUST Principles for digital repositories. *Sci. Data* **2020**, *7*, 144. [CrossRef]

153. Yozwiak, N.L.; Schaffner, S.F.; Sabeti, P.C. Data sharing: Make outbreak research open access. *Nature* **2015**, *518*, 477–479. [CrossRef]
154. Romero, O.; Wrembel, R. Data engineering for data science: Two sides of the same coin. In Proceedings of the Big Data Analytics and Knowledge Discovery: 22nd International Conference, DaWaK 2020, Bratislava, Slovakia, 14–17 September 2020; p. 22.
155. Tamburri, D.; van den Heuvel, W.-J. Big Data Engineering, in Data Science for Entrepreneurship: Principles and Methods for Data Engineering. In *Analytics, Entrepreneurship, and the Society*; Springer: Berlin/Heidelberg, Germany, 2023; pp. 25–35.
156. Schelter, S.; Stoyanovich, J. Taming technical bias in machine learning pipelines. *Bull. Tech. Comm. Data Eng.* **2020**, *43*, 1926250.
157. Gudivada, V.; Apon, A.; Ding, J. Data quality considerations for big data and machine learning: Going beyond data cleaning and transformations. *Int. J. Adv. Softw.* **2017**, *10*, 1–20.
158. Chu, X.; Ilyas, I.F.; Krishnan, S.; Wang, J. Data cleaning: Overview and emerging challenges. In Proceedings of the 2016 International Conference on Management of Data, San Francisco, CA, USA, 26 June–1 July 2016.
159. Martinez, D.; Malyska, N.; Streilein, B.; Caceres, R.; Campbell, W.; Dagli, C.; Gadepally, V.; Greenfield, K.; Hall, R.; King, A. *Artificial Intelligence: Short History, Present Developments, and Future Outlook*; MIT Lincoln Laboratory: Lexington, MA, USA, 2019.
160. Libbrecht, M.W.; Noble, W.S. Machine learning applications in genetics and genomics. *Nat. Rev. Genet.* **2015**, *16*, 321–332. [CrossRef]
161. Shimizu, H.; Nakayama, K.I. Artificial intelligence in oncology. *Cancer Sci.* **2020**, *111*, 1452–1460. [CrossRef]
162. Khodabandehlou, S.; Zivari Rahman, M. Comparison of supervised machine learning techniques for customer churn prediction based on analysis of customer behavior. *J. Syst. Inf. Technol.* **2017**, *19*, 65–93. [CrossRef]
163. Hambarde, K.; Silahtaroğlu, G.; Khamitkar, S.; Bhalchandra, P.; Shaikh, H.; Kulkarni, G.; Tamsekar, P.; Samale, P. Data Analytics Implemented over E-Commerce Data to Evaluate Performance of Supervised Learning Approaches in Relation to Customer Behavior. In *Soft Computing for Problem Solving: SocProS 2018*; Springer: Berlin/Heidelberg, Germany, 2020; Volume 1.
164. Liu, M.; Ylanko, J.; Weekman, E.; Beckett, T.; Andrews, D.; McLaurin, J. Utilizing supervised machine learning to identify microglia and astrocytes in situ: Implications for large-scale image analysis and quantification. *J. Neurosci. Methods* **2019**, *328*, 108424. [CrossRef]
165. Janssens, T.; Antanas, L.; Derde, S.; Vanhorebeek, I.; Van den Berghe, G.; Grandas, F.G. CHARISMA: An integrated approach to automatic H&E-stained skeletal muscle cell segmentation using supervised learning and novel robust clump splitting. *Med. Image Anal.* **2013**, *17*, 1206–1219. [PubMed]
166. Wani, M.A.; Bhat, F.A.; Afzal, S.; Khan, A.I.; Wani, M.A.; Bhat, F.A.; Afzal, S.; Khan, A.I. Supervised deep learning in face recognition. In *Advances in Deep Learning. Studies in Big Data*; Springer: Singapore, 2020; pp. 95–110.
167. Nagaraj, P.; Banala, R.; Prasad, A.K. Real Time Face Recognition using Effective Supervised Machine Learning Algorithms. *J. Physics Conf. Ser.* **2021**, *1998*, 012007. [CrossRef]
168. Han, T.; Yao, H.; Sun, X.; Zhao, S.; Zhang, Y. Unsupervised discovery of crowd activities by saliency-based clustering. *Neurocomputing* **2016**, *171*, 347–361. [CrossRef]
169. Xu, S.; Ho, E.S.; Aslam, N.; Shum, H.P. Unsupervised abnormal behaviour detection with overhead crowd video. In Proceedings of the 2017 11th International Conference on Software, Knowledge, Information Management and Applications (SKIMA), Malabe, Sri Lanka, 6–8 December 2017.
170. He, J.; Baxter, S.L.; Xu, J.; Xu, J.; Zhou, X.; Zhang, K. The practical implementation of artificial intelligence technologies in medicine. *Nat. Med.* **2019**, *25*, 30–36. [CrossRef] [PubMed]
171. Patil, A.; Rane, M. Convolutional neural networks: An overview and Its applications in pattern recognition. In *Information and Communication Technology for Intelligent Systems: Proceedings of ICTIS 2020*; Springer: Singapore, 2020; Volume 1, pp. 21–30.
172. Graves, A.; Liwicki, M.; Bunke, H.; Schmidhuber, J.; Fernández, S. Unconstrained on-line handwriting recognition with recurrent neural networks. *Adv. Neural Inf. Process. Syst.* **2007**, *20*, 1–8.
173. Behnke, S. *Hierarchical Neural Networks for Image Interpretation*; Springer: Berlin/Heidelberg, Germany, 2003; Volume 2766.
174. Clark, E.; Ross, A.S.; Tan, C.; Ji, Y.; Smith, N.A. Creative writing with a machine in the loop: Case studies on slogans and stories. In Proceedings of the 23rd International Conference on Intelligent User Interfaces, Tokyo, Japan, 7–11 March 2018.
175. Hadjeres, G.; Pachet, F.; Nielsen, F. Deepbach: A steerable model for bach chorales generation. In Proceedings of the 34th International Conference on Machine Learning, PMLR, Sydney, Australia, 6–11 August 2017.
176. Guzdial, M.; Liao, N.; Chen, J.; Chen, S.-Y.; Shah, S.; Shah, V.; Reno, J.; Smith, G.; Riedl, M.O. Friend, collaborator, student, manager: How design of an ai-driven game level editor affects creators. In Proceedings of the 2019 CHI Conference on Human Factors in Computing Systems, Glasgow, UK, 4–9 May 2019.
177. Introducing ChatGPT. Available online: https://openai.com/blog/chatgpt (accessed on 10 June 2023).
178. Bard, an Experiment by Google. Available online: https://bard.google.com/ (accessed on 10 June 2023).
179. Teubner, T.; Flath, C.M.; Weinhardt, C.; van der Aalst, W.; Hinz, O. Welcome to the era of chatgpt et al. The prospects of large language models. *Bus. Inf. Syst. Eng.* **2023**, *65*, 95–101. [CrossRef]
180. Mondal, S.; Das, S.; Vrana, V.G. How to Bell the Cat? A Theoretical Review of Generative Artificial Intelligence towards Digital Disruption in All Walks of Life. *Technologies* **2023**, *11*, 44. [CrossRef]
181. Piccolo, S.R.; Denny, P.; Luxton-Reilly, A.; Payne, S.; Ridge, P.G. Many bioinformatics programming tasks can be automated with ChatGPT. *arXiv* **2023**, arXiv:2303.13528.
182. Surameery, N.M.S.; Shakor, M.Y. Use chat GPT to solve programming bugs. *Int. J. Inf. Technol. Comput. Eng.* **2023**, *3*, 17–22. [CrossRef]

183. van Dis, E.A.; Bollen, J.; Zuidema, W.; van Rooij, R.; Bockting, C.L. ChatGPT: Five priorities for research. *Nature* **2023**, *614*, 224–226. [CrossRef]
184. Naser, M.; Ross, B.; Ogle, J.; Kodur, V.; Hawileh, R.; Abdalla, J.; Thai, H.-T. Can AI Chatbots Pass the Fundamentals of Engineering (FE) and Principles and Practice of Engineering (PE) Structural Exams? *arXiv* **2023**, arXiv:2303.18149.
185. Geerling, W.; Mateer, G.D.; Wooten, J.; Damodaran, N. Is ChatGPT Smarter than a Student in Principles of Economics? 2023. Available online: https://ssrn.com/abstract=4356034 (accessed on 10 June 2023).
186. The Brilliance and Weirdness of ChatGPT. Available online: https://www.nytimes.com/2022/12/05/technology/chatgpt-ai-twitter.html (accessed on 10 June 2023).
187. Ram, B.; Pratima Verma, P.V. Artificial intelligence AI-based Chatbot Study of ChatGPT, Google AI Bard and Baidu AI. *World J. Adv. Eng. Technol. Sci.* **2023**, *8*, 258–261.
188. Cascella, M.; Montomoli, J.; Bellini, V.; Bignami, E. Evaluating the feasibility of ChatGPT in healthcare: An analysis of multiple clinical and research scenarios. *J. Med. Syst.* **2023**, *47*, 33. [CrossRef] [PubMed]
189. Kung, T.H.; Cheatham, M.; Medenilla, A.; Sillos, C.; De Leon, L.; Elepaño, C.; Madriaga, M.; Aggabao, R.; Diaz-Candido, G.; Maningo, J. Performance of ChatGPT on USMLE: Potential for AI-assisted medical education using large language models. *PLoS Digit. Health* **2023**, *2*, e0000198. [CrossRef]
190. Vert, J.-P. How will generative AI disrupt data science in drug discovery? *Nat. Biotechnol.* **2023**, *41*, 750–751. [CrossRef]
191. Uprety, D.; Zhu, D.; West, H. ChatGPT—A promising generative AI tool and its implications for cancer care. *Cancer* **2023**, *129*, 2284–2289. [CrossRef] [PubMed]
192. Sakamoto, T.; Furukawa, T.; Lami, K.; Pham, H.H.N.; Uegami, W.; Kuroda, K.; Kawai, M.; Sakanashi, H.; Cooper, L.A.D.; Bychkov, A. A narrative review of digital pathology and artificial intelligence: Focusing on lung cancer. *Transl. Lung Cancer Res.* **2020**, *9*, 2255. [CrossRef] [PubMed]
193. Zheng, J.; Chan, K.; Gibson, I. Virtual reality. *IEEE Potentials* **1998**, *17*, 20–23. [CrossRef]
194. Carmigniani, J.; Furht, B. *Augmented reality: An overview. Handbook of Augmented Reality*; Springer: New York, NY, USA, 2011; pp. 3–46.
195. Berryman, D.R. Augmented reality: A review. *Med. Ref. Serv. Q.* **2012**, *31*, 212–218. [CrossRef] [PubMed]
196. Fuchsova, M.; Korenova, L. Visualisation in Basic Science and Engineering Education of Future Primary School Teachers in Human Biology Education Using Augmented Reality. *Eur. J. Contemp. Educ.* **2019**, *8*, 92–102.
197. Paembonan, T.L.; Ikhsan, J. Supporting Students' Basic Science Process S kills by Augmented Reality Learning Media. *J. Educ. Sci. Technol.* **2021**, *7*, 188–196.
198. Chen, S.-Y.; Liu, S.-Y. Using augmented reality to experiment with elements in a chemistry course. *Comput. Hum. Behav.* **2020**, *111*, 106418. [CrossRef]
199. Li, L.; Yu, F.; Shi, D.; Shi, J.; Tian, Z.; Yang, J.; Wang, X.; Jiang, Q. Application of virtual reality technology in clinical medicine. *Am. J. Transl. Res.* **2017**, *9*, 3867.
200. Pottle, J. Virtual reality and the transformation of medical education. *Future Healthc. J.* **2019**, *6*, 181. [CrossRef]
201. Ayoub, A.; Pulijala, Y. The application of virtual reality and augmented reality in Oral & Maxillofacial Surgery. *BMC Oral Health* **2019**, *19*, 238.
202. McKnight, R.R.; Pean, C.A.; Buck, J.S.; Hwang, J.S.; Hsu, J.R.; Pierrie, S.N. Virtual reality and augmented reality—Translating surgical training into surgical technique. *Curr. Rev. Musculoskelet. Med.* **2020**, *13*, 663–674. [CrossRef]
203. Casari, F.A.; Navab, N.; Hruby, L.A.; Kriechling, P.; Nakamura, R.; Tori, R.; de Lourdes dos Santos Nunes, F.; Queiroz, M.C.; Fürnstahl, P.; Farshad, M. Augmented reality in orthopedic surgery is emerging from proof of concept towards clinical studies: A literature review explaining the technology and current state of the art. *Curr. Rev. Musculoskelet. Med.* **2021**, *14*, 192–203. [CrossRef] [PubMed]
204. Carl, B.; Bopp, M.; Saß, B.; Voellger, B.; Nimsky, C. Implementation of augmented reality support in spine surgery. *Eur. Spine J.* **2019**, *28*, 1697–1711. [CrossRef]
205. Georgescu, R.; Fodor, L.A.; Dobrean, A.; Cristea, I.A. Psychological interventions using virtual reality for pain associated with medical procedures: A systematic review and meta-analysis. *Psychol. Med.* **2020**, *50*, 1795–1807. [CrossRef] [PubMed]
206. Pittara, M.; Matsangidou, M.; Stylianides, K.; Petkov, N.; Pattichis, C.S. Virtual reality for pain management in cancer: A comprehensive review. *IEEE Access* **2020**, *8*, 225475–225489. [CrossRef]
207. Sharifpour, S.; Manshaee, G.R.; Sajjadian, I. Effects of virtual reality therapy on perceived pain intensity, anxiety, catastrophising and self-efficacy among adolescents with cancer. *Couns. Psychother. Res.* **2021**, *21*, 218–226. [CrossRef]
208. Cipresso, P.; Giglioli, I.A.C.; Raya, M.A.; Riva, G. The past, present, and future of virtual and augmented reality research: A network and cluster analysis of the literature. *Front. Psychol.* **2018**, *9*, 2086. [CrossRef]
209. Peddie, J. *Augmented Reality: Where We Will All Live*; Springer: Berlin/Heidelberg, Germany, 2017.
210. Riva, G.; Baños, R.M.; Botella, C.; Mantovani, F.; Gaggioli, A. Transforming experience: The potential of augmented reality and virtual reality for enhancing personal and clinical change. *Front. Psychiatry* **2016**, *7*, 164. [CrossRef] [PubMed]
211. Garcke, J.; Roscher, R. Explainable Machine Learning. *Mach. Learn. Knowl. Extr.* **2023**, *5*, 169–170. [CrossRef]
212. Roscher, R.; Bohn, B.; Duarte, M.F.; Garcke, J. Explainable Machine Learning for Scientific Insights and Discoveries. *IEEE Access* **2019**, *8*, 42200–42216. [CrossRef]

213. Rasheed, K.; Qayyum, A.; Ghaly, M.; Al-Fuqaha, A.I.; Razi, A.; Qadir, J. Explainable, trustworthy, and ethical machine learning for healthcare: A survey. *Comput. Biol. Med.* **2021**, *149*, 106043. [CrossRef]
214. Ratti, E.; Graves, M. Explainable machine learning practices: Opening another black box for reliable medical AI. *AI Ethics* **2022**, *2*, 801–814. [CrossRef]
215. Linardatos, P.; Papastefanopoulos, V.; Kotsiantis, S.B. Explainable AI: A Review of Machine Learning Interpretability Methods. *Entropy* **2020**, *23*, 18. [CrossRef]
216. Wells, L.; Bednarz, T. Explainable AI and Reinforcement Learning—A Systematic Review of Current Approaches and Trends. *Front. Artif. Intell.* **2021**, *4*, 550030. [CrossRef]
217. Nauta, M.; Trienes, J.; Pathak, S.; Nguyen, E.; Peters, M.; Schmitt, Y.; Schlötterer, J.; Keulen, M.V.; Seifert, C. From Anecdotal Evidence to Quantitative Evaluation Methods: A Systematic Review on Evaluating Explainable AI. *ACM Comput. Surv.* **2022**, *55*, 1–42. [CrossRef]
218. Jin, W.; Li, X.; Hamarneh, G. Evaluating explainable AI on a multi-modal medical imaging task: Can existing algorithms fulfill clinical requirements? *Proc. AAAI Conf. Artif. Intelligence.* **2022**, *36*, 11945–11953. [CrossRef]
219. Dwivedi, R.; Dave, D.; Naik, H.; Singhal, S.; Omer, R.; Patel, P.; Qian, B.; Wen, Z.; Shah, T.; Morgan, G. Explainable AI (XAI): Core ideas, techniques, and solutions. *ACM Comput. Surv.* **2023**, *55*, 194. [CrossRef]
220. Jiménez-Luna, J.; Grisoni, F.; Schneider, G. Drug discovery with explainable artificial intelligence. *Nat. Mach. Intell.* **2020**, *2*, 573–584. [CrossRef]
221. de Souza, L.A., Jr.; Mendel, R.; Strasser, S.; Ebigbo, A.; Probst, A.; Messmann, H.; Papa, J.P.; Palm, C. Convolutional Neural Networks for the evaluation of cancer in Barrett's esophagus: Explainable AI to lighten up the black-box. *Comput. Biol. Med.* **2021**, *135*, 104578. [CrossRef]
222. Zeiler, M.D.; Fergus, R. Visualizing and understanding convolutional networks. In Proceedings of the Computer Vision–ECCV 2014: 13th European Conference, Zurich, Switzerland, 6–12 September 2014.
223. Zafar, M.R.; Khan, N.M. DLIME: A deterministic local interpretable model-agnostic explanations approach for computer-aided diagnosis systems. *arXiv* **2019**, arXiv:1906.10263.
224. Palatnik de Sousa, I.; Vellasco, M.M.B.R.; da Silva, E.C. Local interpretable model-agnostic explanations for classification of lymph node metastases. *Sensors* **2019**, *19*, 2969. [CrossRef]
225. Kumarakulasinghe, N.B.; Blomberg, T.; Liu, J.; Leao, A.S.; Papapetrou, P. Evaluating local interpretable model-agnostic explanations on clinical machine learning classification models. In Proceedings of the 2020 IEEE 33rd International Symposium on Computer-Based Medical Systems (CBMS), Rochester, MN, USA, 28–30 July 2020.
226. Binder, A.; Bockmayr, M.; Hägele, M.; Wienert, S.; Heim, D.; Hellweg, K.; Ishii, M.; Stenzinger, A.; Hocke, A.; Denkert, C. Morphological and molecular breast cancer profiling through explainable machine learning. *Nat. Mach. Intell.* **2021**, *3*, 355–366. [CrossRef]
227. Alsinglawi, B.; Alshari, O.; Alorjani, M.; Mubin, O.; Alnajjar, F.; Novoa, M.; Darwish, O. An explainable machine learning framework for lung cancer hospital length of stay prediction. *Sci. Rep.* **2022**, *12*, 607. [CrossRef]
228. Creswell, A.; White, T.; Dumoulin, V.; Arulkumaran, K.; Sengupta, B.; Bharath, A.A. Generative adversarial networks: An overview. *IEEE Signal Process. Mag.* **2018**, *35*, 53–65. [CrossRef]
229. Aggarwal, A.; Mittal, M.; Battineni, G. Generative adversarial network: An overview of theory and applications. *Int. J. Inf. Manag. Data Insights* **2021**, *1*, 100004. [CrossRef]
230. Arjovsky, M.; Bottou, L. Towards principled methods for training generative adversarial networks. *arXiv* **2017**, arXiv:1701.04862.
231. Gui, J.; Sun, Z.; Wen, Y.; Tao, D.; Ye, J. A review on generative adversarial networks: Algorithms, theory, and applications. *IEEE Trans. Knowl. Data Eng.* **2021**, *35*, 3313–3332. [CrossRef]
232. Nam, S.; Kim, Y.; Kim, S.J. Text-adaptive generative adversarial networks: Manipulating images with natural language. *Adv. Neural Inf. Process. Syst.* **2018**, *31*, 1–10.
233. Xu, J.; Ren, X.; Lin, J.; Sun, X. Diversity-promoting GAN: A cross-entropy based generative adversarial network for diversified text generation. In Proceedings of the 2018 Conference on Empirical Methods in Natural Language Processing, Brussels, Belgium, 31 October–4 November 2018.
234. Lai, C.-T.; Hong, Y.-T.; Chen, H.-Y.; Lu, C.-J.; Lin, S.-D. Multiple text style transfer by using word-level conditional generative adversarial network with two-phase training. In Proceedings of the 2019 Conference on Empirical Methods in Natural Language Processing and the 9th International Joint Conference on Natural Language Processing (EMNLP-IJCNLP), Hong Kong, China, 3–7 November 2019.
235. Yinka-Banjo, C.; Ugot, O.-A. A review of generative adversarial networks and its application in cybersecurity. *Artif. Intell. Rev.* **2020**, *53*, 1721–1736. [CrossRef]
236. Toshpulatov, M.; Lee, W.; Lee, S. Generative adversarial networks and their application to 3D face generation: A survey. *Image Vis. Comput.* **2021**, *108*, 104119. [CrossRef]
237. Kusiak, A. Convolutional and generative adversarial neural networks in manufacturing. *Int. J. Prod. Res.* **2020**, *58*, 1594–1604. [CrossRef]
238. Singh, R.; Garg, R.; Patel, N.S.; Braun, M.W. Generative adversarial networks for synthetic defect generation in assembly and test manufacturing. In Proceedings of the 2020 31st Annual SEMI Advanced Semiconductor Manufacturing Conference (ASMC), Saratoga Springs, NY, USA, 24–26 August 2020.

239. Wang, J.; Yang, Z.; Zhang, J.; Zhang, Q.; Chien, W.-T.K. AdaBalGAN: An improved generative adversarial network with imbalanced learning for wafer defective pattern recognition. *IEEE Trans. Semicond. Manuf.* **2019**, *32*, 310–319. [CrossRef]
240. Meng, F.-j.; Li, Y.-q.; Shao, F.-m.; Yuan, G.-h.; Dai, J.-y. Visual-simulation region proposal and generative adversarial network based ground military target recognition. *Def. Technol.* **2022**, *18*, 2083–2096. [CrossRef]
241. Thompson, S.; Teixeira-Dias, F.; Paulino, M.; Hamilton, A. Predictions on multi-class terminal ballistics datasets using conditional Generative Adversarial Networks. *Neural Netw.* **2022**, *154*, 425–440. [CrossRef] [PubMed]
242. Achuthan, S.; Chatterjee, R.; Kotnala, S.; Mohanty, A.; Bhattacharya, S.; Salgia, R.; Kulkarni, P. Leveraging deep learning algorithms for synthetic data generation to design and analyze biological networks. *J. Biosci.* **2022**, *47*, 43. [CrossRef] [PubMed]
243. Yi, X.; Walia, E.; Babyn, P. Generative adversarial network in medical imaging: A review. *Med. Image Anal.* **2019**, *58*, 101552. [CrossRef]
244. Chen, Y.; Yang, X.-H.; Wei, Z.; Heidari, A.A.; Zheng, N.; Li, Z.; Chen, H.; Hu, H.; Zhou, Q.; Guan, Q. Generative adversarial networks in medical image augmentation: A review. *Comput. Biol. Med.* **2022**, *144*, 105382. [CrossRef]
245. von Werra, L.; Schöngens, M.; Uzun, E.D.G.; Eickhoff, C. Generative adversarial networks in precision oncology. In Proceedings of the 2019 ACM SIGIR International Conference on Theory of Information Retrieval, Santa Clara, CA, USA, 2–5 October 2019.
246. Zhan, B.; Xiao, J.; Cao, C.; Peng, X.; Zu, C.; Zhou, J.; Wang, Y. Multi-constraint generative adversarial network for dose prediction in radiotherapy. *Med. Image Anal.* **2022**, *77*, 102339. [CrossRef] [PubMed]
247. Heilemann, G.; Zimmermann, L.; Matthewman, M. Investigating the Potential of Generative Adversarial Networks (GANs) for Autosegmentation in Radiation Oncology. Ph.D. Thesis, Technische Universitä, Viena, Austria, 2021.
248. Nakamura, M.; Nakao, M.; Imanishi, K.; Hirashima, H.; Tsuruta, Y. Geometric and dosimetric impact of 3D generative adversarial network-based metal artifact reduction algorithm on VMAT and IMPT for the head and neck region. *Radiat. Oncol.* **2021**, *16*, 96. [CrossRef]
249. Hersche, M.; Zeqiri, M.; Benini, L.; Sebastian, A.; Rahimi, A. A neuro-vector-symbolic architecture for solving Raven's progressive matrices. *Nat. Mach. Intell.* **2023**, *5*, 363–375. [CrossRef]
250. Serre, T. Deep learning: The good, the bad, and the ugly. *Annu. Rev. Vis. Sci.* **2019**, *5*, 399–426. [CrossRef]
251. Shrestha, A.; Mahmood, A. Review of deep learning algorithms and architectures. *IEEE Access* **2019**, *7*, 53040–53065. [CrossRef]
252. Kleyko, D.; Rachkovskij, D.A.; Osipov, E.; Rahimi, A. A Survey on Hyperdimensional Computing aka Vector Symbolic Architectures, Part I: Models and Data Transformations. *ACM Comput. Surv.* **2021**, *55*, 1–40. [CrossRef]
253. Kleyko, D.; Rachkovskij, D.A.; Osipov, E.; Rahimi, A. A Survey on Hyperdimensional Computing aka Vector Symbolic Architectures, Part II: Applications, Cognitive Models, and Challenges. *ACM Comput. Surv.* **2021**, *55*, 1–52. [CrossRef]
254. This AI Could Likely Beat You at an IQ Test. Available online: https://research.ibm.com/blog/neuro-vector-symbolic-architecture-IQ-test (accessed on 10 June 2023).
255. Widdows, D.; Cohen, T. Reasoning with vectors: A continuous model for fast robust inference. *Log. J. IGPL* **2015**, *23*, 141–173. [CrossRef]
256. Abhijith, M.; Nair, D.R. Neuromorphic High Dimensional Computing Architecture for Classification Applications. In Proceedings of the 2021 IEEE International Conference on Nanoelectronics, Nanophotonics, Nanomaterials, Nanobioscience & Nanotechnology (5NANO), Kottayam, Kerala, India, 29–30 April 2021.
257. Fortunato, L.; Galassi, M. The case for free and open source software in research and scholarship. *Philos. Trans. R. Soc. A Math. Phys. Eng. Sci.* **2021**, *379*, 20200079. [CrossRef]
258. Sahay, S. Free and open source software as global public goods? What are the distortions and how do we address them? *Electron. J. Inf. Syst. Dev. Ctries.* **2019**, *85*, e12080. [CrossRef]
259. Singh, P.; Manure, A.; Singh, P.; Manure, A. Introduction to Tensorflow 2.0. In *Learn TensorFlow 2.0: Implement Machine Learning and Deep Learning Models with Python*; Apress: New York, NY, USA, 2020; pp. 1–24.
260. Stevens, E.; Antiga, L.; Viehmann, T. *Deep Learning with PyTorch*; Manning Publications: Shelter Island, NY, USA, 2020.
261. Pang, B.; Nijkamp, E.; Wu, Y.N. Deep learning with tensorflow: A review. *J. Educ. Behav. Stat.* **2020**, *45*, 227–248. [CrossRef]
262. Sarang, P. *Artificial Neural Networks with TensorFlow 2*; Apress: Berkeley, CA, USA, 2021.
263. Rao, D.; McMahan, B. *Natural Language Processing with PyTorch: Build Intelligent Language Applications Using Deep Learning*; O'Reilly Media, Inc.: Sebastopol, CA, USA, 2019.
264. Wolf, T.; Debut, L.; Sanh, V.; Chaumond, J.; Delangue, C.; Moi, A.; Cistac, P.; Rault, T.; Louf, R.; Funtowicz, M. Huggingface's transformers: State-of-the-art natural language processing. *arXiv* **2019**, arXiv:1910.03771.
265. Ayyadevara, V.K.; Reddy, Y. *Modern Computer Vision with Pytorch: Explore Deep Learning Concepts and Implement Over 50 Real-World Image Applications*; Packt Publishing Ltd.: Birmingham, UK, 2020.
266. Anderson, B.M.; Wahid, K.A.; Brock, K.K. Simple python module for conversions between DICOM images and radiation therapy structures, masks, and prediction arrays. *Pract. Radiat. Oncol.* **2021**, *11*, 226–229. [CrossRef] [PubMed]
267. Norton, I.; Essayed, W.I.; Zhang, F.; Pujol, S.; Yarmarkovich, A.; Golby, A.J.; Kindlmann, G.; Wassermann, D.; Estepar, R.S.J.; Rathi, Y. SlicerDMRI: Open source diffusion MRI software for brain cancer research. *Cancer Res.* **2017**, *77*, e101–e103. [CrossRef]
268. Gutman, D.A.; Khalilia, M.; Lee, S.; Nalisnik, M.; Mullen, Z.; Beezley, J.; Chittajallu, D.R.; Manthey, D.; Cooper, L.A. The digital slide archive: A software platform for management, integration, and analysis of histology for cancer research. *Cancer Res.* **2017**, *77*, e75–e78. [CrossRef]

269. Zhu, Y.; Qiu, P.; Ji, Y. TCGA-assembler: Open-source software for retrieving and processing TCGA data. *Nat. Methods* **2014**, *11*, 599–600. [CrossRef] [PubMed]
270. Jo, A. The Promise and Peril of Generative AI. *Nature* **2023**, *614*, 214–216.
271. Bucknall, B.S.; Dori-Hacohen, S. Current and near-term AI as a potential existential risk factor. In Proceedings of the 2022 AAAI/ACM Conference on AI, Ethics, and Society, Oxford, UK, 19–21 May 2022.
272. Roose, K.A.I. Poses 'Risk of Extinction,' Industry Leaders Warn. *New York Times*, 30 May 2023.
273. Letters, F.O. Pause giant AI Experiments: An Open Letter. Future of Life Institution. 2023. Available online: https://futureoflife.org/open-letter/pause-giant-ai-experiments (accessed on 10 June 2023).
274. Kann, B.H.; Hosny, A.; Aerts, H. Artificial intelligence for clinical oncology. *Cancer Cell* **2021**, *39*, 916–927. [CrossRef] [PubMed]
275. Gha, A.E.; Takov, P.; Shang, N. The Superhuman Born Out of Artificial Intelligence and Genetic Engineering: The Destruction of Human Ontological Dignity. *Horiz. J. Humanit. Artif. Intell.* **2023**, *2*, 56–65.
276. Zhang, J.; Zhang, Z.-M. Ethics and governance of trustworthy medical artificial intelligence. *BMC Med. Inform. Decis. Mak.* **2023**, *23*, 7. [CrossRef]
277. Langlotz, C.P. Will artificial intelligence replace radiologists? *Radiol. Soc. North America.* **2019**, *1*, e190058. [CrossRef]
278. Goldhahn, J.; Rampton, V.; Spinas, G.A. Could artificial intelligence make doctors obsolete? *BMJ* **2018**, *363*, k4563. [CrossRef]
279. Razzaki, S.; Baker, A.; Perov, Y.; Middleton, K.; Baxter, J.; Mullarkey, D.; Sangar, D.; Taliercio, M.; Butt, M.; Majeed, A. A comparative study of artificial intelligence and human doctors for the purpose of triage and diagnosis. *arXiv* **2018**, arXiv:1806.10698.
280. Botha, J.; Pieterse, H. Fake news and deepfakes: A dangerous threat for 21st century information security. In Proceedings of the ICCWS 2020 15th International Conference on Cyber Warfare and Security, Norfolk, VA, USA, 12–13 March 2020.
281. Pantserev, K.A. The Malicious Use of AI-Based Deepfake Technology as the New Threat to Psychological Security and Political Stability. In *Cyber Defence in the Age of AI, Smart Societies and Augmented Humanity*; Springer: Cham, Switzerland, 2020; pp. 37–55.
282. Borji, A. A categorical archive of ChatGPT failures. *arXiv* **2023**, arXiv:2302.03494.
283. Brainard, J. Journals take up arms against AI-written text. *Science* **2023**, *379*, 740–741. [CrossRef]
284. Vaishya; Misra, A.; Vaish, A. ChatGPT: Is this version good for healthcare and research? *Diabetes Metab. Syndr. Clin. Res. Rev.* **2023**, *17*, 102744. [CrossRef]
285. DeGrave, A.; Janizek, J.; Lee, S. AI for radiographic COVID-19 detection selects shortcuts over signal. *Nat. Mach. Intell.* **2021**, *3*, 610–619. [CrossRef]
286. Khullar, D.; Casalino, L.P.; Qian, Y.; Lu, Y.; Krumholz, H.M.; Aneja, S. Perspectives of patients about artificial intelligence in health care. *JAMA Netw. Open* **2022**, *5*, e2210309. [CrossRef]
287. Seyyed-Kalantari, L.; Zhang, H.; McDermott, M.B.; Chen, I.Y.; Ghassemi, M. Underdiagnosis bias of artificial intelligence algorithms applied to chest radiographs in under-served patient populations. *Nat. Med.* **2021**, *27*, 2176–2182. [CrossRef]
288. Kim, E.J.; Woo, H.S.; Cho, J.H.; Sym, S.J.; Baek, J.-H.; Lee, W.-S.; Kwon, K.A.; Kim, K.O.; Chung, J.-W.; Park, D.K. Early experience with Watson for oncology in Korean patients with colorectal cancer. *PLoS ONE* **2019**, *14*, e0213640. [CrossRef]
289. Richardson, J.P.; Smith, C.; Curtis, S.; Watson, S.; Zhu, X.; Barry, B.; Sharp, R.R. Patient apprehensions about the use of artificial intelligence in healthcare. *NPJ Digit. Med.* **2021**, *4*, 140. [CrossRef]
290. Lim, A.K.; Thuemmler, C. Opportunities and challenges of internet-based health interventions in the future internet. In Proceedings of the 2015 12th International Conference on Information Technology-New Generations, Las Vegas, NV, USA, 13–15 April 2015.
291. Hatherley, J.J. Limits of trust in medical AI. *J. Med. Ethics* **2020**, *46*, 478–481. [CrossRef] [PubMed]
292. DeCamp, M.; Tilburt, J.C. Why we cannot trust artificial intelligence in medicine. *Lancet Digit. Health* **2019**, *1*, e390. [CrossRef] [PubMed]
293. Nundy, S.; Montgomery, T.; Wachter, R.M. Promoting trust between patients and physicians in the era of artificial intelligence. *JAMA* **2019**, *322*, 497–498. [CrossRef] [PubMed]
294. Johnson, M.; Albizri, A.; Harfouche, A. Responsible artificial intelligence in healthcare: Predicting and preventing insurance claim denials for economic and social wellbeing. *Inf. Syst. Front.* **2021**, 1–17. [CrossRef]
295. Lysaght, T.; Lim, H.Y.; Xafis, V.; Ngiam, K.Y. AI-assisted decision-making in healthcare: The application of an ethics framework for big data in health and research. *Asian Bioeth. Rev.* **2019**, *11*, 299–314. [CrossRef]
296. Dueno, T. Racist Robots and the Lack of Legal Remedies in the Use of Artificial Intelligence in Healthcare. *Conn. Ins. LJ* **2020**, *27*, 337.
297. Formosa, P.; Rogers, W.; Griep, Y.; Bankins, S.; Richards, D. Medical AI and human dignity: Contrasting perceptions of human and artificially intelligent (AI) decision making in diagnostic and medical resource allocation contexts. *Comput. Hum. Behav.* **2022**, *133*, 107296. [CrossRef]
298. Muthukrishnan, N.; Maleki, F.; Ovens, K.; Reinhold, C.; Forghani, B.; Forghani, R. Brief history of artificial intelligence. *Neuroimaging Clin.* **2020**, *30*, 393–399. [CrossRef] [PubMed]
299. Pan, Y. Heading toward artificial intelligence 2.0. *Engineering* **2016**, *2*, 409–413. [CrossRef]
300. Fan, J.; Campbell, M.; Kingsbury, B. Artificial intelligence research at IBM. *IBM J. Res. Dev.* **2011**, *55*, 16:1–16:4. [CrossRef]
301. Bory, P. Deep new: The shifting narratives of artificial intelligence from Deep Blue to AlphaGo. *Convergence* **2019**, *25*, 627–642. [CrossRef]

302. McCorduck, P.; Minsky, M.; Selfridge, O.; Simon, H.A. History of artificial intelligence. In Proceedings of the IJCAI'77: Proceedings of the 5th International Joint Conference on Artificial Intelligence, Cambridge, MA, USA, 22–25 August 1977.
303. Bernstein, J. *Three Degrees above Zero: Bell Laboratories in the Information Age*; Cambridge University Press: Cambridge, UK, 1987.
304. Horowitz, M.C. Artificial intelligence, international competition, and the balance of power. *Texas National Security Review* **2018**, *2018*, 22.
305. Dick, S. Artificial intelligence. *Harv. Data Sci. Rev.* **2019**, *1*, 1–9.
306. Wasilow, S.; Thorpe, J.B. Artificial intelligence, robotics, ethics, and the military: A Canadian perspective. *AI Mag.* **2019**, *40*, 37–48. [CrossRef]
307. Bistron, M.; Piotrowski, Z. Artificial intelligence applications in military systems and their influence on sense of security of citizens. *Electronics* **2021**, *10*, 871. [CrossRef]
308. Lohr, S. MIT Plans College for Artificial Intelligence, Backed by $1 Billion. *The New York Times*, 15 October 2018; 15.
309. Kubassova, O.; Shaikh, F.; Melus, C.; Mahler, M. History, current status, and future directions of artificial intelligence. In *Precision Medicine and Artificial Intelligence*; Academic Press: Cambridge, MA, USA, 2021; pp. 1–38.
310. Fulbright, R. A Brief History of Artificial Intelligence. In *Democratization of Expertise*; Routledge: London, UK, 2020.
311. Marvin, M.; Seymour, A.P. *Perceptrons*; MIT Press: Cambridge, MA, USA, 1969; pp. 318–362.
312. McCarthy, J.; Minsky, M.; Rochester, N.; Shannon, C. *Dartmouth Artificial Intelligence (AI) Conference*; Dartmouth College: Hanover, NH, USA, 1956.
313. Lim, H.-o.; Takanishi, A. Biped walking robots created at Waseda University: WL and WABIAN family. *Philos. Trans. R. Soc. A Math. Phys. Eng. Sci.* **2007**, *365*, 49–64. [CrossRef] [PubMed]
314. Güzel, M.S. Autonomous vehicle navigation using vision and mapless strategies: A survey. *Adv. Mech. Eng.* **2013**, *5*, 234747. [CrossRef]
315. Fei-Fei, L.; Deng, J.; Li, K. ImageNet: Constructing a large-scale image database. *J. Vis.* **2009**, *9*, 1037. [CrossRef]
316. Chang, A. History of Artificial Intelligence in Medicine. *Gastrointest. Endosc.* **2020**, *92*, 807–812.
317. Beam, A.L.; Drazen, J.M.; Kohane, I.S.; Leong, T.Y.; Manrai, A.K.; Rubin, E.J. Artificial Intelligence in Medicine. *N. Engl. J. Med.* **2023**, *388*, 1220–1221. [CrossRef]
318. Kumar, V.; Gu, Y.; Basu, S.; Berglund, A.; Eschrich, S.A.; Schabath, M.B.; Forster, K.; Aerts, H.J.; Dekker, A.; Fenstermacher, D. Radiomics: The process and the challenges. *Magn. Reson. Imaging* **2012**, *30*, 1234–1248. [CrossRef]
319. Tang, A.; Tam, R.; Cadrin-Chênevert, A.; Guest, W.; Chong, J.; Barfett, J.; Chepelev, L.; Cairns, R.; Mitchell, J.R.; Cicero, M.D. Canadian Association of Radiologists white paper on artificial intelligence in radiology. *Can. Assoc. Radiol. J.* **2018**, *69*, 120–135. [CrossRef]
320. Johnson, K.W.; Soto, J.T.; Glicksberg, B.S.; Shameer, K.; Miotto, R.; Ali, M.; Ashley, E.; Dudley, J.T. Artificial intelligence in cardiology. *J. Am. Coll. Cardiol.* **2018**, *71*, 2668–2679. [CrossRef]
321. Lopez-Jimenez, F.; Attia, Z.; Arruda-Olson, A.M.; Carter, R.; Chareonthaitawee, P.; Jouni, H.; Kapa, S.; Lerman, A.; Luong, C.; Medina-Inojosa, J.R. Artificial intelligence in cardiology: Present and future. *Mayo Clin. Proc.* **2020**, *95*, 1015–1039. [CrossRef]
322. Miyazawa, A.A. Artificial intelligence: The future for cardiology. *Heart* **2019**, *105*, 1214. [CrossRef]
323. Le Berre, C.; Sandborn, W.J.; Aridhi, S.; Devignes, M.-D.; Fournier, L.; Smail-Tabbone, M.; Danese, S.; Peyrin-Biroulet, L. Application of artificial intelligence to gastroenterology and hepatology. *Gastroenterology* **2020**, *158*, 76–94.e2. [CrossRef]
324. Christou, C.D.; Tsoulfas, G. Challenges and opportunities in the application of artificial intelligence in gastroenterology and hepatology. *World J. Gastroenterol.* **2021**, *27*, 6191. [CrossRef]
325. Kröner, P.T.; Engels, M.M.; Glicksberg, B.S.; Johnson, K.W.; Mzaik, O.; van Hooft, J.E.; Wallace, M.B.; El-Serag, H.B.; Krittanawong, C. Artificial intelligence in gastroenterology: A state-of-the-art review. *World J. Gastroenterol.* **2021**, *27*, 6794. [CrossRef] [PubMed]
326. Kaplan, A.; Cao, H.; FitzGerald, J.M.; Iannotti, N.; Yang, E.; Kocks, J.W.; Kostikas, K.; Price, D.; Reddel, H.K.; Tsiligianni, I. Artificial intelligence/machine learning in respiratory medicine and potential role in asthma and COPD diagnosis. *J. Allergy Clin. Immunol. Pract.* **2021**, *9*, 2255–2261. [CrossRef] [PubMed]
327. Mekov, E.; Miravitlles, M.; Petkov, R. Artificial intelligence and machine learning in respiratory medicine. *Expert Rev. Respir. Med.* **2020**, *14*, 559–564. [CrossRef]
328. Ferrante, G.; Licari, A.; Marseglia, G.L.; La Grutta, S. Artificial intelligence as an emerging diagnostic approach in paediatric pulmonology. *Respirology* **2020**, *25*, 1029–1030. [CrossRef]
329. Hunter, B.; Hindocha, S.; Lee, R.W. The role of artificial intelligence in early cancer diagnosis. *Cancers* **2022**, *14*, 1524. [CrossRef] [PubMed]
330. Kenner, B.; Chari, S.T.; Kelsen, D.; Klimstra, D.S.; Pandol, S.J.; Rosenthal, M.; Rustgi, A.K.; Taylor, J.A.; Yala, A.; Abul-Husn, N. Artificial intelligence and early detection of pancreatic cancer: 2020 summative review. *Pancreas* **2021**, *50*, 251. [CrossRef]
331. Ballester, P.J.; Carmona, J. Artificial intelligence for the next generation of precision oncology. *NPJ Precis. Oncol.* **2021**, *5*, 79. [CrossRef]
332. Windisch, P.; Hertler, C.; Blum, D.; Zwahlen, D.; Förster, R. Leveraging advances in artificial intelligence to improve the quality and timing of palliative care. *Cancers* **2020**, *12*, 1149. [CrossRef]
333. Periyakoil, V.S.; Gunten, C.F.v. Palliative Care Is Proven. *J. Palliat. Med.* **2023**, *26*, 2–4. [CrossRef]
334. Courdy, S.; Hulse, M.; Nadaf, S.; Mao, A.; Pozhitkov, A.; Berger, S.; Chang, J.; Achuthan, S.; Kancharla, C.; Kunz, I. The City of Hope POSEIDON enterprise-wide platform for real-world data and evidence in cancer. *J. Clin. Oncol.* **2021**, *39*, e18813. [CrossRef]

335. Melstrom, L.G.; Rodin, A.S.; Rossi, L.A.; Fu, P., Jr.; Fong, Y.; Sun, V. Patient generated health data and electronic health record integration in oncologic surgery: A call for artificial intelligence and machine learning. *J. Surg. Oncol.* **2021**, *123*, 52–60. [CrossRef] [PubMed]
336. Dadwal, S.; Eftekhari, Z.; Thomas, T.; Johnson, D.; Yang, D.; Mokhtari, S.; Nikolaenko, L.; Munu, J.; Nakamura, R. A Machine-Learning Sepsis Prediction Model for Patients Undergoing Hematopoietic Cell Transplantation. *Blood* **2018**, *132*, 711. [CrossRef]
337. Deng, H.; Eftekhari, Z.; Carlin, C.; Veerapong, J.; Fournier, K.F.; Johnston, F.M.; Dineen, S.P.; Powers, B.D.; Hendrix, R.; Lambert, L.A. Development and Validation of an Explainable Machine Learning Model for Major Complications After Cytoreductive Surgery. *JAMA Netw. Open* **2022**, *5*, e2212930. [CrossRef]
338. Zachariah, F.J.; Rossi, L.A.; Roberts, L.M.; Bosserman, L.D. Prospective comparison of medical oncologists and a machine learning model to predict 3-month mortality in patients with metastatic solid tumors. *JAMA Netw. Open* **2022**, *5*, e2214514. [CrossRef]
339. Rossi, L.A.; Shawber, C.; Munu, J.; Zachariah, F. Evaluation of embeddings of laboratory test codes for patients at a cancer center. *arXiv* **2019**, arXiv:1907.09600.
340. Achuthan, S.; Chang, M.; Shah, A. SPIRIT-ML: A machine learning platform for deriving knowledge from biomedical datasets. In Proceedings of the Data Integration in the Life Sciences: 11th International Conference, DILS 2015, Los Angeles, CA, USA, 9–10 July 2015.
341. Karolak, A.; Branciamore, S.; McCune, J.S.; Lee, P.P.; Rodin, A.S.; Rockne, R.C. Concepts and applications of information theory to immuno-oncology. *Trends Cancer* **2021**, *7*, 335–346. [CrossRef] [PubMed]
342. Rosen, S.T. Why precision medicine continues to be the future of health care. *Oncol. Times UK* **2017**, *39*, 24. [CrossRef]
343. Budhraja, K.K.; McDonald, B.R.; Stephens, M.D.; Contente-Cuomo, T.; Markus, H.; Farooq, M.; Favaro, P.F.; Connor, S.; Byron, S.A.; Egan, J.B. Genome-wide analysis of aberrant position and sequence of plasma DNA fragment ends in patients with cancer. *Science Transl. Med.* **2023**, *15*, eabm6863. [CrossRef] [PubMed]
344. Liu, A.; Germino, E.; Han, C.; Watkins, W.; Amini, A.; Wong, J.; Williams, T. Clinical Validation of Artificial Intelligence Based Auto-Segmentation of Organs-at-Risk in Total Marrow Irradiation Treatment. *Int. J. Radiat. Oncol. Biol. Phys.* **2021**, *111*, e302–e303. [CrossRef]
345. Watkins, W.; Du, D.; Qing, K.; Ladbury, C.; Liu, J.; Liu, A. Validation of Automated Segmentation Algorithms. *Int. J. Radiat. Oncol. Biol. Phys.* **2021**, *111*, e152. [CrossRef]
346. Watkins, W.; Liu, J.; Hui, S.; Liu, A. Clinical Efficiency Gains with Artificial-Intelligence Auto-Segmentation in the Entire Human Body. *Int. J. Radiat. Oncol. Biol. Phys.* **2022**, *114*, e558. [CrossRef]
347. Jossart, J.; Kenjić, N.; Perry, J. Structural-Based Drug Discovery Targeting PCNA: A Novel Cancer Therapeutic. *FASEB J.* **2021**, *35*. [CrossRef]
348. Djulbegovic, B.; Teh, J.B.; Wong, L.; Hozo, I.; Armenian, S.H. Diagnostic Predictive Model for Diagnosis of Heart Failure after Hematopoietic Cell Transplantation (HCT): Comparison of Traditional Statistical with Machine Learning Modeling. *Blood* **2019**, *134*, 5799. [CrossRef]
349. Ladbury, C.; Amini, A.; Govindarajan, A.; Mambetsariev, I.; Raz, D.J.; Massarelli, E.; Williams, T.; Rodin, A.; Salgia, R. Integration of artificial intelligence in lung cancer: Rise of the machine. *Cell Rep. Med.* **2023**, *4*, 100933. [CrossRef] [PubMed]
350. Kothari, R.; Jones, V.; Mena, D.; Reyes, V.B.; Shon, Y.; Smith, J.P.; Schmolze, D.; Cha, P.D.; Lai, L.; Fong, Y. Raman spectroscopy and artificial intelligence to predict the Bayesian probability of breast cancer. *Sci. Rep.* **2021**, *11*, 6482. [CrossRef]
351. Pozhitkov, A.; Seth, N.; Kidambi, T.D.; Raytis, J.; Achuthan, S.; Lew, M.W. Machine learning algorithm to perform ASA Physical Status Classification. *medRxiv* **2021**. [CrossRef]
352. Han, C.; Liu, A.; Wong, J. Application of Machine Learning for Prediction of Normal Organ Dose: Feasibility Study in Treatment Planning for Total Marrow Irradiation. *Int. J. Radiat. Oncol. Biol. Phys.* **2020**, *108*, e782–e783. [CrossRef]
353. Two Studies for Patients with High Risk Prostate Cancer Testing Less Intense Treatment for Patients with a Low Gene Risk Score and Testing a More Intense Treatment for Patients With a High Gene Risk Score, The PREDICT-RT Trial. Available online: https://clinicaltrials.gov/ct2/show/NCT04513717 (accessed on 10 June 2023).
354. Ngiam, K.Y.; Khor, W. Big data and machine learning algorithms for health-care delivery. *Lancet Oncol.* **2019**, *20*, e262–e273. [CrossRef]
355. Bosserman, L.D.; Cianfrocca, M.; Yuh, B.; Yeon, C.; Chen, H.; Sentovich, S.; Polverini, A.; Zachariah, F.; Deaville, D.; Lee, A.B. Integrating academic and community cancer care and research through multidisciplinary oncology pathways for value-based care: A review and the City of Hope experience. *J. Clin. Med.* **2021**, *10*, 188. [CrossRef] [PubMed]

Journal of
Clinical Medicine

MDPI

Article

Building Team Medicine in the Management of CNS Metastases

Archit B. Baskaran [1], Robin A. Buerki [2], Osaama H. Khan [3], Vinai Gondi [4], Roger Stupp [5], Rimas V. Lukas [6,†] and Victoria M. Villaflor [7,*,†]

1 Department of Neurology, The University of Chicago, Chicago, IL 60637, USA
2 Health System Clinician of Neurology (Neuro-Oncology), Northwestern Medicine Regional Medical Group, Warrenville, IL 60555, USA
3 Surgical Neuro-Oncology, Northwestern Medicine Central DuPage Hospital, Winfield, IL 60190, USA
4 Department of Radiation Oncology, Nothwestern Medicine West Region, Lou & Jean Malnati Brain Tumor Institute, Northwestern University, Warrenville, IL 60555, USA
5 Neuro-Oncology Division, Neurological Surgery, Medicine (Hematology and Oncology), Neurology, Department of Neurology, Lou & Jean Malnati Brain Tumor Institute Northwestern University, Chicago, IL 60611, USA
6 Neuro-Oncology Division, Department of Neurology, Lou & Jean Malnati Brain Tumor Institute, Northwestern University, Chicago, IL 60611, USA
7 Department of Medical Oncology and Therapeutics Research, City of Hope, Duarte, CA 91010, USA
* Correspondence: vvillaflor@coh.org
† These authors contributed equally to this work.

Abstract: CNS metastases are often terminal for cancer patients and occur at an approximately 10-fold higher rate than primary CNS tumors. The incidence of these tumors is approximately 70,000–400,000 cases annually in the US. Advances that have occurred over the past two decades have led to more personalized treatment approaches. Newer surgical and radiation techniques, as well as targeted and immune therapies, have enanled patient to live longer, thus increasing the risk for the development of CNS, brain, and leptomeningeal metastases (BM and LM). Patients who develop CNS metastases have often been heavily treated, and options for future treatment could best be addressed by multidisciplinary teams. Studies have indicated that patients with brain metastases have improved survival outcomes when cared for in high-volume academic institutions using multidisciplinary teams. This manuscript discusses a multidisciplinary approach for both parenchymal brain metastases as well as leptomeningeal metastases implemented in three academic institutions. Additionally, with the increasing development of healthcare systems, we discuss optimizing the management of CNS metastases across healthcare systems and integrating basic and translational science into our clinical care to further improve outcomes. This paper summarizes the existing therapeutic approaches to the treatment of BM and LM and discusses novel and emerging approaches to optimizing access to neuro-oncologic care while simultaneously integrating multidisciplinary teams in the care of patients with BM and LM.

Keywords: brain metastases; delivery of health care; leptomeningeal metastases; tumor board

Citation: Baskaran, A.B.; Buerki, R.A.; Khan, O.H.; Gondi, V.; Stupp, R.; Lukas, R.V.; Villaflor, V.M. Building Team Medicine in the Management of CNS Metastases. *J. Clin. Med.* **2023**, *12*, 3901. https://doi.org/10.3390/jcm12123901

Academic Editor: Morgan Broggi

Received: 13 April 2023
Revised: 30 May 2023
Accepted: 3 June 2023
Published: 7 June 2023

SUMMARY TABLE

• The treatment of brain metastases (BM) and leptomeningeal metastases (LM) includes a combination of surgical intervention, radiation therapy, systemically administered therapies, and therapies directly administered to the CNS.
• Surgical interventions for BM include craniotomy and emerging modalities such as laser interstitial thermal therapy (LITT), which are often utilized with large tumor metastases that produce neurologic symptoms and increase intracranial pressure; for LM, options are more limited to the placement of intraventricular reservoirs for intrathecal chemotherapy or shunt placement for communicating hydrocephalus.
• Radiation therapies for BM include whole-brain radiation therapy (WBRT) and stereotactic radiosurgery (SRS), of which neurocognitive toxicity remains a pressing concern; for LM, options include WBRT and craniospinal irradiation (CSI).
• Systemically administered therapies for BM include tyrosine kinase inhibitors (TKI), such as osimertinib and tucatinib, and immune checkpoint inhibitors (ICI), such as cytotoxic T-lymphocyte-associated antigen 4 (CTLA4) inhibitors and programmed death 1 (PD1) pathway inhibitors; for LM, there is not a standard-of-care systemic therapy, with overall prognosis remaining poor.
• Intrathecal (or intra-CSF) chemotherapy is an ongoing area of study for the treatment of LM, with novel therapies demonstrating some promise.
• Hospital-system-wide tumor boards, also known as multidisciplinary cancer meetings (MCM), are emerging settings through which providers representing multidisciplinary neuro-oncology subspecialties meet to discuss complex cases and facilitate information sharing for the improvement of patient outcomes. These tumor boards may be in person or virtual, which remains an ongoing area of interest.
• Telemedicine has been a growing technology since COVID-19, with widespread potential for improving access and quality of care in the field of neuro-oncology.

1. Background

In this review, we discuss the clinical management of brain metastases (BM) and leptomeningeal metastases (LM), which are diseases managed by a multidisciplinary team of medical oncologists in conjunction with neuro-oncologists, radiation oncologists, and neurosurgeons [1]. There remains an unmet need to provide specific, central nervous system (CNS)-metastases-directed treatment in metastatic solid tumors. A multidisciplinary-team-based medicine model may add value for patients, practitioners, and hospital systems following such an approach. For example, the survival outcomes in brain metastases patients cared for at academic institutions appear superior [2]. This benefit of care in high-volume centers is in line with what has been observed for patients with glioblastoma [3,4]. While numerous non-clinical factors, including selection bias, may contribute to this benefit, we propose that specific factors related to the coordinated multispecialty oncology care model may positively impact patient outcomes. For example, a multidisciplinary approach has been shown to decrease treatment-associated clinical encounters and decrease the time to the initiation of radiation therapy for patients with BM [5]. However, available data to indicate which are the important elements and how to structure a coordinated, interdisciplinary care model are sparse and often lack granularity [6].

The various therapeutic modalities chosen and delivered by assorted sub-specialists in the care of patients with BM and LM will be briefly reviewed. These interventions include surgical intervention, radiation therapy, systemically administered therapies, and therapies directly administered to the CNS. How individual physicians can use these therapeutic modalities within a team model and in the context of a multidisciplinary care plan developed across a healthcare system will be highlighted. Finally, the integration of basic and translational science discoveries into the clinical management of BM and LM points toward future areas of focus to continue to benefit this complex oncology patient population.

2. Multi-Disciplinary Clinical Management of Brain Metastases

2.1. Epidemiology and Prognosis of Brain Metastases

The exact incidence of BM from solid tumors is not as certain as that of primary CNS tumors. It is thought there are ~70,000–400,000 new cases in the United States per annum. If correct, this makes them ~10-fold more common than primary CNS malignancies [1,7]. In turn, the scope of the problem presented by BM is substantial. Coordinated, high-touch multidisciplinary management may be essential for all patients with BM. This may be the case when the burden of progressive systemic disease supersedes CNS disease in driving morbidity and mortality, a concept that can be quantified as brain metastases velocity. Patients who develop progressive BM at a rapid annualized rate are at greater risk of dying from CNS disease [8,9]. Logistical and organizational factors also impact the feasibility of team-based multidisciplinary care for BM patients, including a limited workforce of sub-specialty providers, BM-specific physical and cognitive patient debility, and the complexities presented by health system infrastructure for well-coordinated care.

Until recently, BM was viewed as being associated with a universally dismal prognosis [10]. We now have a more nuanced perspective of outcomes for these patients. For patients with BM from non-small cell lung, breast, melanoma, gastrointestinal, and renal cancers, median survival ranges from 7–47 months, 3–36 months, 5–34 months, 3–17 months, and 4–35 months, respectively per the Disease Specific Graded Prognostic Assessment (DS-GPA). Numerous factors comprise GPA score, including histology, molecular characteristics, patient age, performance status, and the number of metastases [11]. Considering these various points is important when developing an optimal plan tailored to specific patient characteristics and circumstances.

2.2. Surgical Therapies

Surgical resection has demonstrated efficacy in prolonging survival in patients with solitary BM. The role of surgical resection is predominantly in patients with large BM that produce either neurologic symptoms or those associated with increased intracranial pressure due to mass effect. It also serves a diagnostic role, including when a patient may benefit from more detailed molecular characterization of their BM. The role of surgical resection in oligometastatic disease is less clear, although it can be considered when there is a dominant symptomatic BM in the setting of other smaller lesions. One key study was the pivotal phase 3 randomized controlled trial comparing the effectiveness of surgical resection plus post-operative whole-brain radiotherapy (WBRT) vs. WBRT alone. This trial demonstrated that surgical resection followed by radiotherapy was superior to WBRT for the treatment of a solitary BM. The median overall survival (OS) was over 15 months in the surgical group vs. over 6 months in the RT-only group ($p < 0.0009$) [12]. Superiority was also observed across numerous other clinically relevant endpoints.

MRI-guided laser interstitial thermal therapy (LITT) is a minimally invasive surgical alternative for metastases up to 3 cm in diameter in locations where surgical resection may not be readily feasible [13]. The technology delivers laser light through a stereotactically navigated fiber optic probe to create thermal damage, leading to cellular death within the target lesion [14]. In patients with BM and radiation necrosis, who may be sicker and thereby unable to tolerate resection of the necrotic lesion, LITT may present itself as a viable alternative; in patients with radiation necrosis, it has been shown to decrease lesion size and improve progression-free survival [15]. Some studies suggest that LITT may even be superior to bevacizumab in the management of select patients with radiation necrosis [16]. Similarly, other studies have suggested comparable outcomes between LITT and other surgical resection approaches for patients with brain metastases [17]. However, to date, the evidence comparing LITT to craniotomy or medical management for the management of radiation necrosis or brain metastases is largely retrospective (e.g., LITT vs. craniotomy [18], LITT vs. bevacizumab [16]), with one multicenter prospective study demonstrating benefits to laser ablation following SRS in patients with brain metastases and radiation necrosis [19]. One additional phase I clinical trial is underway, exploring the efficacy of a combination therapy of LITT with Pembrolizumab for recurrent BM (NCT04187872).

2.3. Radiation Therapies

Radiation therapy (RT) has formed the backbone of the treatment of BM. The most common approaches, WBRT and stereotactic radiosurgery (SRS), have different benefits in terms of acute and late toxicity, ease of implementation, and treatment duration, without clear differences in survival from CNS failure in oligometastatic disease (OMD). OMD has been defined as up to five metastatic lesions [20]. Prophylactic cranial irradiation (PCI) is another therapeutic approach using WBRT that has demonstrated success in reducing the incidence of symptomatic BM in patients with limited- and extensive-stage small cell lung cancer (ES-SCLC), as well as improving OS [21]. However, significant knowledge gaps surrounding its neurotoxicity and unique effects in different subgroups of patients have made its continued routine use somewhat controversial [22,23]. With ongoing investigations evaluating the growing role of SRS in this disease, it is possible that this current approach may be supplanted.

WBRT has the advantages of ubiquity in being able to be implemented quickly, its ability to be administered across healthcare systems with varying resources, and the ability to treat both macro- (i.e., radiographically visible) and micrometastatic CNS disease. A key limitation of WBRT is the neurocognitive toxicity from irradiating healthy brain tissue unaffected by tumors and the resulting deterioration of patient independence and quality of life (QOL) [24]. In contrast, SRS can provide local control of BM without the risk of the potential late neurotoxicity associated with larger volumes of radiation and at the expense of not treating micrometastatic disease.

There have been efforts to avoid the neurocognitive toxicity of WBRT [25]. Two recent interventions have been demonstrated to reduce the risk of delayed cognitive decline: (i) the addition of the N-methyl-D-aspartate receptor (NMDAR) antagonist memantine, a drug marketed to treat Alzheimer's disease, and (ii) hippocampal avoidance WBRT (HA-WBRT) [26], i.e., a careful method of treatment planning that protects the hippocampi from high-dose radiation exposure when delivering WBRT [26]. The addition of 6 months of memantine to WBRT in the randomized RTOG 0614 trial resulted in delayed cognitive decline in the investigational arm [27]. The phase 3 NRG CC-001 trial used this same approach and investigated the addition of HA-WBRT. The trial compared the neurocognitive decline in patients with BM treated with either standard WBRT or HA-WBRT [28]. The employment of HA-WBRT was associated with lesser deterioration of executive function as well as learning and memory without any detriment to overall survival or intracranial progression-free survival. This benefit was observed across all patient ages and was first noted at 4 months, persisting out to at least 12 months.

There has been continual work on optimizing the role of SRS for BM. SRS has been shown to be both safe and efficacious for treating several "oligo" BM. Absolute tumor number, cumulative tumor volume, and tumor location are all factors that may influence the efficacy, safety, and tolerability of SRS for BM, making the discussion more complex than simply defining an optimal numeric cut-off for the number of brain metastases. In patients with 1–3 brain metastases, SRS has already replaced WBRT as the standard of care [29]. A more recent randomized phase III trial demonstrated level 1 evidence for the superiority of SRS over WBRT in non-melanoma patients with 4–15 brain metastases [30], providing support for the expansion of the number of BM that can feasibly be treated. The latest ASCO-SNO-ASTRO guidelines recommend SRS for patients with 1–4 unresected, asymptomatic brain metastases, excluding small cell lung cancer (SCLC) [31]. Two additional phase III clinical trials (NRG Oncology CC009 and NRG BN009) are currently underway. NRG-CC009 is comparing SRS to HA-WBRT plus memantine for 10 or fewer BMs from small cell lung cancer with a primary endpoint of cognitive function failure; NRG BN009 is comparing salvage SRS to salvage HA-WBRT for distant brain relapse following upfront SRS with high brain metastasis velocity (BMV). BMV is defined as the total number of new brain metastases a patient develops over time since their first treatment with SRS; it is a newer metric that has been associated with neurologic death and overall survival and is a predictor for determining a patient's need for salvage WBRT after initial distant brain failure following upfront SRS alone [9].

For large BM and limited intracranial disease burden, post-operative SRS to the surgical resection cavity may be employed as a component of multimodal therapy [32]. Over time, prospective trials have validated the superiority of post-operative SRS over WBRT for this indication [33,34]. However, post-operative SRS has been associated in some studies with radiation necrosis, leptomeningeal metastatic spread, and local failure at the SRS treatment site, thereby prompting the exploration of the use of pre-operative SRS [35]. One retrospective multi-institutional analysis of 279 patients found lower rates of radiation necrosis, leptomeningeal spread, and local failure in patients undergoing multimodal therapy for large BM with limited intracranial disease [36]. A recently initiated phase III trial is exploring the efficacy of pre-operative versus post-operative SRS for patients with surgically resected BM (NRG BN012; NCT05438212). Currently, the ASCO-SNO-ASTRO guidelines recommend SRS to the surgical cavity in patients with 1–2 resected BM and SRS vs. WBRT vs. combination therapy for other patients [31].

2.4. Systemic Therapies

Traditionally, systemic chemotherapy had a limited role in the management of BM due to the CNS penetration of available agents, yet growing evidence suggests that systemically effective therapies of newer small molecules are also demonstrating responses in the CNS. The intracranial failure rate has been reduced in randomized trials, likely impacted by the low CNS penetration of the agents used. A wide range of therapeutic agents are now considered in the management of BM. The most frequently used agents are systemic targeted therapies (e.g., tyrosine kinase inhibitors (TKI)) and immune checkpoint inhibitors (ICI).

One example within the targeted therapy class of systemic therapy is the TKI osimertinib, a third-generation agent for the treatment of epidermal growth factor receptor (EGFR) mutant non-small cell lung cancer (NSCLC). Osimertinib has demonstrated efficacy in the treatment of T790M-positive advanced NSCLC BM [37]. Another TKI is tucatinib, a human epidermal growth factor receptor 2 (HER-2) inhibitor often used for the treatment of HER2(+) breast cancer. A post hoc exploratory analysis of patients with BM treated on the HER2-CLIMB trial with tucatinib in combination with capecitabine and trastuzumab further demonstrated a role in preventing intracranial disease progression, showing a 68% reduced risk of intracranial progression or death versus capecitabine and trastuzumab alone [38]. Similar results have been seen for a variety of other solid tumor types. BRAF/MEK targeted therapies, such as dabrafenib and trametinib, have become the standard of care for the treatment of melanoma brain metastases with a BRAF-MEK pathway driver mutation [39] and currently have histology-agnostic approval for tumors harboring the *BRAF* V600E mutation. Retrospective cohort data have demonstrated the efficacy and acceptable safety profiles of another TKI, cabozantinib, for the treatment of renal cell carcinoma (RCC) BM [40].

Drugs within the ICI class of systemic therapies used for the treatment of BM often target the cytotoxic T-lymphocyte-associated antigen 4 (CTLA4) and programmed death 1 (PD1) pathways. Theoretical benefits of ICI therapies include their lower toxicity profiles compared to traditional chemotherapeutic agents as well as their potential for efficacy both outside of and within the CNS [41]. Combination ICI therapies have also often been investigated for the treatment of BM. One phase II trial exploring the combination of nivolumab with ipilumab for patients with mainly asymptomatic melanoma BM demonstrated a response rate of 57% (95% CI 47–67%) and a three-year intracranial PFS of 54% (95%CI 43–64%) in 101 patients with asymptomatic melanoma BM [42]. Several ongoing trials are further investigating the incorporation of ICI for the treatment of BM secondary to multiple primary solid tumors (NCT05704647, NCT04511013, NCT05012254, NCT03873818, NCT04187872, NCT05341349, and NCT04348747).

3. Multi-Disciplinary Clinical Management of Leptomeningeal Metastases

3.1. Epidemiology and Prognosis of LM

Leptomeningeal spread is increasingly observed in advanced metastatic cancer, with a relative increase as systemic treatments and disease control have substantially improved over the last decade. LM occurs in approximately 5% of advanced solid tumors, yet autopsy series reveal a prevalence of up to 20% for asymptomatic or undiagnosed LM [43,44]. The observed incidence of LM diagnoses may be increasing due to advances in the diagnosis of LM and treatment strategies that prolong survival from systemic disease [45,46]. While cancer prognoses have broadly improved, the prognosis of LM remains poor, with survival times typically quoted at only 2 to 3 months [47,48]. The most important (and consistent) prognostic factor across numerous studies remains the patients' performance status [49,50]. This serves as a readily assessable quantifiable measure that can help guide the aggressiveness of the management approach.

Recently, progress has been made in the understanding of patients at risk for and the pathogenesis of LM [51–53]. The presence of a blood–CSF barrier (distinct from the blood–brain barrier (BBB)) presents unique therapeutic challenges for the management of LM compared to BM. The blood–CSF barrier is the space between the choroid plexus and the CSF. Unlike the BBB capillaries, which form the endothelial layer of the brain parenchyma to separate brain interstitial fluid from the blood, the choroid plexus forms the epithelial layer to separate CSF from the blood; it has no tight junctions, contains fenestrations, and thereby utilizes distinct active transport mechanisms such as bulk CSF flow and transcapillary exchange to determine the concentration of molecules in the CSF [54]. Given its relative leakiness compared to the BBB, the blood–CSF barrier permits the crossing of certain substances that would not otherwise cross the BBB [55]. LM have been shown to upregulate complement component 3 (C3), activating the C3a receptor on the choroid plexus, thereby disrupting the blood–CSF barrier and enabling the passage of growth factors such as amphiregulin into the CSF that enable LM to adapt and spread in the CSF microenvironment [56]. This preclinical work raises the opportunity for further study in inhibiting C3 signaling pathways to therapeutically manipulate the blood–CSF barrier to improve the penetration of systemic chemotherapeutic agents.

3.2. Surgical Therapies

The role of surgical intervention in LM is limited to the placement of an intraventricular reservoir (Ommaya, Rickham) to facilitate the delivery of intrathecal chemotherapy or the placement of a shunt in patients with communicating hydrocephalus due to flow obstruction by metastatic cells in the cerebrospinal fluid [57,58]. One critical consideration for safety in proposing direct IT chemotherapy is establishing unobstructed CSF circulation flowing through the ventricular system. Chemotherapy can be delivered via lumbar puncture; an intraventricular reservoir allows not only for less burdensome repeat drug delivery but also appears to be associated with superior survival when compared to delivery via lumbar puncture based on a small retrospective series [59]. Although not as common, however, placement of a reservoir in continuity with a ventriculoperitoneal shunt with the ability to temporarily turn "off" the shunt provides an opportunity to deliver therapy and divert CSF in the same system, thereby avoiding multiple surgeries or procedures (recurrent lumbar punctures).

3.3. Radiation Therapies

Radiation has been used to treat LM, although practice patterns vary between institutions [48]. Different approaches for RT in this setting include focal radiation to address symptomatic bulky or obstructive leptomeningeal metastases, WBRT to treat a substantial but incomplete portion of the target space, and full craniospinal irradiation (CSI).

The benefits of CSI have traditionally been limited by significant potential CNS and extra-CNS toxicity. However, in comparison to standard photon radiation techniques, the application of protons for CSI may be a means to overcome some of the toxicities, particularly myelosuppression and radiation esophagitis. Proton RT is a type of radiation treatment [60]. A recent phase I study of proton CSI for LM patients demonstrated a favorable safety profile with self-resolution of any dose-limiting toxicities and improved overall survival compared to historical norms [61], and a subsequent small randomized phase II trial supported the superiority of proton CSI (both in terms of increased meaningful overall survival as well as reduced neurologic symptom burden) when compared to involved field photon radiation; this is the only intervention trial to have ever shown an overall survival benefit in LM [62]. This concept requires further validation, and a larger multi-institution phase 3 trial is currently in development. Another phase I clinical study is currently underway exploring an intrathecal rhenium-186 nanoliposome for the treatment of LM (NCT05034497).

3.4. Systemic Therapies

The role of systemic therapies in the treatment of LM is evolving. One study compared the efficacy of systemically administered high-dose methotrexate (HD-MTX) with IT MTX for patients with LM, finding that systemic MTX may be superior for the cytologic clearing of CSF and for OS in patients (mOS 13.8 months for the IV HD-MTX group vs. 2.3 months for the IT-MTX group, $p = 0.003$) [63]. However, more recent case series have demonstrated the potential viability of integrating HD-MTX into broader multimodal treatment plans for LM [64].

Following up on prior data demonstrating the benefit of combination therapy with tucatinib and capecitabine plus trastuzumab for parenchymal BM, a randomized phase II study is currently ongoing to further explore this treatment combination in LM (NCT03501979). Furthermore, a phase II trial of ANG1005, a novel taxane derivative designed to penetrate the blood–brain and blood–CSF barriers more effectively, showed that 79% of patients with LM had intracranial disease control with a median OS of 8 months [65]. Despite these findings, there is no standard-of-care systemic therapy in this LM setting, with the overall prognosis remaining poor [66].

3.5. CSF-Administered Therapies

The direct administration of antineoplastic agents into the CSF is a means to circumvent the blood–CSF barrier [67]. Therapeutic trends investigated within the context of clinical trials have recently been reviewed [68].

Intrathecal (or intra-CSF) chemotherapy may be delivered via surgically implanted ventricular Ommaya reservoirs or via lumbar puncture [69]. The theoretical benefit of direct intrathecal delivery is the ability of agents to achieve homogenous distribution within the subarachnoid space [70]. A recent review of intrathecal therapies for LM over a 39-year period showed that the most commonly administered intrathecal regimens consisted of a combination of singular therapy of methotrexate (MTX), cytarabine (Ara-C), and thiotepa, with the additional limited use of topotecan, pemetrexed, and systemic antibodies or immunotherapies such as trastuzumab and interleukin-2 (IL-2) [68,71,72]. The rationale for choosing each agent depends largely on their pharmacokinetic properties, namely their half-life, clearance from the CSF space, and lipophilicity [68,69]. An additional agent, sustained-release liposomal cytarabine (DepoCyt), was initially utilized but discontinued in 2017 following increased adverse events secondary to neurotoxicity [73]. Ultimately, while there are no specific guideline-based therapies for determining which intrathecal agent to use for which primary tumor, data suggest a benefit of (1) MTX in LM due to solid neoplasms, (2) trastuzumab for LM due to HER2-positive malignancies, and (3) pemetrexed for LM secondary to lung adenocarcinoma [68].

Of particular interest has been the use of antibodies for specific targets, such as HER2(+). These have proven to be very well tolerated and associated with favorable survival outcomes [74,75]. One phase I/II trial of IT trastuzumab for patients with HER2-positive breast cancer patients demonstrated an mOS of 10.5 months for the HER2-positive breast cancer population with LM versus 3.3–4.4 months for historical controls [49,74]. Another phase II study exploring IT trastuzumab for HER2-positive breast cancer demonstrated a mOS of 7.9 months [75]. Numerous other small single-arm studies of a variety of IT approaches have been explored, with modest results observed thus far [62–64].

4. Optimizing Management of CNS Metastases across a Health Care System (and Beyond)

4.1. Tumor Boards

The integrated, interdisciplinary team management of BM and LM necessitates seamless cross-coordination between different subspecialty providers. One such platform that has enabled this includes hospital-system-wide tumor boards, also known as multidisciplinary cancer meetings (MCM), specifically directed for CNS metastases. These can bring together experts from medical oncology, neurosurgery, neuro-oncology, radiation oncology, neuroradiology, neuropathology, epilepsy neurology, and cancer genetic counseling to review complex cases.

In oncology, tumor boards have been established as a means of improving accuracy in diagnosis, bettering adherence to clinical guidelines, advancing the integration of novel research and clinical trials in clinical management, and improving patient outcomes [76]. Prior studies have demonstrated that multidisciplinary tumor boards improve both the quality of medical services offered to patients as well as OS rates [77–80]. One retrospective study of an institution's neuro-oncology tumor board at a large tertiary care academic medical center in Italy suggested that the multidisciplinary management of challenging cases improved overall effectiveness in managing brain tumors [76]. At this institution, complex cases often involving gliomas and brain metastases were discussed on a weekly basis between neurosurgeons, neuro-oncologists, neuro-radiologists, neuropathologists, and other key multidisciplinary teams; neuro-oncologists presented most cases, with neuroradiologists providing interpretations of possible image reassessments, with ongoing cross-team communication between all providers on updated therapies utilized for each patient's care along with interval reassessments of previous exams [76]. The discussions enabled not only the broadening of differential diagnoses but also treatment plan changes [76]. Additional benefits described in the literature of institutions with neuro-oncologic multidisciplinary tumor boards included improved resident education, increased clinical trial access for patients, and increased published guideline adherence [81].

4.2. Telemedicine

Since the COVID-19 pandemic, telemedicine has rapidly expanded across the globe to facilitate both more seamless interdisciplinary care and increased access to oncologic care, particularly within neuro-oncology [82]. One review posits that certain types of neuro-oncology encounters, including chemotherapy monitoring visits, chemotherapy consent and education, second opinion visits, or discussions around new laboratory or imaging results, may all prove viable in a telemedicine setting [83]. One study of an institution's telemedicine program for neuro-oncology visits demonstrated equivalent patient satisfaction, with time and cost savings for patients [84]. A review of the telemedicine programs at Barrow Neurological Institute and Geisinger Health during the early COVID-19 pandemic period in 2020 demonstrated that the neuro-oncology divisions were able to return to 90% or greater capacity within 6 weeks of initial closure due to both systems' effective implementation of telehealth programs [85]. Although further research is needed, the promise of telehealth for patients and providers is the largest when considering its potential to expand access to care in remote areas, particularly for patients with financial or transportation barriers to care.

Furthermore, the ease of telemedicine platforms in enabling multidisciplinary discussions regardless of geographic location may help facilitate virtual tumor boards in the future [82]. Recently, multi-institutional virtual tumor boards (VTBs) have been an emerging platform for case discussions; a review of three VTBs in the United States demonstrated that neuro-oncology VTBs provide for faster expert analysis of clinical cases with an average response time under 24 h [86]. Thus, VTBs represent an effective means of conducting multidisciplinary care for patients with neuro-oncologic disease burdens without the inhibition of institutional barriers.

5. Integration of Science into Clinical Care

The molecular pathological characterization of CNS tumors is becoming increasingly important in diagnosis and clinical management [87]. Practically, there may be greater value in performing molecular pathology evaluations earlier versus later in a patient's treatment course. The application of next-generation sequencing to detect mutations and oncogenic fusions can also be applied to brain metastases [88]. Genomic analysis of BM has compared them to systemic tumor metastases to distal extracranial and regional lymph node sites, finding that BM harbor several distinct genomic alterations compared to primary tumors, particularly in the CDK and PI3K/AKT/mTOR gene pathways [89]. As such, BM may demonstrate sensitivity to inhibitors targeting these pathways when those aberrancies are present. An ongoing National Cancer Institute (NCI) cooperative group phase II randomized clinical trial (NCT03994796) is investigating agents with known CNS/blood–brain barrier penetrance, including abemaciclib, paxalisib and entrectinib, to selectively target these pathways, as well as other pathways, such as NTRK and ROS1. In this trial, 136 patients with new, recurrent, or progressive BM will be divided into three experimental arms: the first arm with CDK mutations, the second with PI3K mutations, and the third with NTRK/ROS1 mutations. The groups will, respectively, be administered twice daily oral abemaciclib, daily oral paxalisib, and daily oral entrectinib, each for a total of 28 days, with cycles repeating every 28 days. The primary outcome measure will be the objective response rate in the brain as determined by Response Assessment in Neuro-Oncology (RANO) criteria; the secondary endpoints will consist of systemic response and clinical benefit rates, duration of response, PFS, OS, and adverse rate incidence. The goal of the trial is to determine the efficacy of targeted therapies in patients harboring specific BM gene mutations. Although this targeted, personalized approach for patients with BM will require tissue from the CNS for analysis, it may represent a new paradigm for the clinical management of new, recurrent, or progressive BM.

6. Conclusions

The management of CNS metastases is an important component of oncologic care. It is optimal if this is integrated into the other aspects of the care of this patient population. This may be complex as some patients benefit from surgical, radiation, systemic, and/or CNS-delivered therapies delivered by a panoply of subspecialty providers. There are numerous models on how to best accomplish this high-quality multi-disciplinary care, with no single approach demonstrating definitive superiority over others. Considerations for implementation include resources, including sub-specialty clinical care providers with adequate availability and interest. We present a broad overview of how this care can be delivered.

Author Contributions: V.M.V. and R.V.L. contributed to the manuscript's conception and design. A.B.B. wrote the manuscript. All authors (A.B.B.; R.A.B.; O.H.K.; V.G.; R.S.; R.V.L.; V.M.V.) critically reviewed and revised the manuscript. All authors have read and agreed to the published version of the manuscript.

Funding: The authors declare that no funds, grants, or other support were received during the preparation of this manuscript.

Institutional Review Board Statement: Not applicable.

Informed Consent Statement: Not applicable.

Data Availability Statement: Not applicable.

Conflicts of Interest: Archit B Baskaran has no disclosures to report. Robin A. Buerki has no competing interests. Osaama H Khan has no competing interests. Vinai Gondi reports ownership or investment interests in Radiation Oncology Consultants, Ltd. (ROC) and honoraria from UpToDate. He is supported by a National Institute of Health R42 Small Business Technology Transfer grant through Immunochem Therapeutics. Roger Stupp reports consulting/related activities for AstraZeneca, Boston Scientific Corporation, CarThera, Celularity Inc, CranioVation, Elsevier Inc., GT Medical Technologies, Inc., Insightec, Lookwood/BlackDiamond, NorthWest Biotherapeutics Inc., Novocure, Inc. and Syneos Health, as well as ownership or investment interests in CranioVation. Rimas Lukas reports membership to a speaker's bureau for Merck, a speaker's bureau for Novocure, and a scientific advisory board for Merck, as well as research support (drug only) from BMS. Victoria Villaflor reports a position in consulting/advising for Astra Zeneca, as well as stock ownership in Johnson & Johnson.

References

1. Lamba, N.; Wen, P.Y.; Aizer, A.A. Epidemiology of brain metastases and leptomeningeal disease. *Neuro Oncol.* **2021**, *23*, 1447–1456. [CrossRef]
2. Amin, S.; Baine, M.; Meza, J.; Lin, C. The impact of treatment facility type on the survival of brain metastases patients regardless of the primary cancer type. *BMC Cancer* **2021**, *21*, 387. [CrossRef] [PubMed]
3. Koshy, M.; Sher, D.J.; Spiotto, M.; Husain, Z.; Engelhard, H.; Slavin, K.; Nicholas, M.K.; Weichselbaum, R.R.; Rusthoven, C. Association between hospital volume and receipt of treatment and survival in patients with glioblastoma. *J. Neurooncol.* **2017**, *135*, 529–534. [CrossRef] [PubMed]
4. Lukas, R.V.; Lesniak, M.S.; Stupp, R. Hospital volume and group expertise in newly diagnosed glioblastoma management. *J. Neurooncol.* **2018**, *136*, 213–214. [CrossRef]
5. Moss, N.S.; El Ahmadieh, T.Y.; Brown, S.; Chen, J.; Imber, B.S.; Pike, L.; Reiner, A.S.; Panageas, K.S.; Brennan, C.; Tabar, V.; et al. Integrated Multidisciplinary Brain Metastasis Care Reduces Patient Visits and Shortens Time to Adjuvant Irradiation. *JCO Oncol. Pract.* **2022**, *18*, e1732–e1738. [CrossRef]
6. Bajwa, M.H.; Bakhshi, S.K.; Shamim, M.S. Role of multidisciplinary neuro-oncology tumour boards in cancer management. *J. Pak. Med. Assoc.* **2021**, *71*, 2285–2286.
7. Ostrom, Q.T.; Wright, C.H.; Barnholtz-Sloan, J.S. Brain metastases: Epidemiology. *Handb. Clin. Neurol.* **2018**, *149*, 27–42. [CrossRef] [PubMed]
8. Abdulhaleem, M.; Ruiz, J.; Cramer, C.; Xing, F.; Lo, H.W.; Su, J.; Chan, M.D. Brain metastasis prognostic nomograms and brain metastasis velocity: A narrative review. *Chin. Clin. Oncol.* **2022**, *11*, 10. [CrossRef] [PubMed]
9. Farris, M.; McTyre, E.R.; Cramer, C.K.; Hughes, R.; Randolph, D.M.; Ayala-Peacock, D.N.; Bourland, J.D.; Ruiz, J.; Watabe, K.; Laxton, A.W.; et al. Brain Metastasis Velocity: A Novel Prognostic Metric Predictive of Overall Survival and Freedom From Whole-Brain Radiation Therapy After Distant Brain Failure Following Upfront Radiosurgery Alone. *Int. J. Radiat. Oncol. Biol. Phys.* **2017**, *98*, 131–141. [CrossRef]
10. Gaspar, L.; Scott, C.; Rotman, M.; Asbell, S.; Phillips, T.; Wasserman, T.; McKenna, W.G.; Byhardt, R. Recursive partitioning analysis (RPA) of prognostic factors in three Radiation Therapy Oncology Group (RTOG) brain metastases trials. *Int. J. Radiat. Oncol. Biol. Phys.* **1997**, *37*, 745–751. [CrossRef]
11. Sperduto, P.W.; Mesko, S.; Li, J.; Cagney, D.; Aizer, A.; Lin, N.U.; Nesbit, E.; Kruser, T.J.; Chan, J.; Braunstein, S.; et al. Survival in Patients With Brain Metastases: Summary Report on the Updated Diagnosis-Specific Graded Prognostic Assessment and Definition of the Eligibility Quotient. *J. Clin. Oncol.* **2020**, *38*, 3773–3784. [CrossRef] [PubMed]
12. Patchell, R.A.; Tibbs, P.A.; Walsh, J.W.; Dempsey, R.J.; Maruyama, Y.; Kryscio, R.J.; Markesbery, W.R.; Macdonald, J.S.; Young, B. A randomized trial of surgery in the treatment of single metastases to the brain. *N. Engl. J. Med.* **1990**, *322*, 494–500. [CrossRef] [PubMed]

13. Kamath, A.A.; Friedman, D.D.; Hacker, C.D.; Smyth, M.D.; Limbrick, D.D., Jr.; Kim, A.H.; Hawasli, A.H.; Leuthardt, E.C. MRI-Guided Interstitial Laser Ablation for Intracranial Lesions: A Large Single-Institution Experience of 133 Cases. *Stereotact. Funct. Neurosurg.* **2017**, *95*, 417–428. [CrossRef] [PubMed]
14. Luther, E.; Mansour, S.; Echeverry, N.; McCarthy, D.; Eichberg, D.G.; Shah, A.; Nada, A.; Berry, K.; Kader, M.; Ivan, M.; et al. Laser Ablation for Cerebral Metastases. *Neurosurg. Clin. N. Am.* **2020**, *31*, 537–547. [CrossRef]
15. Luther, E.; McCarthy, D.; Shah, A.; Semonche, A.; Borowy, V.; Burks, J.; Eichberg, D.G.; Komotar, R.; Ivan, M. Radical Laser Interstitial Thermal Therapy Ablation Volumes Increase Progression-Free Survival in Biopsy-Proven Radiation Necrosis. *World Neurosurg.* **2020**, *136*, e646–e659. [CrossRef]
16. Sujijantarat, N.; Hong, C.S.; Owusu, K.A.; Elsamadicy, A.A.; Antonios, J.P.; Koo, A.B.; Baehring, J.M.; Chiang, V.L. Laser interstitial thermal therapy (LITT) vs. bevacizumab for radiation necrosis in previously irradiated brain metastases. *J. Neurooncol.* **2020**, *148*, 641–649. [CrossRef]
17. Srinivasan, E.S.; Grabowski, M.M.; Nahed, B.V.; Barnett, G.H.; Fecci, P.E. Laser interstitial thermal therapy for brain metastases. *Neurooncol. Adv.* **2021**, *3*, v16–v25. [CrossRef]
18. Hong, C.S.; Deng, D.; Vera, A.; Chiang, V.L. Laser-interstitial thermal therapy compared to craniotomy for treatment of radiation necrosis or recurrent tumor in brain metastases failing radiosurgery. *J. Neurooncol.* **2019**, *142*, 309–317. [CrossRef]
19. Ahluwalia, M.; Barnett, G.H.; Deng, D.; Tatter, S.B.; Laxton, A.W.; Mohammadi, A.M.; Leuthardt, E.; Chamoun, R.; Judy, K.; Asher, A.; et al. Laser ablation after stereotactic radiosurgery: A multicenter prospective study in patients with metastatic brain tumors and radiation necrosis. *J. Neurosurg. JNS* **2019**, *130*, 804–811. [CrossRef]
20. Lievens, Y.; Guckenberger, M.; Gomez, D.; Hoyer, M.; Iyengar, P.; Kindts, I.; Méndez Romero, A.; Nevens, D.; Palma, D.; Park, C.; et al. Defining oligometastatic disease from a radiation oncology perspective: An ESTRO-ASTRO consensus document. *Radiother. Oncol.* **2020**, *148*, 157–166. [CrossRef] [PubMed]
21. Slotman, B.; Faivre-Finn, C.; Kramer, G.; Rankin, E.; Snee, M.; Hatton, M.; Postmus, P.; Collette, L.; Musat, E.; Senan, S. Prophylactic cranial irradiation in extensive small-cell lung cancer. *N. Engl. J. Med.* **2007**, *357*, 664–672. [CrossRef] [PubMed]
22. Manapov, F.; Käsmann, L.; Roengvoraphoj, O.; Dantes, M.; Schmidt-Hegemann, N.S.; Belka, C.; Eze, C. Prophylactic cranial irradiation in small-cell lung cancer: Update on patient selection, efficacy and outcomes. *Lung Cancer* **2018**, *9*, 49–55. [CrossRef] [PubMed]
23. Crockett, C.; Belderbos, J.; Levy, A.; McDonald, F.; Le Péchoux, C.; Faivre-Finn, C. Prophylactic cranial irradiation (PCI), hippocampal avoidance (HA) whole brain radiotherapy (WBRT) and stereotactic radiosurgery (SRS) in small cell lung cancer (SCLC): Where do we stand? *Lung Cancer* **2021**, *162*, 96–105. [CrossRef] [PubMed]
24. Pospisil, P.; Kazda, T.; Hynkova, L.; Bulik, M.; Dobiaskova, M.; Burkon, P.; Laack, N.N.; Slampa, P.; Jancalek, R. Post-WBRT cognitive impairment and hippocampal neuronal depletion measured by in vivo metabolic MR spectroscopy: Results of prospective investigational study. *Radiother. Oncol.* **2017**, *122*, 373–379. [CrossRef] [PubMed]
25. Lehrer, E.J.; Jones, B.M.; Dickstein, D.R.; Green, S.; Germano, I.M.; Palmer, J.D.; Laack, N.; Brown, P.D.; Gondi, V.; Wefel, J.S.; et al. The Cognitive Effects of Radiotherapy for Brain Metastases. *Front. Oncol.* **2022**, *12*, 893264. [CrossRef]
26. Gondi, V.; Pugh, S.L.; Tome, W.A.; Caine, C.; Corn, B.; Kanner, A.; Rowley, H.; Kundapur, V.; DeNittis, A.; Greenspoon, J.N.; et al. Preservation of memory with conformal avoidance of the hippocampal neural stem-cell compartment during whole-brain radiotherapy for brain metastases (RTOG 0933): A phase II multi-institutional trial. *J. Clin. Oncol.* **2014**, *32*, 3810–3816. [CrossRef]
27. Brown, P.D.; Pugh, S.; Laack, N.N.; Wefel, J.S.; Khuntia, D.; Meyers, C.; Choucair, A.; Fox, S.; Suh, J.H.; Roberge, D.; et al. Memantine for the prevention of cognitive dysfunction in patients receiving whole-brain radiotherapy: A randomized, double-blind, placebo-controlled trial. *Neuro Oncol.* **2013**, *15*, 1429–1437. [CrossRef]
28. Brown, P.D.; Gondi, V.; Pugh, S.; Tome, W.A.; Wefel, J.S.; Armstrong, T.S.; Bovi, J.A.; Robinson, C.; Konski, A.; Khuntia, D.; et al. Hippocampal Avoidance During Whole-Brain Radiotherapy Plus Memantine for Patients With Brain Metastases: Phase III Trial NRG Oncology CC001. *J. Clin. Oncol.* **2020**, *38*, 1019–1029. [CrossRef]
29. Shinde, A.; Akhavan, D.; Sedrak, M.; Glaser, S.; Amini, A. Shifting paradigms: Whole brain radiation therapy versus stereotactic radiosurgery for brain metastases. *CNS Oncol.* **2019**, *8*, Cns27. [CrossRef]
30. Li, J.; Ludmir, E.; Wang, Y.; Guha-Thakurta, N.; McAleer, M.; Settle, S.; Yeboa, D.; Ghia, A.; McGovern, S.; Chung, C. Stereotactic radiosurgery versus whole-brain radiation therapy for patients with 4-15 brain metastases: A phase III randomized controlled trial. *Int. J. Radiat. Oncol. Biol. Phys.* **2020**, *108*, S21–S22. [CrossRef]
31. Vogelbaum, M.A.; Brown, P.D.; Messersmith, H.; Brastianos, P.K.; Burri, S.; Cahill, D.; Dunn, I.F.; Gaspar, L.E.; Gatson, N.T.N.; Gondi, V.; et al. Treatment for Brain Metastases: ASCO-SNO-ASTRO Guideline. *J. Clin. Oncol.* **2022**, *40*, 492–516. [CrossRef]
32. Akanda, Z.Z.; Hong, W.; Nahavandi, S.; Haghighi, N.; Phillips, C.; Kok, D.L. Post-operative stereotactic radiosurgery following excision of brain metastases: A systematic review and meta-analysis. *Radiother. Oncol.* **2020**, *142*, 27–35. [CrossRef] [PubMed]
33. Brown, P.D.; Ballman, K.V.; Cerhan, J.H.; Anderson, S.K.; Carrero, X.W.; Whitton, A.C.; Greenspoon, J.; Parney, I.F.; Laack, N.N.I.; Ashman, J.B.; et al. Postoperative stereotactic radiosurgery compared with whole brain radiotherapy for resected metastatic brain disease (NCCTG N107C/CEC·3): A multicentre, randomised, controlled, phase 3 trial. *Lancet Oncol.* **2017**, *18*, 1049–1060. [CrossRef] [PubMed]
34. Kayama, T.; Sato, S.; Sakurada, K.; Mizusawa, J.; Nishikawa, R.; Narita, Y.; Sumi, M.; Miyakita, Y.; Kumabe, T.; Sonoda, Y.; et al. Effects of Surgery With Salvage Stereotactic Radiosurgery Versus Surgery With Whole-Brain Radiation Therapy in Patients With

One to Four Brain Metastases (JCOG0504): A Phase III, Noninferiority, Randomized Controlled Trial. *J. Clin. Oncol.* **2018**, *36*, 3282–3289. [CrossRef]

35. Palmer, J.D.; Perlow, H.K.; Matsui, J.K.; Ho, C.; Prasad, R.N.; Liu, K.; Upadhyay, R.; Klamer, B.; Wang, J.; Damante, M.; et al. Fractionated pre-operative stereotactic radiotherapy for patients with brain metastases: A multi-institutional analysis. *J. Neurooncol.* **2022**, *159*, 389–395. [CrossRef] [PubMed]
36. Perlow, H.K.; Ho, C.; Matsui, J.K.; Prasad, R.N.; Klamer, B.G.; Wang, J.; Damante, M.; Upadhyay, R.; Thomas, E.; Blakaj, D.M.; et al. Comparing pre-operative versus post-operative single and multi-fraction stereotactic radiotherapy for patients with resectable brain metastases. *Clin. Transl. Radiat. Oncol.* **2023**, *38*, 117–122. [CrossRef] [PubMed]
37. Wu, Y.L.; Ahn, M.J.; Garassino, M.C.; Han, J.Y.; Katakami, N.; Kim, H.R.; Hodge, R.; Kaur, P.; Brown, A.P.; Ghiorghiu, D.; et al. CNS Efficacy of Osimertinib in Patients With T790M-Positive Advanced Non-Small-Cell Lung Cancer: Data From a Randomized Phase III Trial (AURA3). *J. Clin. Oncol.* **2018**, *36*, 2702–2709. [CrossRef]
38. Lin, N.U.; Borges, V.; Anders, C.; Murthy, R.K.; Paplomata, E.; Hamilton, E.; Hurvitz, S.; Loi, S.; Okines, A.; Abramson, V.; et al. Intracranial Efficacy and Survival With Tucatinib Plus Trastuzumab and Capecitabine for Previously Treated HER2-Positive Breast Cancer With Brain Metastases in the HER2CLIMB Trial. *J. Clin. Oncol.* **2020**, *38*, 2610–2619. [CrossRef]
39. Saleem, K.; Davar, D. The role of systemic therapy in melanoma brain metastases: A narrative review. *Chin. Clin. Oncol.* **2022**, *11*, 24. [CrossRef]
40. Hirsch, L.; Martinez Chanza, N.; Farah, S.; Xie, W.; Flippot, R.; Braun, D.A.; Rathi, N.; Thouvenin, J.; Collier, K.A.; Seront, E.; et al. Clinical Activity and Safety of Cabozantinib for Brain Metastases in Patients With Renal Cell Carcinoma. *JAMA Oncol.* **2021**, *7*, 1815–1823. [CrossRef]
41. Darvin, P.; Toor, S.M.; Sasidharan Nair, V.; Elkord, E. Immune checkpoint inhibitors: Recent progress and potential biomarkers. *Exp. Mol. Med.* **2018**, *50*, 1–11. [CrossRef] [PubMed]
42. Tawbi, H.A.; Forsyth, P.A.; Hodi, F.S.; Algazi, A.P.; Hamid, O.; Lao, C.D.; Moschos, S.J.; Atkins, M.B.; Lewis, K.; Postow, M.A.; et al. Long-term outcomes of patients with active melanoma brain metastases treated with combination nivolumab plus ipilimumab (CheckMate 204): Final results of an open-label, multicentre, phase 2 study. *Lancet Oncol.* **2021**, *22*, 1692–1704. [CrossRef] [PubMed]
43. Mantovani, C.; Gastino, A.; Cerrato, M.; Badellino, S.; Ricardi, U.; Levis, M. Modern Radiation Therapy for the Management of Brain Metastases From Non-Small Cell Lung Cancer: Current Approaches and Future Directions. *Front. Oncol.* **2021**, *11*, 772789. [CrossRef]
44. Le Rhun, E.; Taillibert, S.; Chamberlain, M.C. Carcinomatous meningitis: Leptomeningeal metastases in solid tumors. *Surg. Neurol. Int.* **2013**, *4*, S265–S288. [CrossRef] [PubMed]
45. Le Rhun, E.; Galanis, E. Leptomeningeal metastases of solid cancer. *Curr. Opin. Neurol.* **2016**, *29*, 797–805. [CrossRef] [PubMed]
46. Ahn, J.H.; Lee, S.H.; Kim, S.; Joo, J.; Yoo, H.; Lee, S.H.; Shin, S.H.; Gwak, H.S. Risk for leptomeningeal seeding after resection for brain metastases: Implication of tumor location with mode of resection. *J. Neurosurg.* **2012**, *116*, 984–993. [CrossRef]
47. Clarke, J.L.; Perez, H.R.; Jacks, L.M.; Panageas, K.S.; Deangelis, L.M. Leptomeningeal metastases in the MRI era. *Neurology* **2010**, *74*, 1449–1454. [CrossRef]
48. Rinehardt, H.; Kassem, M.; Morgan, E.; Palettas, M.; Stephens, J.A.; Suresh, A.; Ganju, A.; Lustberg, M.; Wesolowski, R.; Sardesai, S.; et al. Assessment of Leptomeningeal Carcinomatosis Diagnosis, Management and Outcomes in Patients with Solid Tumors Over a Decade of Experience. *Eur. J. Breast Health* **2021**, *17*, 371–377. [CrossRef]
49. Le Rhun, E.; Preusser, M.; van den Bent, M.; Andratschke, N.; Weller, M. How we treat patients with leptomeningeal metastases. *ESMO Open* **2019**, *4*, e000507. [CrossRef]
50. Lukas, R.V. Leptomeningeal metastases—What outcomes should we measure and how? *Neuro-Oncology* **2022**, *24*, 1736–1737. [CrossRef]
51. Glover, R.L.; Brook, A.L.; Welch, M.R. Teaching NeuroImages: Leptomeningeal lung carcinoma. *Neurology* **2014**, *82*, e183–e184. [CrossRef] [PubMed]
52. Kokkoris, C.P. Leptomeningeal carcinomatosis. How does cancer reach the pia-arachnoid? *Cancer* **1983**, *51*, 154–160. [CrossRef] [PubMed]
53. Boyle, R.; Thomas, M.; Adams, J.H. Diffuse involvement of the leptomeninges by tumour--a clinical and pathological study of 63 cases. *Postgrad. Med. J.* **1980**, *56*, 149–158. [CrossRef] [PubMed]
54. Chamberlain, M.C.; Baik, C.S.; Gadi, V.K.; Bhatia, S.; Chow, L.Q. Systemic therapy of brain metastases: Non-small cell lung cancer, breast cancer, and melanoma. *Neuro Oncol.* **2017**, *19*, i1–i24. [CrossRef]
55. Pardridge, W.M. CSF, blood-brain barrier, and brain drug delivery. *Expert. Opin. Drug. Deliv.* **2016**, *13*, 963–975. [CrossRef]
56. Boire, A.; Zou, Y.; Shieh, J.; Macalinao, D.G.; Pentsova, E.; Massagué, J. Complement Component 3 Adapts the Cerebrospinal Fluid for Leptomeningeal Metastasis. *Cell* **2017**, *168*, 1101–1113.e13. [CrossRef]
57. Karschnia, P.; Le Rhun, E.; Vogelbaum, M.A.; van den Bent, M.; Grau, S.J.; Preusser, M.; Soffietti, R.; von Baumgarten, L.; Westphal, M.; Weller, M.; et al. The evolving role of neurosurgery for central nervous system metastases in the era of personalized cancer therapy. *Eur. J. Cancer* **2021**, *156*, 93–108. [CrossRef]
58. Murakami, Y.; Ichikawa, M.; Bakhit, M.; Jinguji, S.; Sato, T.; Fujii, M.; Sakuma, J.; Saito, K. Palliative shunt surgery for patients with leptomeningeal metastasis. *Clin. Neurol. Neurosurg.* **2018**, *168*, 175–178. [CrossRef]

59. Montes de Oca Delgado, M.; Cacho Díaz, B.; Santos Zambrano, J.; Guerrero Juárez, V.; López Martínez, M.S.; Castro Martínez, E.; Avendaño Méndez-Padilla, J.; Mejía Pérez, S.; Reyes Moreno, I.; Gutiérrez Aceves, A.; et al. The Comparative Treatment of Intraventricular Chemotherapy by Ommaya Reservoir vs. Lumbar Puncture in Patients With Leptomeningeal Carcinomatosis. *Front. Oncol.* **2018**, *8*, 509. [CrossRef]
60. Tian, X.; Liu, K.; Hou, Y.; Cheng, J.; Zhang, J. The evolution of proton beam therapy: Current and future status. *Mol. Clin. Oncol.* **2018**, *8*, 15–21. [CrossRef]
61. Yang, T.J.; Wijetunga, N.A.; Yamada, J.; Wolden, S.; Mehallow, M.; Goldman, D.A.; Zhang, Z.; Young, R.J.; Kris, M.G.; Yu, H.A.; et al. Clinical trial of proton craniospinal irradiation for leptomeningeal metastases. *Neuro Oncol.* **2021**, *23*, 134–143. [CrossRef] [PubMed]
62. Yang, J.T.; Wijetunga, N.A.; Pentsova, E.; Wolden, S.; Young, R.J.; Correa, D.; Zhang, Z.; Zheng, J.; Steckler, A.; Bucwinska, W.; et al. Randomized Phase II Trial of Proton Craniospinal Irradiation Versus Photon Involved-Field Radiotherapy for Patients With Solid Tumor Leptomeningeal Metastasis. *J. Clin. Oncol.* **2022**, *40*, 3858–3867. [CrossRef]
63. Glantz, M.J.; Cole, B.F.; Recht, L.; Akerley, W.; Mills, P.; Saris, S.; Hochberg, F.; Calabresi, P.; Egorin, M.J. High-dose intravenous methotrexate for patients with nonleukemic leptomeningeal cancer: Is intrathecal chemotherapy necessary? *J. Clin. Oncol.* **1998**, *16*, 1561–1567. [CrossRef] [PubMed]
64. Kapke, J.T.; Schneidewend, R.J.; Jawa, Z.A.; Huang, C.-C.; Connelly, J.M.; Chitambar, C.R. High-dose intravenous methotrexate in the management of breast cancer with leptomeningeal disease: Case series and review of the literature. *Hematol./Oncol. Stem Cell. Ther.* **2019**, *12*, 189–193. [CrossRef] [PubMed]
65. Kumthekar, P.; Tang, S.C.; Brenner, A.J.; Kesari, S.; Piccioni, D.E.; Anders, C.; Carrillo, J.; Chalasani, P.; Kabos, P.; Puhalla, S.; et al. ANG1005, a Brain-Penetrating Peptide-Drug Conjugate, Shows Activity in Patients with Breast Cancer with Leptomeningeal Carcinomatosis and Recurrent Brain Metastases. *Clin. Cancer Res.* **2020**, *26*, 2789–2799. [CrossRef]
66. Prakadan, S.M.; Alvarez-Breckenridge, C.A.; Markson, S.C.; Kim, A.E.; Klein, R.H.; Nayyar, N.; Navia, A.W.; Kuter, B.M.; Kolb, K.E.; Bihun, I.; et al. Genomic and transcriptomic correlates of immunotherapy response within the tumor microenvironment of leptomeningeal metastases. *Nat. Commun.* **2021**, *12*, 5955. [CrossRef]
67. Cohen, S.P.; Dragovich, A. Intrathecal analgesia. *Anesthesiol. Clin.* **2007**, *25*, 863–882, viii. [CrossRef]
68. Palmisciano, P.; Watanabe, G.; Conching, A.; Ogasawara, C.; Vojnic, M.; D'Amico, R.S. Intrathecal therapy for the management of leptomeningeal metastatic disease: A scoping review of the current literature and ongoing clinical trials. *J. Neurooncol.* **2022**, *160*, 79–100. [CrossRef]
69. Khang, M.; Bindra, R.S.; Mark Saltzman, W. Intrathecal delivery and its applications in leptomeningeal disease. *Adv. Drug. Deliv. Rev.* **2022**, *186*, 114338. [CrossRef]
70. Beauchesne, P. Intrathecal chemotherapy for treatment of leptomeningeal dissemination of metastatic tumours. *Lancet Oncol.* **2010**, *11*, 871–879. [CrossRef]
71. Pan, Z.; Yang, G.; He, H.; Cui, J.; Li, W.; Yuan, T.; Chen, K.; Jiang, T.; Gao, P.; Sun, Y.; et al. Intrathecal pemetrexed combined with involved-field radiotherapy as a first-line intra-CSF therapy for leptomeningeal metastases from solid tumors: A phase I/II study. *Ther. Adv. Med. Oncol.* **2020**, *12*, 1758835920937953. [CrossRef] [PubMed]
72. Jaeckle, K.A.; Dixon, J.G.; Anderson, S.K.; Moreno-Aspitia, A.; Colon-Otero, G.; Hebenstreit, K.; Patel, T.A.; Reddy, S.L.; Perez, E.A. Intra-CSF topotecan in treatment of breast cancer patients with leptomeningeal metastases. *Cancer Med.* **2020**, *9*, 7935–7942. [CrossRef] [PubMed]
73. Chamberlain, M.C. Neurotoxicity of intra-CSF liposomal cytarabine (DepoCyt) administered for the treatment of leptomeningeal metastases: A retrospective case series. *J. Neuro-Oncol.* **2012**, *109*, 143–148. [CrossRef]
74. Kumthekar, P.U.; Avram, M.J.; Lassman, A.B.; Lin, N.U.; Lee, E.; Grimm, S.A.; Schwartz, M.; Bell Burdett, K.L.; Lukas, R.V.; Dixit, K.; et al. A Phase I/II Study of Intrathecal Trastuzumab in HER-2 Positive Cancer with Leptomeningeal Metastases: Safety, Efficacy, and Cerebrospinal Fluid Pharmacokinetics. *Neuro Oncol.* **2022**, *25*, 557–565. [CrossRef] [PubMed]
75. Oberkampf, F.; Gutierrez, M.; Trabelsi Grati, O.; Le Rhun, É.; Trédan, O.; Turbiez, I.; Kadi, A.; Dubot, C.; Taillibert, S.; Vacher, S.; et al. Phase II study of intrathecal administration of trastuzumab in patients with HER2-positive breast cancer with leptomeningeal metastasis. *Neuro Oncol.* **2022**, *25*, 365–374. [CrossRef]
76. Gaudino, S.; Giordano, C.; Magnani, F.; Cottonaro, S.; Infante, A.; Sabatino, G.; La Rocca, G.; Della Pepa, G.M.; D'Alessandris, Q.G.; Pallini, R.; et al. Neuro-Oncology Multidisciplinary Tumor Board: The Point of View of the Neuroradiologist. *J. Pers. Med.* **2022**, *12*, 135. [CrossRef]
77. El Saghir, N.S.; Charara, R.N.; Kreidieh, F.Y.; Eaton, V.; Litvin, K.; Farhat, R.A.; Khoury, K.E.; Breidy, J.; Tamim, H.; Eid, T.A. Global Practice and Efficiency of Multidisciplinary Tumor Boards: Results of an American Society of Clinical Oncology International Survey. *J. Glob. Oncol.* **2015**, *1*, 57–64. [CrossRef]
78. Lamb, B.W.; Green, J.S.; Benn, J.; Brown, K.F.; Vincent, C.A.; Sevdalis, N. Improving decision making in multidisciplinary tumor boards: Prospective longitudinal evaluation of a multicomponent intervention for 1421 patients. *J. Am. Coll. Surg.* **2013**, *217*, 412–420. [CrossRef]
79. Liu, J.C.; Kaplon, A.; Blackman, E.; Miyamoto, C.; Savior, D.; Ragin, C. The impact of the multidisciplinary tumor board on head and neck cancer outcomes. *Laryngoscope* **2020**, *130*, 946–950. [CrossRef]

80. Quero, G.; Salvatore, L.; Fiorillo, C.; Bagalà, C.; Menghi, R.; Maria, B.; Cina, C.; Laterza, V.; Di Stefano, B.; Maratta, M.G.; et al. The impact of the multidisciplinary tumor board (MDTB) on the management of pancreatic diseases in a tertiary referral center. *ESMO Open.* **2021**, *6*, 100010. [CrossRef]
81. Schäfer, N.; Bumes, E.; Eberle, F.; Fox, V.; Gessler, F.; Giordano, F.A.; Konczalla, J.; Onken, J.; Ottenhausen, M.; Scherer, M.; et al. Implementation, relevance, and virtual adaptation of neuro-oncological tumor boards during the COVID-19 pandemic: A nationwide provider survey. *J. Neurooncol.* **2021**, *153*, 479–485. [CrossRef] [PubMed]
82. Wasilewski, A.; Mohile, N. Tele-neuro-oncology: Current Practices and Future Directions. *Curr. Oncol. Rep.* **2022**, *24*, 99–103. [CrossRef]
83. Strowd, R.E.; Dunbar, E.M.; Gan, H.K.; Kurz, S.; Jordan, J.T.; Mandel, J.J.; Mohile, N.A.; Nevel, K.S.; Taylor, J.W.; Ullrich, N.J.; et al. Practical guidance for telemedicine use in neuro-oncology. *Neurooncol. Pract.* **2022**, *9*, 91–104. [CrossRef]
84. Liu, J.K.C.; Kang, R.; Bilenkin, A.; Prorok, R.; Whiting, J.; Patel, K.B.; Beer-Furlan, A.; Naso, C.; Rogers, A.; Castro, X.B.; et al. Patient satisfaction and cost savings analysis of the telemedicine program within a neuro-oncology department. *J. Neurooncol.* **2022**, *160*, 517–525. [CrossRef] [PubMed]
85. Fonkem, E.; Gatson, N.T.N.; Tadipatri, R.; Cole, S.; Azadi, A.; Sanchez, M.; Stefanowicz, E. Telemedicine review in neuro-oncology: Comparative experiential analysis for Barrow Neurological Institute and Geisinger Health during the 2020 COVID-19 pandemic. *Neurooncol. Pract.* **2021**, *8*, 109–116. [CrossRef] [PubMed]
86. Ekhator, C.; Kesari, S.; Tadipatri, R.; Fonkem, E.; Grewal, J. The Emergence of Virtual Tumor Boards in Neuro-Oncology: Opportunities and Challenges. *Cureus* **2022**, *14*, e25682. [CrossRef] [PubMed]
87. Horbinski, C.; Ligon, K.L.; Brastianos, P.; Huse, J.T.; Venere, M.; Chang, S.; Buckner, J.; Cloughesy, T.; Jenkins, R.B.; Giannini, C.; et al. The medical necessity of advanced molecular testing in the diagnosis and treatment of brain tumor patients. *Neuro Oncol.* **2019**, *21*, 1498–1508. [CrossRef] [PubMed]
88. Farago, A.F.; Azzoli, C.G. Beyond ALK and ROS1: RET, NTRK, EGFR and BRAF gene rearrangements in non-small cell lung cancer. *Transl. Lung Cancer Res.* **2017**, *6*, 550–559. [CrossRef] [PubMed]
89. Brastianos, P.K.; Carter, S.L.; Santagata, S.; Cahill, D.P.; Taylor-Weiner, A.; Jones, R.T.; Van Allen, E.M.; Lawrence, M.S.; Horowitz, P.M.; Cibulskis, K.; et al. Genomic Characterization of Brain Metastases Reveals Branched Evolution and Potential Therapeutic Targets. *Cancer Discov.* **2015**, *5*, 1164–1177. [CrossRef] [PubMed]

Journal of
Clinical Medicine

MDPI

Article

Overcoming Barriers to Tobacco Cessation and Lung Cancer Screening among Racial and Ethnic Minority Groups and Underserved Patients in Academic Centers and Community Network Sites: The City of Hope Experience

Cary A. Presant *, Kimlin Ashing, Dan Raz, Sophia Yeung, Brenda Gascon, Alexis Stewart, Jonjon Macalintal, Argelia Sandoval, Loretta Ehrunmwunsee, Tanyanika Phillips, Ravi Salgia, Amartej Merla, Shanmuga Subbiah, Michelle El-Hajjouie, Jeffrey Staley, Heather Graves, Ranjan Pathak, Shaira Dingal, Sagus Sampath, Beverly Laksana, Thomas Joseph, Tricia Eugenio, Veronica Degoma, Kathleen Burns, Sarah Phillips, Tingting Tan, Kelly Tarkshian, Virginia Sun, Arya Amini, Khristie Davy, Janet Cronkhite, Mary Cianfrocca, Susan Brown, Yuman Fong and Steven Rosen

City of Hope Medical Center, 1500 East Duarte Rd, Duarte, CA 91010, USA
* Correspondence: cpresant@coh.org; Tel.: +1-818-406-1344

Abstract: Background: Tobacco control is important for cancer patient health, but delivering effective low-dose CT (LDCT) screening and tobacco cessation is more difficult in underserved and patients from racial and ethnic minority groups. At City of Hope (COH), we have developed strategies to overcome barriers to the delivery of LDCT and tobacco cessation. Methods: We performed a needs assessment. New tobacco control program services were implemented focusing on patients from racial and ethnic minority groups. Innovations included Whole Person Care with motivational counseling, placing clinician and nurse champions at points of care, training module and leadership newsletters, and a patient-centric personalized medicine Personalized Pathways to Success (PPS) program. Results: Emphasis on patients from racial and ethnic minority groups was implemented by training cessation personnel and lung cancer control champions. LDCT increased. Tobacco use assessment increased and abstinence was 27.2%. The PPS pilot program achieved 47% engagement in cessation, with self-reported abstinence at 3 months of 38%, with both results slightly higher in patients from racial and ethnic minority groups than in Caucasian patients. Conclusions: Tobacco cessation barrier-focused innovations can result in increased lung cancer screening and tobacco cessation reach and effectiveness, especially among patients from racial and ethnic minority groups. The PPS program is promising as a personalized medicine patient-centric approach to cessation and lung cancer screening.

Keywords: tobacco control; smoking cessation; cancer prevention; lung cancer screening; low dose Ct scans; LDCT; cancer disparities; minority health; personalized medicine; pathways to success

Citation: Presant, C.A.; Ashing, K.; Raz, D.; Yeung, S.; Gascon, B.; Stewart, A.; Macalintal, J.; Sandoval, A.; Ehrunmwunsee, L.; Phillips, T.; et al. Overcoming Barriers to Tobacco Cessation and Lung Cancer Screening among Racial and Ethnic Minority Groups and Underserved Patients in Academic Centers and Community Network Sites: The City of Hope Experience. *J. Clin. Med.* **2023**, *12*, 1275. https://doi.org/10.3390/jcm12041275

Academic Editors: David Barnes and Luca Bertolaccini

Received: 21 December 2022
Revised: 18 January 2023
Accepted: 1 February 2023
Published: 6 February 2023

1. Introduction

Tobacco control is important for cancer patient health. Tobacco is the primary or contributing cause of 30% of cancers in the United States [1] and 80–90% of lung cancers are caused by or associated with tobacco use. Diagnosing lung cancer early can increase the 5-year cancer cure rate to 90.8% [2], emphasizing the benefit of effective implementation of low-dose CT (LDCT) for lung cancer screening (LCS). Since 5-year survival rates in lung cancer are 26% higher in patients undergoing tobacco cessation compared to patients who continue to use tobacco [3], implementing tobacco cessation programs is important as a fourth pillar of cancer care [4].

Delivering effective LCS and tobacco cessation is more difficult in underserved populations of patients from racial and ethnic minority groups [5,6]. This is likely due to poor access to care, insurance, language barriers, lack of trust, cultural beliefs or prejudices, and/or a lack of education about the benefits of tobacco cessation. At City of Hope (COH), we have developed strategies to overcome barriers to the delivery of LDCT and tobacco cessation across our network of academic and community sites. This communication summarizes our experience and recommendations.

2. Methods

We performed a needs assessment by conducting 193 interviews and surveys with clinicians, patients, tobacco treatment specialists, nurses, and administrators, in order to determine the barriers to tobacco control delivery. New tobacco control program services were implemented [7].

After the identification of a patient with current tobacco use, physicians were prompted to refer the patient to tobacco cessation. A multilingual tobacco treatment specialist (TTS) conducted a culturally sensitive motivational interview, gave educational materials, and referred the consenting patient for a tobacco cessation consultation. Patients with a qualifying tobacco use history were referred by the physicians themselves or with TTS prompting to LDCT screening. The use of multilingual support overcame a significant barrier to LCS in patients from racial and ethnic minority groups. Physicians considered patient life expectancy and willingness to have screening and potentially curative therapy before referral to LDCT screening.

We reviewed the tobacco use assessments and engagement with the tobacco control program by race and ethnicity across the City of Hope southern California treatment sites in 2021 (1 academic center and 40 community centers). We analyzed LCS rates and tobacco cessation referral rates, and cessation effectiveness after program implementation.

The COH tobacco control program consisted of quality improvement projects. This was submitted to the COH investigational review board, which concluded the program was deemed non-human-subject research. Therefore, no patient informed consent was required (IRB number 19201).

These new services were implemented in 2019 and continued until the present. Observational evaluations began in 2019 pre- and post-implementation.

3. Results

3.1. The City of Hope Catchment Area

In 2020, 137,125 new patients were seen in the entire enterprise of 1 academic center and 40 community sites in southern California (this does not include any patients seen in Cancer Treatment Centers of America in Chicago, Phoenix, and Atlanta, which were acquired by City of Hope in 2022, and does not include any new patients seen in the second academic center Lennar Hospital in Irvine, CA, USA, which opened in 2022). The details of these patients are presented in Table 1. In community centers, there were more patients with fewer Caucasians, Hispanics, and Asian/Pacific Islanders, and more Black, mixed, and other patients from racial and ethnic minority groups, mostly middle eastern and eastern European/Armenian. Across the entire enterprise, there were 51.4% of minority patients (39.8% in Duarte academic center and 59.5% in the community centers).

3.2. Barriers to Tobacco Cessation and City of Hope Solutions

Among patients, there was frequent refusals to participate in screening and/or cessation programs, especially among patients from racial and ethnic minority groups and patients of low socioeconomic status (SES). Stress in general or due to cancer therapies, anxiety and depression, work problems, and family problems contributed to patient reluctance. A lack of knowledge about tobacco-use-associated reduced cancer outcomes, increased side effects of cancer treatments, the high cost of cessation medications not covered by insurance, non-English speaking, and cultural obstacles all reduced patient agreement to initiate our cessation program, particularly in patients from racial and ethnic minority groups and low SES cancer patients. The cultural context was an important barrier for cessation staff to evaluate. For example, counseling was often stigmatized in Chinese patients because of the misinterpretation of cessation counseling as a mental health problem. Because our network of community clinical sites and our catchment area was so vast in southern California, travel to the academic center in Duarte was not convenient or possible for many tobacco users. Easy access to tobacco products particularly in low-SES neighborhoods and the use of tobacco products by family and friends served as ubiquitous triggers to continue tobacco use [8]. LDCT screening sites were more limited in low-SES areas.

Table 1. Characteristics of City of Hope Patients 2020.

	Duarte	Community Centers
Number of New Patients	51,099	86,026
% Tobacco Users	4.5%	7.4%
Race and Ethnicity		
Caucasian	60.2%	40.5%
Hispanic	21.9%	16.2%
Asian, Pacific Islander	8.2%	5.4%
Black, African American	8.8%	29.7%
Mixed, Other	0.9%	8.2%

In response to these observations in our needs assessment, City of Hope used implementation science to introduce solutions to improve tobacco-use screening and tobacco cessation [7], as outlined in Table 2.

Physician barriers were a lack of education about tobacco cessation, lack of time, and lack of reimbursement. These were addressed by educational training modules, newsletters from leadership, and improvements in the electronic medical record system.

Institutional barriers were the lack of leadership and personnel. These were addressed by hiring personnel (TTS full-time individuals), commitments from COH enterprise leaders, leadership newsletters published monthly, and naming physician and nurse champions (leaders) in academic center clinics and in network clinical sites.

To overcome patient barriers, the first key COH innovation was to include a Whole Person Care approach including training of tobacco treatment specialists to conduct motivational counseling of each patient, providing introductory tobacco use educational videos and brochures to patients even before asking for participation in individual cessation consultations, and TTS assistance to physicians to prompt LDCT screening of eligible patients.

A second COH innovation was to motivate clinicians to refer every patient with current tobacco use to the cessation program by leadership sending out a monthly newsletter (the Moonshot Shoutout) to every staff member reminding them about the importance of referral to cessation and ordering LDCT for eligible patients. This included training modules for physicians, nurses, and tobacco treatment specialists.

A third COH innovation was taking the promotion of cessation and screening to the patient point of care by training and deploying both clinician and nurse champions in selected clinics in the academic center and selected community centers [9]. A fourth COH innovation was providing patients with a choice of 36 distinct services from which a patient could select which services they initially wanted to use to begin their own personalized tobacco cessation and screening journey, which we named the Personalized Pathway to Success (PPS) program. This innovation was patient-centric and encouraged the continued engagement of the patient with the tobacco cessation specialist. A list of the services offered to patients in the PPS is seen in Table 3.

Table 2. Barriers and City of Hope Solutions for Tobacco Cessation and LDCT Screening.

Barrier	City of Hope Solution
Initial patient refusal to consider cessation	Motivational interviewing by TTS TTS Offers PPS Involve champion leader or attending oncologist in Duarte clinic or community center
Lack of knowledge about the risks of tobacco use and benefits of screening and cessation	Educational videos and brochures
Lack of LDCT access in low SES areas	Transportation to Duarte LDCT Collaborate with local hospitals to provide LDCT City of Hope builds LDCT services in community sites
Tobacco use triggers are prevalent with family, friends, and community smoke shops	Consultation and counseling about triggers Free Tobacco use support groups for patients, family, and friends Referrals to psychological counseling Collaborate with local city councils about reducing access to Tobacco use
Cultural barriers and lack of trust in Caucasian cessation providers	Staff program with TTS and clinicians representing patients from racial and ethnic minority groups (Hispanic, Chinese, Filipino, Black, Middle Eastern/Armenian) Training staff in cultural and societal values in patients from racial and ethnic minority groups
Non-English-speaking patients	Staff with bilingual TTS Online or in-person translation services
High cost of cessation medications	Pharmacists assist in finding affordable medications Availability of charity funds
Inconvenient or lack of travel to face-to-face consultations or LDCT screening	Telehealth services for consultations and counseling Transportation services by local and community service providers
Lack of referral by clinicians	Newsletter from leadership Opt out, opt in, and Best Practice Advisory referrals in electronic medical records
Lack of reimbursement for clinical cessation services	Train clinicians in using CPT codes 99406 and 99407 Provide "smart-phrase" to assist documentation of cessation services Up-code E/M code selection appropriately if cessation services provided For managed care, refer to the contracted in-network cessation provider

Abbreviations: TTS: Tobacco treatment specialist; LDCT: Low-dose computerized tomography; PPS: Personalized pathway for success; CPT: current procedural terminology.

Specific tobacco control program and PPS program characteristics were introduced to target patients from racial and ethnic minority groups and/or underserved patients. Educational resources were multilingual, and the staff was diverse and representative of the patients from racial and ethnic minority groups to be emphasized. Monthly meetings with champions emphasized feedback about barriers encountered in screening and cessation. The use of video counseling allowed the inclusion of patients from racial and ethnic minority groups who were reluctant to proceed with face-to-face counseling. Offering family and friend support through the free tobacco-use support groups increased the trust of the patients in the program.

Table 3. Menu of Personalized Pathway to Success (PPS) Services.

Introductory Services
Motivational interview
"Tobacco Cessation as recovery enhancement" video
Tobacco risks and cessation benefits brochure
Essential Elements
Consultation with Tobacco Cessation provider
Nicotine replacement therapy (NRT) prescription
Varenicline or Bupropion eligibility evaluation and prescription
Combination NRT and oral cessation medication prescription and
Initial Tobacco Use Assessment for type of tobacco and doses
City of Hope (COH) tobacco use support group intake form and registration
Individual counseling and support by the Tobacco Treatment Specialist
Rapid action plan for relapse or slips
Educational video, brochure and counseling on preventing tobacco use
Follow up Tobacco Use Assessments
Support Resources
Identify and recruit a "buddy" support friend or a participant peer from
Identify and recruit an "angel" support family member
Offer non-COH support groups: Nicotine Anonymous, Celebrate
Phone support: Kick It CA Quitline
Phone Apps: SmokeFree, QuitGuide, QuitStart
Text Support: SmokeFree Text, Kick It CA text, "DITCHJUUL"
Online live chat: kickitca.org, cancer.gov
Web Resources: smokefree.com, becomeanex.org, trughinitiative.org,
Commitment Adherence Tools
Commitment Agreement
Quit Plan: https://smokefree.gov/build-your-quit-plan (accessed 31 January 2023)
Daily Coping Strategies
5Ds: Distraction, Delaying, Drinking Water, Deep Breathing, and
7 self care behaviors: health coping, healthy eating, regular exercise,
Cinnamon stick/bubble blowing
COH Relaxation Video with Guided Imagery:
COH tobacco Cessation Hypnosis Video: Hypnosis for Smoking Cessation-You Tube (accessed 31 January 2023)
Audiobooks accessible through *Overdrive* and *Audiobooks* apps
Stress management apps: *Headspace* and *Insight Timer*
Spiritual care and support
Educational Tools
Educational handouts
Kick It https://www.youtube.com/c/KickItCa/videos (accessed 31 January 2023)
The CDC E-Cig cessation factsheet
Modular videos
NCCN information sheets

Abbreviations: CDC: Center for Disease Control; E-Cig: electronic cigarettes; NCCN: National Comprehensive Center Network.

3.3. Teachable Moments for Referral to Tobacco Cessation and Lung Cancer Screening

Patients may be more agreeable to accepting referral to screening and tobacco cessation if they are approached at teachable moments. These include the visit at which the diagnosis of cancer is given to the patient; the visit at which the treatment plan is discussed with the patient (cessation is the fourth pillar of cancer care and should be in every treatment plan of a currently tobacco-using patient); a visit at which progression of the cancer is discussed with the patient; a preoperative visit for planning surgery; a visit for discussing cellular therapy with stem cell transplant or CAR-T cell therapy; and/or a visit for pulmonary consultation, pulmonary function testing, or respiratory therapy.

Meanwhile, tobacco cessation is a teachable moment for LCS. LCS is severely underutilized due to a variety of patient, provider, and system barriers [10,11]. The integration of LCS into tobacco cessation workflows is an important strategy for overcoming some of these barriers, including the identification of eligible participants and the lack of knowledge about LCS among people who use tobacco. At our institution, a comprehensive cancer center without affiliated primary care clinics, tobacco cessation services, and LCS are integrated, and the majority of patients screened for lung cancer are identified through referral for tobacco cessation [12]. We have trained tobacco cessation counselors to provide education about LCS, and there is one nurse practitioner and program coordinator for both programs. We have found that training tobacco cessation staff to provide LCS education is feasible and that tobacco cessation staff embrace providing LCS education as part of their role [13].

3.4. Results of Screening and Cessation Services

We have utilized several strategies to improve the utilization of LCS in our catchment areas, particularly in underserved communities with patients from racial and ethnic minority groups, rural populations, and non-English speakers. First, we have expanded our ability to provide LDCT to eligible patients by opening LCS programs at various community sites. Since the start of the COVID-19 pandemic, a face-to-face encounter for shared-decision making is no longer required, and telehealth visits have been expanded. This has allowed us to expand our screening program using a centralized staffing model from one site in 2019 to six sites throughout Los Angeles County and Orange County by 2023. Expanding the reach of our screening program helps to provide quality LCS to patients closer to their homes. We have also performed educational outreach to primary care providers in these areas, as well as in partnership with Federally Qualified Health Clinics (FQHCs) with large numbers of patients from racial and ethnic minority groups, resulting in improved utilization of LCS [14].

Educational outreach directed to patients about LCS, including community health fairs and educational material translated into Spanish, Chinese, Vietnamese, Korean, and Armenian languages has also been an important strategy.

Finally, we recently started an Early Lung Cancer Navigation program in the Antelope Valley, a community with the highest rates of lung cancer mortality and tobacco use in the Los Angeles area as well as large populations of African Americans, Latinos, and rural populations, to improve lung cancer screening and expeditious treatment of lung cancer. This program is led by a Community Navigator who has lived and worked in the Antelope Valley community, and the program is guided by a Community Advisory Board comprised of clinicians, lung cancer patients, and community stakeholders. The navigator performs a needs assessment of patients in our LCS program, patients with imaging findings that are suspicious for lung cancer, and patients with newly diagnosed lung cancer, and then helps guide the patient so that these potential barriers to care can be overcome. The program accesses funds for transportation and even internet-enabled tablets to allow patients easier access to see and communicate with their providers. In 2021, our program screened 153 new patients, which included 59% non-Hispanic Whites, 18% Hispanics, 12% Asians, 4% Blacks, and 7% other or declined to answer.

Tobacco cessation services have increased as a result of our expanded tobacco control initiatives after the needs assessment. Before the COH quality improvement project was implemented, documentation of Tobacco use was only 80.8%. Referrals of patients to tobacco cessation counselors were low at 1.4% of all patients. After the implementation of the project, the assessment of tobacco use increased to 96.6%. Referrals of patients increased to 6.2%.

Counseling of referred patients for consultation for cessation by a nurse practitioner increased to 98% in the Duarte academic center. At a 6-month follow-up of those patients, self-reported abstinence was 27.2%.

This was compared to the experience in the Antelope Valley community site (which had the highest tobacco use in the COH enterprise). Counseling by a TTS or consultation with a nurse practitioner increased to 83.2%. At a 6-month follow-up, self-reported abstinence was 22.5%.

The reach of this program was evaluated as part of the Cancer Center Cessation Initiative (C3I). Engagement of patients who used tobacco with at least one element of our tobacco treatment program was 93.0% of patients in the Duarte academic center and 59.6% of patients in our community site Antelope Valley with the highest smoking rate. The reach of participants measured by active participation in the tobacco treatment program was different for patients from minority groups 45.3%, compared to 66.6% of Caucasian patients.

LCS services have also increased as a result of our expanded tobacco control activities. Compared to the pre-expansion year of 2018, total LDCT referrals increased by 24.4%. Importantly, while referrals of Caucasian patients increased by 11%, referrals of patients from racial and ethnic minority groups increased by 59%.

To extend these previous results, we piloted the PPS program with tobacco treatment specialists and champions in the preoperative anesthesia testing clinic. Fifty-four patients were evaluated in the PPS project. We observed a 47% engagement of patients to initiate cessation. This was higher than the enterprise-wide historical engagement rate of counseling of only 6.2% before the cessation program innovations described above. The engagement rate of patients from racial and ethnic minority groups was slightly higher than that of Caucasian patients (not statistically significant). Self-reported abstinence from tobacco use (at 3 months after engaging in cessation) was achieved by 38% of patients and was slightly higher in patients from racial and ethnic minority groups than in Caucasian patients.

4. Discussion

Tobacco use is a major cause of human infirmity and reduced quality of life. Tobacco control is a major public health goal [15]. National guidelines for tobacco cessation [16] and LDCT screening [17] are widely accepted, but the effectiveness of tobacco cessation and LCS utilization remains low [18,19].

The population of patients from racial and ethnic minority groups in City of Hope is a large proportion of the enterprise's cancer patients (51.4%), and the communication of methods and results of our interventions may be helpful to other institutions. Our focus has been on implementing personalized and patient-centric services, which have been actively supported by senior leadership and well accepted by clinical staff. Our program has been dependent on high staff resources and long-term support from the institution. Efforts must be continued to utilize methods that are likely to ensure the sustainability of the program [20] and attempt to minimize staffing needs [21,22].

It is remarkable that the pilot study results of the PPS cessation program achieved similar results in patients from racial and ethnic minority groups and Caucasian patients. If confirmed in our future studies, these preliminary results suggest that specific innovative strategies can possibly overcome barriers to tobacco cessation in patients from racial and ethnic minority groups, but broader implementation is needed to be certain of the success of these innovations. The observation that patients from racial and ethnic minority groups have higher tobacco usage does not necessarily imply that tobacco cessation is always more difficult, as our results with the PPS program suggest.

This experience was a quality improvement project. As such, it was observational and not a randomized research study. Thus, the methods used were based on combining separate principles that had been developed using implementation science at COH and other institutions. Our project showed that using these principles with trained multilingual support personnel can result in valuable benefits for both patients from racial and ethnic minority groups and Caucasian patients. The PPS pilot study is extending these principles in a novel fashion to academic and community sites, focusing on teachable moments in the cancer patient experience.

Limitations of this communication include that this was performed in only one institutional enterprise limited to southern California and should be extended to other institutions and locations. The PPS program was a pilot, and an extension to other clinics and centers is in progress. The number of patients in the pilot PPS program is low and larger numbers will be required to determine specific services that are most effectively utilized by specific patients from racial and ethnic minority groups.

5. Conclusions

Using innovative strategies and implementation science, lung cancer screening, and tobacco cessation programs can improve outcomes and may reduce disparities in patients from racial and ethnic minority groups.

Author Contributions: Conceptualization: C.A.P., K.A., D.R., R.S., S.S. (Shanmuga Subbiah), J.M. and S.R.; methodology: C.A.P., K.A., D.R., R.S., A.M., V.S. and M.C.; investigation: C.A.P., K.A., D.R., S.Y., B.G., A.S. (Alexis Stewart), J.M., A.S. (Argelia Sandoval), L.E., T.P., A.M., S.S. (Shanmuga Subbiah), M.E.-H., J.S., H.G., R.P., S.D., S.S. (Sagus Sampath), B.L., T.J., T.E., V.D., K.B., S.P., T.T., K.T., A.A., K.D. and J.C.; data evaluation: C.A.P., K.A., D.R., B.G., A.S. (Alexis Stewart), J.M. and A.S. (Argelia Sandoval); writing—initial draft: C.A.P.; writing—review and editing: all authors; supervision: C.A.P., K.A. and D.R.; project administration: C.A.P., K.A., D.R., R.S., S.Y., S.B., Y.F. and S.R. All authors have read and agreed to the published version of the manuscript.

Funding: This work was funded in part by NIH Grants P30 CA033572 and P30 CA033572-37S5.

Institutional Review Board Statement: The COH tobacco control program consisted of quality improvement projects. This was submitted to the COH investigational review board, which concluded the program was deemed non-human-subject research. Therefore, no patient informed consent was required (IRB number 19201).

Informed Consent Statement: Informed consent for treatment was obtained from all patients involved in the project. The study was a Quality Improvement project.

Data Availability Statement: The data are available for review at City of Hope Medical Center, Department of Population Science.

Conflicts of Interest: The authors declare no conflict of interest.

References

1. Lortet-Tieulent, J.; Sauer, A.G.; Siegel, R.L.; Miller, K.D.; Islami, F.; Fedewa, S.A.; Jacobs, E.J.; Jemal, A. State-Level Cancer Mortality Attributable to Cigarette Smoking in the United States. *JAMA Intern. Med.* **2016**, *176*, 1792–1798. [CrossRef]
2. Yang, C.-Y.; Lin, Y.-T.; Lin, L.-J.; Chang, Y.-H.; Chen, H.-Y.; Wang, Y.-P.; Shih, J.-Y.; Yu, C.-J.; Yang, P.-C. Stage Shift Improves Lung Cancer Survival: Real-World Evidence. *J. Thorac. Oncol.* **2022**, *18*, 47–56. [CrossRef]
3. Caini, S.; Del Riccio, M.; Vettori, V.; Scotti, V.; Martinoli, C.; Raimondi, S.; Cammarata, G.; Palli, D.; Banini, M.; Masala, G.; et al. Quitting Smoking At or Around Diagnosis Improves the Overall Survival of Lung Cancer Patients: A Systematic Review and Meta-Analysis. *J. Thorac. Oncol.* **2022**, *17*, 623–636. [CrossRef] [PubMed]
4. Fiore, M.C.; D'Angelo, H.; Baker, T. Effective Cessation Treatment for Patients With Cancer Who Smoke—The Fourth Pillar of Cancer Care. *JAMA Netw. Open* **2019**, *2*, e1912264. [CrossRef] [PubMed]
5. Sheppard, R.S.; Beale, S.; Joseph, J.; Achi, S.S.; Showunmi, A.; Ayinla, R.; Ahluwalia, M. Potential barriers to lung cancer screening in a minority population: Assessing attitudes, beliefs, and values towards screening with low-dose computed tomography amongst a high-risk minority population. *J. Clin. Oncol.* **2021**, *39* (Suppl. 28), 17. [CrossRef]
6. Leventhal, A.M.; Dai, H.; Higgins, S.T. Smoking Cessation Prevalence and Inequalities in the United States: 2014-2019. *J. Natl. Cancer Inst.* **2022**, *114*, 381–390. [CrossRef]
7. Presant, C.A.; Salgia, R.; Kulkarni, P.; Tiep, B.L.; Sanani, S.; Leach, B.; Ashing, K.; Sandoval, J.; Sedrak, M.S.; Landau, S.; et al. Implementing Lung Cancer Screening and Prevention in Academic Centers, Affiliated Network Offices and Collaborating Care Sites. *J. Clin. Med.* **2020**, *9*, 1820. [CrossRef]

8. Ashing, K.T.; Tiep, B.; Macalintal, J.; Yeung, S.; O'Connor, T.; Xie, B.; Obodo, U.; Tsou, M.-H.; Song, G.; Abuan, F.; et al. A geospacial approach to explore the socioecological context of tobacco and vape shop location. *Cancer Res.* **2020**, *80*, CT087. [CrossRef]
9. Presant, C.A.; Ashing, K.; Yeung, S.; Macalintal, J.; Tiep, B.; Sandoval, A.; Brown, S.; Cianfrocca, M.; Erhunmwunsee, L.; Raz, D.; et al. Increasing clinician participation in tobacco cessation by an implementation science-based tobacco cessation champion program. *Cancer Causes Control* **2022**, *34*, 81–88. [CrossRef]
10. Raz, D.J.; Wu, G.; Nelson, R.A.; Sun, V.; Wu, S.; Alem, A.; Haupt, E.C.; Ismail, M.H.; Gould, M.K. Perceptions and Utilization of Lung Cancer Screening Among Smokers Enrolled in a Tobacco Cessation Program. *Clin. Lung Cancer* **2019**, *20*, e115–e122. [CrossRef]
11. Raz, D.J.; Wu, G.X.; Consunji, M.; Nelson, R.; Sun, C.; Erhunmwunsee, L.; Ferrell, B.; Sun, V.; Kim, J.Y. Perceptions and Utilization of Lung Cancer Screening Among Primary Care Physicians. *J. Thorac. Oncol.* **2016**, *11*, 1856–1862. [CrossRef]
12. Raz, D.J.; Dunham, R.; Tiep, B.; Sandoval, A.; Grannis, F.; Rotter, A.; Kim, J.Y. Augmented Meaningful Use Criteria to Identify Patients Eligible for Lung Cancer Screening. *Ann. Thorac. Surg.* **2014**, *98*, 996–1002. [CrossRef]
13. Raz, D.J.; Ismail, M.H.; Sun, V.; Park, S.; Alem, A.C.; Haupt, E.C.; Gould, M.K. Incorporating lung cancer screening education into tobacco cessation group counseling. *Tob. Prev. Cessat.* **2020**, *6*, 12. [CrossRef] [PubMed]
14. Akhtar, A.; Sosa, E.; Castro, S.; Sur, M.; Lozano, V.; D'Souza, G.; Yeung, S.; Macalintal, J.; Patel, M.; Zou, X.; et al. A Lung Cancer Screening Education Program Impacts both Referral Rates and Provider and Medical Assistant Knowledge at Two Federally Qualified Health Centers. *Clin. Lung Cancer* **2022**, *23*, 356–363. [CrossRef] [PubMed]
15. Redfield, R.R.; Hahn, S.M.; Sharpless, N.E. Redoubling Efforts to Help Americans Quit Smoking—Federal Initiatives to Tackle the Country's Longest-Running Epidemic. *N. Engl. J. Med.* **2020**, *383*, 1606–1609. [CrossRef] [PubMed]
16. The US Preventive Services Task Force. Interventions for Smoking Cessation in adults, including pregnant persons: US Preventive Services Task Force recommendation statement. *JAMA* **2021**, *325*, 265–279. [CrossRef]
17. The US Preventive Services Task Force. Screening for lung cancer: US Preventive Services Task Force recommendation statement; CCN Guidelines®Insights: Lung Cancer Screening, Version 1.2022. *JAMA* **2021**, *325*, 962–970.
18. Hohl, S.D.; A Shoenbill, K.; Taylor, K.L.; Minion, M.; E Bates-Pappas, G.; Hayes, R.B.; Nolan, M.B.; Simmons, V.N.; Steinberg, M.B.; Park, E.R.; et al. The Impact of the COVID-19 Pandemic on Tobacco Treatment Program Implementation at National Cancer Institute-Designated Cancer Centers. *Nicotine Tob. Res.* **2022**, *25*, 345–349. [CrossRef]
19. Pham, D.; Bhandari, S.; Pinkston, C.; Oechsli, M.; Kloecker, G. Lung Cancer Screening Registry Reveals Low-dose CT Screening Remains Heavily Underutilized. *Clin. Lung Cancer* **2020**, *21*, e206–e211. [CrossRef]
20. Cancer Center Cessation Initiative Sustainability Working Group. Sustainability of tobacco treatment programs in the cancer center cessation initiative. *J. Natl. Compr. Cancer Netw.* **2022**, *19* (Suppl. 1), S16–S20.
21. Hohl, S.D.; Matulewicz, R.S.; Salloum, R.G.; Ostroff, J.S.; Baker, T.B.; Schnoll, R.; Warren, G.; Bernstein, S.L.; Minion, M.; Lenhoff, K.; et al. Integrating tobacco treatment into oncology care: Reach and effectiveness of evidence-based tobacco treatment across NCI-designated cancer centers. *J. Clin. Oncol.* 2022, in press. [CrossRef]
22. Ramsey, A.T.; Baker, T.B.; Stoneking, F.; Smock, N.; Chen, J.; Pham, G.; James, A.S.; Colditz, G.A.; Govindan, R.; Bierut, L.J.; et al. Increased Reach and Effectiveness With a Low-Burden Point-of-Care Tobacco Treatment Program in Cancer Clinics. *J. Natl. Compr. Cancer Netw.* **2022**, *20*, 488–495. [CrossRef] [PubMed]

Journal of
Clinical Medicine

MDPI

Article

Patient Care Satisfaction and Emergency Room Utilization among Young Adult Colorectal Cancer Survivors during the SARS-CoV-2 Pandemic: Lessons Learned

Dalia Kagramanov [1,*], Kimberly A. Miller [1,2], Phuong Gallagher [3], David R. Freyer [1,4,5,6,7], Joel E. Milam [8], Heinz-Josef Lenz [7,9] and Afsaneh Barzi [10]

1 Department of Population and Public Health Sciences, Keck School of Medicine, University of Southern California, Los Angeles, CA 90033, USA
2 Department of Dermatology, Keck School of Medicine of the University of Southern California, Los Angeles, CA 90033, USA
3 The Colon Club, Pasadena, CA 91105, USA
4 Cancer and Blood Disease Institute, Children's Hospital Los Angeles, Los Angeles, CA 90027, USA
5 Department of Pediatrics, Keck School of Medicine of the University of Southern California, Los Angeles, CA 90033, USA
6 Department of Medicine, Keck School of Medicine of the University of Southern California, Los Angeles, CA 90033, USA
7 USC Norris Comprehensive Cancer Center, Los Angeles, CA 90033, USA
8 Department of Epidemiology and Biostatistics, Chao Family, Comprehensive Cancer Center, School of Medicine, University of California, Irvine, CA 92868, USA
9 Department of Oncology, University of Southern California, Norris Comprehensive Cancer Center, Los Angeles, CA 90033, USA
10 Department of Medical Oncology and Therapeutics Research, City of Hope National Medical Center, Duarte, CA 91010, USA
* Correspondence: dkagrama@usc.edu

Abstract: Introduction: Survivors of colorectal cancer (CRC) are at risk for late effects of therapy and recurrence of cancer. With recurrence rates ranging between 30–40%, follow-up care is needed for both early detection and management of late effects. Cancer care delivery for CRC patients was significantly disrupted by the SARS-CoV-2 pandemic, with decreases of 40% in such services in the United States between April 2020 and 2019. Survivors were left with fewer options for care, potentially causing increases in emergency room (ER) utilization. Methods: This cross-sectional study examined the patterns of ER utilization during the SARS-CoV-2 pandemic among young adult CRC survivors and assessed the relationship between self-reported care satisfaction and ER use. Eligible participants were colon or rectal cancer survivors diagnosed between 18–39 years of age, 6–36 months from diagnosis/relapse, English speaking and residing in the United States. Multivariable logistic regression assessed the association between patient care satisfaction and ER utilization, adjusting for pandemic factors. Covariates were chosen by significance of $p < 0.1$ at the univariate level and perceived clinical significance. Results: The overall sample (N = 196) had mean age (SD) 32.1 (4.5); 59% were male. Tumor location was colon or rectal in 42% and 57%, respectively, and the majority (56%) were diagnosed with stage 2 disease; 42.6% reported relapsed disease, and 20% had an ostomy. Most survivors (72.5%) had between 1–4 visits to an ER in the last 12 months and were categorized as normal utilizers. Approximately 24.7% of the sample had greater than 4 visits to the ER in the last 12 months and were categorized as super-utilizers. CRC survivors that reported a delay in their follow-up care as a result of the pandemic were two times (OR: 2.05, 95% CI 0.99, 4.24) more likely to be super-utilizers of the ER. Higher self-reported satisfaction with care was associated with a 13.7% lower likelihood of being a super-utilizer (OR: 0.86, 95%CI: −0.68, 1.09). Conclusions: This study found strong associations between delays in care, self-reported care satisfaction, and being a super-utilizer of the ER during the pandemic among young adult CRC survivors off treatment. Increasing patient satisfaction and minimizing care interruptions amongst this vulnerable population may aid in mitigating over-utilization in the ER during an ongoing pandemic.

Keywords: colorectal cancer survivor; emergency utilization; SARS-CoV-2; care satisfaction; pandemic

Citation: Kagramanov, D.; Miller, K.A.; Gallagher, P.; Freyer, D.R.; Milam, J.E.; Lenz, H.-J.; Barzi, A. Patient Care Satisfaction and Emergency Room Utilization among Young Adult Colorectal Cancer Survivors during the SARS-CoV-2 Pandemic: Lessons Learned. *J. Clin. Med.* **2023**, *12*, 469. https://doi.org/10.3390/jcm12020469

Academic Editor: Enrico Capobianco

Received: 8 December 2022
Revised: 24 December 2022
Accepted: 28 December 2022
Published: 6 January 2023

1. Introduction

Colorectal cancer (CRC) is the second cause of cancer-related death and the third most common cancer in the United States [1]. Upon diagnosis, treatment plans are often multimodal, involving a combination of chemotherapy, radiation, and surgery. With advancements in early detection and improved treatment modalities, nearly two-thirds of CRC survivors are living more than 5 years post diagnosis [2]. However, survivors of this malignancy are at risk for late effects of therapy, as well as recurrence of cancer. Side effects may include both physical ailments and/or psychosocial complications [3]. Additionally, overall recurrence rates for colorectal cancer remain high, ranging between 30–40% among survivors [2].

The incidence of colorectal cancer has continued to rise among young adults aged 18–39 years [4]. These are individuals who face unique challenges navigating cancer care during the transformative stages of young adulthood which commonly include changes in residence, career development, starting a family, etc. Post-treatment concerns characteristic of young adulthood include reproductive health, genetics, social impacts, and future employment [5].

Consistent follow-up care is critical for identifying late-effects and recurrence [6], thereby increasing one's chances for long-term survival. Follow up care is also a venue for assisting survivors to develop coping strategies and access resources for ongoing issues [7]. Common survivorship care models where follow-up care is conducted include multidisciplinary care, characterized by a dedicated team of healthcare professionals who provide a range of health services to survivors, and shared care models which include collaborative care between the oncologist and primary care provider (PCP) [8]. Considering the high rates of recurrence among CRC survivors, regular cancer-directed screening and physical examinations via any survivorship care model are of high importance. Both the National Comprehensive Cancer Network (NCCN) and the American Society of Clinical Oncology (ASCO) have developed survivorship care guidelines for clinicians to utilize when working with cancer survivors to prevent and support late-effects of treatment. However, attendance rates to survivorship care models overall among young adult survivors remain low [9]. Further, the literature on CRC has shown that higher care density, the extent to which a patient's providers share patients with one another, and lower care fragmentation are associated with a reduced likelihood of hospitalization and emergency room visits [10].

The SARS-CoV-2 pandemic had a drastic impact on health care globally, including cancer treatment care and survivorship care. CRC-related care delivery was significantly disrupted with decreases of up to 40% in CRC services in the United States between April 2020 and April 2019 [11]. As a result, survivors may have been left with few options for cancer-related care. With the feeling of uncertainty that accompanied the global pandemic, disruptions in available survivor care services may have played a role in potentially driving emergency room utilization higher in this population. Additionally, delays in receiving care as a result of the pandemic may have also impacted patients' care satisfaction with cancer specialists or general care providers, further impacting the utilization patterns in this vulnerable population with potential long-term consequences. SARS-CoV-2 created an unprecedented disruption in care, social norms, and health care expectations, including the impactful changes in the experiences of cancer survivors. The impact of disruptions in care during SARS-CoV-2 (especially the earlier period) may be viewed as a social experiment to understand the interaction between perception of access to care by patients and utilization of services. The lessons of this social experiment can pave the way to design more robust systems and to equip future patients and survivors with tools to address their unmet needs. Although the effects of the COVID-19 pandemic were unprecedented, delays in care can happen for a variety of reasons and are likely to occur in the future. Deeper understanding of the needs of CRC survivors during these times will help to reduce negative impacts.

We set to explore the patterns of emergency room utilization among young adult CRC survivors in the United States during SARS-CoV-2 pandemic. Using survey data from a national society for young adult survivors of colorectal cancer, we explored the patterns of ER utilization during this time and assessed the relationship between self-reported satisfaction with care and emergency-care use. We hypothesized that lower self-reported satisfaction measures would be associated with greater emergency-room utilization.

2. Materials and Methods

This cross-sectional study was conducted using an online survey administered on the Facebook page of a national Colorectal Cancer advocacy group between 31 August and 3 September 2020. Participants were eligible if they were colon or rectal cancer survivors, aged 18–39 at time of diagnosis, between 6–36 months from diagnosis or relapse, English speaking and based in the United States. Study procedures have been detailed elsewhere [4]. An electronic gift card valued at USD20 was provided to participants who completed the survey. The study was approved by the University of Southern California Institutional Review Board (IRB).

2.1. Data Verification

The data cleaning process aimed to ensure validity and reduce fraudulent responses inherent within social media recruitment. Participants were asked questions regarding eligibility at the start of the survey to eliminate automated software or "bots." Additionally, duplicate email use was prohibited. This was monitored by removing respondents whose survey completion time was less than five minutes, given an average completion time of 17-min. Lastly, respondent data was removed if reporting included "highly improbable" medical treatment patterns as reviewed by a medical oncologist [4].

2.2. Variables

Participants were surveyed regarding general demographic information, cancer-related treatment data, five self-reported satisfaction with care questions, scaled 0–10, and the number of emergency room visits in the last 12 months. The collection of survey questions was created from previously validated scales or measures widely used in cancer research. For example, CRC survivors self-reported gender, race/ethnicity, age, stage at diagnosis, whether they had experienced relapsed disease and the year of most recent relapse. Questions on care satisfaction were based on the Consumer Assessment of Healthcare Providers and Systems (CAHPS) patient experience survey (https://www.ahrq.gov/cahps/surveys-guidance/index.html URL accessed on 1 August 2020) [12]. Additionally, survivors were asked about a number of pandemic-related questions such as delays in access to care, financial impacts, psychological and emotional distress, and job loss [13]. These were based on questions from The Pandemic Stress Index [14]. The pandemic-related variables were then also used as co-factors in the statistical modelling process. The full scope of the survey was also pilot tested and reviewed by a patient advocate (P.G.) to ensure the questions used were acceptable and comprehensible to the target group of young adult CRC survivors.

2.3. Statistical Analysis

Frequencies and percentages were calculated for sample demographics, emergency care utilization, and self-reported satisfaction of care measures. A multivariate logistic regression was conducted on the overall sample to assess the association between patient satisfaction and emergency room utilization, adjusting for the influence of the COVID-19 global pandemic. Covariates for this analysis were chosen based on a significance of $p < 0.05$ at the univariate level, as well as general clinical significance. The demographic/clinical characteristic covariates included in the multivariable model were sex, race/ethnicity, and age at diagnosis. Treatment intensity was assessed as a potential covariate in the relationship between prior cancer therapy and the outcomes of ER utilization and patient care satisfaction. Treatment intensity was calculated as the sum of the self-reported treatment modalities received (chemotherapy, radiation, surgery, and/or immunotherapy) and scored on a scale from 0 (defined as no therapy received) to 4 (defined as receiving all four modalities). This covariate was not statistically significant at the univariate level and was therefore not included in the final multivariable model.

Based on similar prior literature on emergency room utilization, the outcome of emergency room utilization was dichotomized. Survivors who visited an emergency room greater than 4 times in the last 12 months were termed "super-utilizers" while survivors who visited an emergency room less than or equal to 4 times in the last 12 months were considered "normal utilizers" [15]. All statistical analyses were performed using SAS (Version 9.4).

3. Results

A total of 371 survey responses were received, of which 196 (53%) were retained after screening eligibility criteria and removing responses that were identified as potentially fraudulent based on our previous algorithm. Sample characteristics are presented in Table 1. Overall mean age (SD) was 32.1 years (4.5), and 116 survivors (59%) were male. Diagnosis tumor location was colon or rectal in 39% and 61%, respectively, and the majority (56%) were diagnosed with stage 2 disease. Relapsed disease was reported by 58% of respondents, and 30% had an ostomy. Lastly, the majority of respondents were non-Latino white (79%).

Table 1. Characteristic of the sample overall and by age category n = 196.

	Current Age	
	20–29 (*n* = 56)	**30–42 (*n* = 140)**
Sex		
Male	33 (61.1)	83 (59.3)
Female	21 (38.9)	57 (40.7)
Race/Ethnicity		
Hispanic/Latino	8 (14.5)	12 (8.7)
Non-Hispanic White	41 (74.5)	112 (81.2)
Black/African American	3 (5.5)	10 (7.3)
Asian/Pacific Islander/Other	3 (5.5)	4 (2.8)
Region		
Midwest	6 (10.9)	31 (22.1)
Northeast	8 (14.6)	20 (14.3)
South	27 (49.1)	43 (30.7)
West	14 (25.4)	46 (32.9)
Income Per Year		
<USD35,000	14 (25.0)	18 (12.9)
USD35,000–USD74,999	26 (46.4)	92 (65.7)
USD75,000–USD149,999	15 (26.8)	28 (20.0)
>USD150,000	1 (1.8)	2 (1.4)
Cancer Type		
Colon	23 (42.6)	52 (38.0)
Rectal	31 (57.4)	85 (62.0)
Stage At Diagnosis		
Stage 1	18 (32.1)	25 (18.0)
Stage 2	23 (41.1)	87 (62.6)
Stage 3	13 (23.2)	23 (16.5)
Stage 4	2 (3.6)	4 (2.9)
Relapse		
Yes	23 (42.6)	89 (63.6)
Ostomy		
Yes	11 (20.0)	46 (33.6)

Approximately one quarter of the sample were super-utilizers of the emergency room (24.7%) (Figure 1). The majority of survivors (72.5%) had between 1–4 visits to an emergency room in the last 12 months.

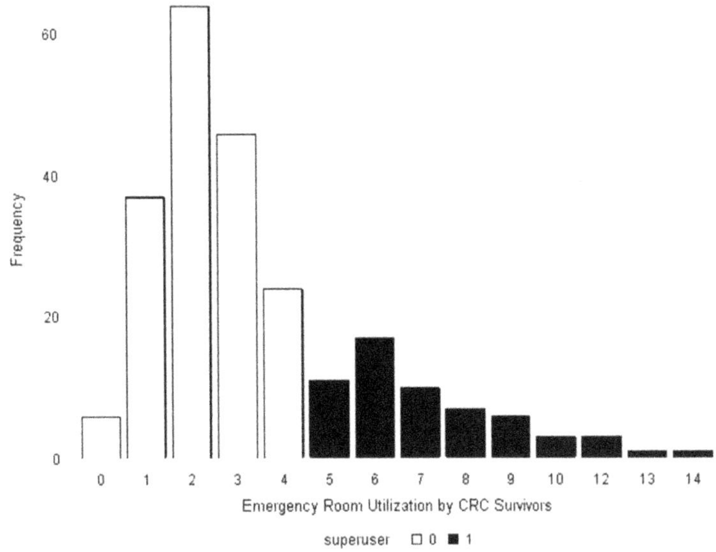

Figure 1. Distribution of "super-utilizers" of the emergency room among adolescent and young adult colorectal cancer survivors.

Participants who had experienced a delay in their cancer care as a result of the pandemic were two times (OR: 2.05, 95% CI 0.99, 4.24) more likely to be super-utilizers of the emergency room. Additionally, those that experienced a delay in general care as a result of the pandemic were 92% more likely to be super-utilizers (OR: 1.92, 95% CI: 0.95, 3.86). However, this result was marginally significant at the $p < 0.05$ level. Survivors that had a higher self-reported care satisfaction rating for their primary provider were 23.5% (OR: 0.76, 95%CI: −0.60, 0.97) less likely to be super-utilizers of the emergency room. Similarly, higher self-reported satisfaction with overall care was associated with a 13.7% (OR: 0.86, 95%CI: −0.68, 1.09) less likely to be a super-utilizer. However, this result was not statistically significant. The described results are presented in Table 2.

Table 2. Multivariable regression of ER super-utilization (>4 visits) and patient care satisfaction.

	OR	SE	95% CI	*p*
Gender				
Female	1.26	0.36	0.61, 2.57	0.53
Male (ref)				
Age at diagnosis	0.96	0.03	0.90, 1.02	0.19
Race/Ethnicity				
Hispanic/Latino	1.67	0.56	0.55, 5.02	0.36
Black/African American	0.65	0.69	0.17, 2.51	0.53
Asian/Pacific Islander/Other	0.47	1.25	0.04, 5.45	0.55
Non-Hispanic White (ref)				
Overall Healthcare Satisfaction Rating	0.86	0.12	0.68, 1.09	0.22
Primary Healthcare Provider Satisfaction Rating	0.77	0.12	0.60, 0.97	0.03
Specialist Healthcare Satisfaction Rating	1.00	0.11	0.81, 1.24	0.98
General Delays in Care (Past 12 months)	0.64	0.49	0.24, 1.69	0.37
Delays in Cancer Care Due to Pandemic	2.05	0.37	0.99, 4.24	0.05
Delays in General Care Due to Pandemic	1.92	0.36	0.95, 3.86	0.07

4. Discussion

The patient care experience is an important aspect of health care quality and is associated with health care utilization and health outcomes. The results found in this study indicate that higher patient satisfaction with care was associated with lower use of the emergency room, which may be a result of perceived or actual changes in one's access to care, as well as their experienced care. For example, clinic closures, delays in getting an appointment, or long wait times during scheduled appointments may have resulted in a perceived negative health care experience. Survivors as a result may be more likely to choose care from the emergency room for health-related concerns if they are unable or unwilling (due to poor experience) to access care elsewhere. Delays in care with one' regular cancer care provider because of the pandemic may play a role in influencing a survivor's future health behaviors. While the results of the cross-sectional study only look at one point in time, the highlighted associations shed light on areas for further study.

Emergency room utilization is a subject of interest in cancer care [16], with ongoing quality measures in development to reduce unnecessary use of this valuable resource [17]. The issue of higher use of ER due to lack of timely and proper access to outpatient services has been reported before [18], however, population-based research access has predominantly viewed and measured this though the channel of health insurance [19]. On the individual level, satisfaction with care can be a major determinant of perceived access to care. Therefore, poor satisfaction with care or sudden changes in an ongoing relationship with the care provider are barriers for access to appropriate care and thus increased use of ER.

In support of our findings, research on the general population has shown that higher patient satisfaction is associated with less emergency department use[7]. More so, previous literature also presented differences in care satisfaction and delays in care among Hispanic communities in the United States. Particularly, a study done in 2012 showed that non-Hispanic Black patient experience in the Los Angeles County may have an even greater impact on disease outcomes as a result of worse patient experiences with care being strongly associated with patient reports of discrimination [20]. This can further negatively impact health care utilization, driving individuals to seek care only in urgent cases through the emergency room. Due to the lack of diversity in the sample of this study, it is possible to have missed capturing even greater associations between self-reported care satisfaction and emergency room utilization among different race/ethnicity groups. Future research would benefit from obtaining data on a more diverse sample and stratifying analyses outcome measures by race/ethnicity groups.

Young adult colorectal cancer survivors commonly experience delays in care, financial hardship, and a reduced quality of life [15]. These components were further exacerbated for this vulnerable population during the SARS-CoV-2 pandemic when access to healthcare drastically changed and life for most was put on pause. With the knowledge that in-person care was greatly disrupted as a result of the pandemic [11], it is important to gain insight on the barriers and facilitators of this population's health care utilization during such global events to aid in preparation of future care disruptions. This study found moderate to high associations between delays in care as a result of the SARS-CoV-2 pandemic and emergency room utilization, as well as self-reported satisfaction measures and emergency room utilization. It is possible that increased emergency care use by this population may be indicative of increased late effect symptomology during the pandemic, as well as a lack of obtaining recommended survivor care screening and physical assessments. Such changes in health behavior can have negative consequences on the health status of young adult CRC survivors.

The insight of our results provides valuable information on the potential drivers of this population's health care patterns during the current SARS-CoV-2 pandemic. It is clear that CRC cancer care was disrupted in some form for this population of at-risk survivors, and it is important to recognize that changes in their 'typical' survivor-focused care can have great implications for their long-term outcomes. Added knowledge in the field can help to inform leaders on how to best support this vulnerable group in future health care disruptions. As a future direction, added exploration into the reasons for and nature of each emergency room visit will be beneficial towards understanding how to best care for these patients.

Some limitations of the interpretation of this study include the limited ability to infer causality based on the cross-sectional design, as well as the self-reported data being subject to bias. Despite rigorous attempts to reduce fraudulent responses, social media sampling prevented full verification of respondents' patient status. Moreover, a social media sample may not be representative of the overall patient population as respondents were connected to an online resource and may represent a more motivated sample. Use of a social media survey also limits clinical verification of disease status. Lastly, considering the survey was conducted during the time of an unexpected pandemic, we do not have information on pre-pandemic ER usage patterns in this group, and are therefore unable to draw comparisons. Further information on methodological limitations is described in more detail in the parent study [4].

5. Conclusions

This study found strong associations between delays in care, self-reported care satisfaction, and emergency room utilization during the SARS-CoV-2 pandemic. The identification of such relationships adds valuable insight to the barriers and facilitators of care utilization during periods of extreme health care disruption. Importantly, survivors opting for emergency room use as opposed to regular follow-up care from their specialist or general care provider may be at risk for long-term consequences. Knowledge of these health behavior changes can help health care professionals recognize the impact of their individual approach and interactions on patient choice and facilitate interventions to improve such interactions. Undoubtedly, this learning can better prepare us for future global events, aiming to minimize cancer survivor impacts. Future research should aim to better characterize the relationship between patient satisfaction and with care and emergency room use in this at-risk population, as well as examine other common survivor populations, such as breast cancer survivors in the United States.

Author Contributions: Conceptualization, D.K., K.A.M. and A.B.; Formal analysis, D.K. and K.A.M.; Funding acquisition, K.A.M.; Methodology, D.K., K.A.M. and A.B., Writing—original draft, D.K.; Writing—review and editing, K.A.M., A.B., P.G., D.R.F., H.-J.L. and J.E.M. All authors have read and agreed to the published version of the manuscript.

Funding: This research was funded by The Aflac Archie Bleyer Young Investigator Award in Adolescent and Young Adult Oncology from the Children's Oncology Group. Additional support was provided by the NCI Cancer Center Support Grant P30 CA014089 from the USC Norris Comprehensive Cancer Center.

Institutional Review Board Statement: The study was conducted according to the guidelines of the Declaration of Helsinki and approved by the Institutional Review Board of the University of Southern California (protocol HS-19-00288 approved 17 July 2020).

Informed Consent Statement: Informed consent was obtained from all subjects involved in the study.

Data Availability Statement: The data presented in this study are available on request as a limited use dataset from the corresponding author. The data are not publicly available due to privacy considerations for participants.

Acknowledgments: The authors would also like to extend appreciation to all of the Colon Club survivors who participated in the study.

Conflicts of Interest: The authors declare no conflict of interest. The funders had no role in the design of the study; in the collection, analyses, or interpretation of data; in the writing of the manuscript; or in the decision to publish the results.

References

1. Curtin, S.C.; Heron, M.; Miniño, A.M.; Warner, M. Recent increases in injury mortality among children and adolescents aged 10-19 years in the United States: 1999-2016. *Natl. Vital Stat. Rep.* **2018**, *67*, 1–16. [PubMed]
2. Siegel, R.L.; Miller, K.D.; Goding Sauer, A.; Fedewa, S.A.; Butterly, L.F.; Anderson, J.C.; Cercek, A.; Smith, R.A.; Jemal, A. Colorectal cancer statistics, 2020. *CA A Cancer J. Clin.* **2020**, *70*, 145–164. [CrossRef] [PubMed]
3. Olver, I. (Ed.) *The MASCC Textbook of Cancer Supportive Care and Survivorship*; Springer: Berlin/Heidelberg, Germany, 2018.

4. Miller, K.A.; Stal, J.; Gallagher, P.; Weng, Z.; Freyer, D.R.; Kaslander, J.N.; Marin, P.; Lenz, H.J.; Milam, J.E.; Govaerts, L.; et al. Time from diagnosis and correlates of health-related quality of life among young adult colorectal cancer survivors. *Cancers* **2021**, *13*, 4045. [CrossRef] [PubMed]
5. El-Shami, K.; Oeffinger, K.C.; Erb, N.L.; Willis, A.; Bretsch, J.K.; Pratt-Chapman, M.L.; Cannady, R.S.; Wong, S.L.; Rose, J.; Barbour, A.L.; et al. American Cancer Society colorectal cancer survivorship care guidelines. *CA A Cancer J. Clin.* **2015**, *65*, 427–455. [CrossRef] [PubMed]
6. Jeffery, M.; Hickey, B.E.; Hider, P.N. Follow-up strategies for patients treated for non-metastatic colorectal cancer. *Cochrane Database Syst. Rev.* **2019**, *9*, CD002200. [CrossRef] [PubMed]
7. Elit, L.; Reade, C.J. Recommendations for follow-up care for gynecologic cancer survivors. *Obstet. Gynecol.* **2015**, *126*, 1207–1214. [CrossRef] [PubMed]
8. Oeffinger, K.C.; McCabe, M.S. Models for delivering survivorship care. *J. Clin. Oncol.* **2006**, *24*, 5117–5124. [CrossRef] [PubMed]
9. Zheng, D.J.; Sint, K.; Mitchell, H.R.; Kadan-Lottick, N.S. Patterns and predictors of survivorship clinic attendance in a population-based sample of pediatric and young adult childhood cancer survivors. *J. Cancer Surviv.* **2016**, *10*, 505–513. [CrossRef] [PubMed]
10. Pollack, C.E.; Frick, K.D.; Herbert, R.J.; Blackford, A.L.; Neville, B.A.; Wolff, A.C.; Carducci, M.A.; Earle, C.C.; Snyder, C.F. It's who you know: Patient-sharing, quality, and costs of cancer survivorship care. *J. Cancer Surviv.* **2014**, *8*, 156–166. [CrossRef] [PubMed]
11. Jammu, A.S.; Chasen, M.R.; Lofters, A.K.; Bhargava, R. Systematic rapid living review of the impact of the COVID-19 pandemic on cancer survivors: Update to August 27, 2020. *Support. Care Cancer* **2020**, *29*, 2841–2850. [CrossRef] [PubMed]
12. CAHPS Patient Experience Surveys and Guidance. *Content Last Reviewed May 2022*; Agency for Healthcare Research and Quality: Rockville, MD, USA, 2022. Available online: https://www.ahrq.gov/cahps/surveys-guidance/index.html/ (accessed on 1 August 2020).
13. Miller, K.A.; Kagramanov, D.; Cohen-Cutler, S.; Nadim, B.; Weng, Z.; Gallagher, P.; Kaslander, J.N.; Freyer, D.R.; Barzi, A.; Lenz, H.J. Impacts of the SARS-CoV-2 pandemic on young adult colorectal cancer survivors. *J. Adolesc. Young Adult Oncol.* **2022**, *11*, 229–233. [CrossRef] [PubMed]
14. Harkness, A.; Behar-Zusman, V.; Safren, S.A. Understanding the Impact of COVID-19 on Latino Sexual Minority Men in a US HIV Hot Spot. *AIDS Behav.* **2020**, *24*, 2017–2023. [CrossRef] [PubMed]
15. Johnson, T.L.; Rinehart, D.J.; Durfee, J.; Brewer, D.; Batal, H.; Blum, J.; Oronce, C.I.; Melinkovich, P.; Gabow, P. For Many Patients Who Use Large Amounts of Health Care Services, The Need Is Intense Yet Temporary. *Health Aff.* **2015**, *34*, 1312–1319. [CrossRef] [PubMed]
16. Brown, J.; Grudzen, C.; Kyriacou, D.N.; Obermeyer, Z.; Quest, T.; Rivera, D.; Stone, S.; Wright, J.; Shelburne, N. The Emergency Care of Patients with Cancer: Setting the Research Agenda. *Ann. Emerg. Med.* **2016**, *68*, 706–711. [CrossRef] [PubMed]
17. Centers for Medicare & Medicaid Services. *Evaluation of the Oncology Care Model: Performance Period One*; Centers for Medicare & Medicaid Services: Baltimore, MD, USA, 2021.
18. Agarwal, P.; Bias, T.K.; Vasile, E.; Moore, L.; Davis, S.; Davidov, D. Exploring health insurance status and emergency department utilization. *Health Serv. Res. Manag. Epidemiol.* **2015**, *2*, 2333392815606094. [CrossRef] [PubMed]
19. Parsons, H.M.; Harlan, L.C.; Lynch, C.F.; Hamilton, A.S.; Wu, X.-C.; Kato, I.; Schwartz, S.; Smith, A.W.; Keel, G.; Keegan, T.H. Impact of Cancer on Work and Education Among Adolescent and Young Adult Cancer Survivors. *J. Clin. Oncol.* **2012**, *30*, 2393–2400. [CrossRef] [PubMed]
20. Weech-Maldonado, R.; Hall, A.; Bryant, T.; Jenkins, K.A.; Elliott, M.N. The relationship between perceived discrimination and patient experiences with health care. *Med. Care* **2012**, *50*, S62–S68. [CrossRef]

Journal of
Clinical Medicine

MDPI

Article

Pyramidal Decision Support Framework Leverages Subspecialty Expertise across Enterprise to Achieve Superior Cancer Outcomes and Personalized, Precision Care Plans

Linda D. Bosserman [1], Isa Mambetsariev [2], Colton Ladbury [3], Afsaneh Barzi [1], Deron Johnson [4], Denise Morse [5], Debbie Deaville [6], Wade Smith [7], Swapnil Rajurkar [8], Amartej Merla [9], George Hajjar [10], Daniel Kim [2], Jeremy Fricke [2], Vijay Trisal [11] and Ravi Salgia [2,*]

1 Department of Medical Oncology and Therapeutics Research, City of Hope, Irwindale, CA 91706, USA
2 Department of Medical Oncology and Therapeutics Research, City of Hope, Duarte, CA 91010, USA
3 Department of Radiation Oncology, City of Hope, Duarte, CA 91010, USA
4 Department of Clinical Informatics, City of Hope, Duarte, CA 91010, USA
5 Department of Quality, Risk and Regulatory Management, City of Hope, Duarte, CA 91010, USA
6 Department of Enterprise Business Intelligence, City of Hope, Irwindale, CA 91706, USA
7 Department of Medical Oncology and Therapeutics Research, City of Hope, Newport Beach, CA 92660, USA
8 Department of Medical Oncology and Therapeutics Research, City of Hope, Upland, CA 91784, USA
9 Department of Medical Oncology and Therapeutics Research, City of Hope, Antelope Valley, CA 93534, USA
10 Department of Medical Oncology and Therapeutics Research, City of Hope, Mission Hills, CA 91345, USA
11 Department of Medicine, City of Hope, Duarte, CA 91010, USA
* Correspondence: rsalgia@coh.org

Citation: Bosserman, L.D.; Mambetsariev, I.; Ladbury, C.; Barzi, A.; Johnson, D.; Morse, D.; Deaville, D.; Smith, W.; Rajurkar, S.; Merla, A.; et al. Pyramidal Decision Support Framework Leverages Subspecialty Expertise across Enterprise to Achieve Superior Cancer Outcomes and Personalized, Precision Care Plans. *J. Clin. Med.* **2022**, *11*, 6738. https://doi.org/10.3390/jcm11226738

Academic Editor: Maria Lina Tornesello

Received: 29 September 2022
Accepted: 9 November 2022
Published: 14 November 2022

Abstract: The complexity of cancer care requires integrated and continuous support to deliver appropriate care. An expert network with complementary expertise and the capability of multidisciplinary care is an integral part of contemporary oncology care. Appropriate infrastructure is necessary to empower this network to deliver personalized precision care to their patients. Providing decision support as cancer care becomes exponentially more complex with new diagnostic and therapeutic choices remains challenging. City of Hope has developed a Pyramidal Decision Support Framework to address these challenges, which were exacerbated by the COVID pandemic, health plan restrictions, and growing geographic site diversity. Optimizing efficient and targeted decision support backed by multidisciplinary cancer expertise can improve individual patient treatment plans to achieve improved care and survival wherever patients are treated.

Keywords: complex case discussions; decision support; oncology pathways; personalized medicine; subspecialty expertise

1. Introduction

The complexity of oncology care continues to increase across cancer types with discoveries of new germline and somatic mutations; new diagnostic, prognostic, and predictive testing; and new systemic, radiation, surgical and supportive therapies [1]. Luckily, this increasing complexity of diagnostic and therapeutic options can provide better outcomes for patients just as oncology care is consolidating into larger network enterprises where multidisciplinary research-focused academic oncology experts partner with their network of oncology clinicians to offer personalized precision cancer care to each patient [2–5]. For both academic and network oncologists, an increasing number of cancer patients along with the increasing complexity of diagnostics, treatments, supportive care, survival, and end-of-life care has also increased the time pressure to fully engage patients and their support systems in understanding these complexities and developing individual care plans through shared decision making [6,7].

Personalized Precision Medicine (PPM) for cancer patients means getting the correct diagnosis and therapy reviewed, ordered, and delivered for each cancer patient to achieve

academic specialists for informal and formal consultations, offering network clinicians participation with respect to their patients in disease-specific and precision oncology tumor boards, and providing regional COCDs. These COCDs allow the community practice leads to be at the forefront of selecting patient cases and requesting experts as needed without straining the geographic hub's operations. To this end, the experts chosen to attend the complex case discussions are selected based on the individual cases that are challenging and require their expertise. Unlike traditional tumor boards where a large majority of cases are evaluated, our model allows the community practice oncologists to select only the complex cases. Our experience has shown that this heuristic approach to evidence-based learning can improve outcomes and overall network practice care [57]. Furthermore, our model helps address the time constraints of academic center oncologists that have experienced significant disruptions in consultations and care due to COVID-19 [58,59]. Our model limits the strain by lowering the number of consultations from network practice oncologists, which in turn provides value to the patient and lowers their costs without sacrificing care expertise. Network and academic oncologists who participate in complex case discussions are interested in obtaining academic and continuing medical education credits. The work towards such credit is under discussion. Both patients and network clinicians have reported a high degree of satisfaction with respect to knowing that the care plan for an individual patient has the best chance of offering the patient the best health outcome for their cancer diagnosis.

Value-based medicine is another key factor in our model that assesses not only the survival, toxicities, and financial costs to the patient from additional consultations and treatments but also takes into consideration personal values when establishing their plan of care. While costs of precision oncology continue to rise, with significant contributions from expensive genomic testing that ranges between USD 3000–6500 from commercially available sequencing platforms, the solution to this problem may be in implementing our approach in network practices where molecular data and genetic testing are performed based on granular evidence or clinical trials that are shared with the payer to justify the costs [60]. Such a model was slated to be adopted nationally through the Oncology Care First model and incorporated into CMS's 2023 enhanced Oncology Medical Home (eOMH) model, where cost-savings may be dependent on the data presented to the payers for higher reimbursement [61]. Our model enhances the precision oncology promise in our network practices by allowing network physicians direct access and consultation to nationally and internationally recognized expertise without a requirement for a traditional consultation. This time- and cost-saving solution also allows patients to receive the latest available information and care, as many patients treated in the community do not obtain a second opinion and rely on their primary oncologist for the entirety of their cancer care [62].

The aim of this study is to detail and describe City of Hope's pyramidal decision support framework for providing efficient and targeted support to busy clinicians in collaboration with expert faculty and to understand how any gaps in patient care can be further improved through academic and network practice collaboration.

2. Materials and Methods

A four-tiered pyramid of decision support was developed at the City of Hope to better serve a growing regional, national, and international network of cancer practices in this time of continuous rapid expansion of cancer diagnostics and therapy options. The 4 tiers are (1) evidence-based pathways via ClinPath, (2) formal and informal 1:1 faculty consultations, (3) 13 regular subspecialty or precision oncology tumor boards, and the newer (4) Complex Oncology Case Discussions (COCD).

Evidence-Based Pathways (EBP): Evidence-based pathways are a key component of City of Hope's digital strategy for value-based care [5]. COH implemented the VIA—now Elsevier ClinPath—evidence-based pathways in January of 2017. Currently, ClinPath pathways provide standard-of-care treatment pathways for medical oncology, hematology, and radiation oncology for 29 diseases (22 solid tumors—breast, neuro, anal, colorectal, gastroesophageal, neuroendocrine, pancreatic adenocarcinoma, bladder, prostate, renal cell, testicular, ovarian, uterine, head and neck, thyroid, mesothelioma, non-small cell lung, small cell lung, melanoma, squamous and basal cell skin cancers, and sarcoma, and 7 hematologic—chronic myelogenous leukemia, immune thrombocytopenia, lymphomas, chronic leucocytic leukemia, myelodysplastic syndromes, multiple myeloma, and other plasma cell dyscrasias). The sarcoma pathways were added in April 2019 (Supplemental Figure S1). The pathways are determined by 19 disease committees of which COH faculty co-chair 4 (Breast, CNS, Gastroesophageal, and Bladder/Renal) and faculty with disease specialties participate in most committee meetings. Disease committees oversee navigations for common and, depending on committee consensus, add guidance or pathways for rarer tumors or germline mutations within pathways to provide deeper navigational guidance to clinicians. The initial goals were to provide guidance on common diseases with a goal of 80% pathway compliance. As complexities of molecular mutations and sequential therapies have evolved, there is a growing consensus regarding the addition of specific navigations for all evidence-based care to help busy clinicians, most of whom treat multiple types of cancer patients in a day.

At COH, the clinical informatics team has had a Pathway and Protocol Informatics Pharmacist (DJ) oversee monthly meetings with the academic disease leads and their specialty PharmDs, along with Epic Beacon builders and the value-based care medical director (LB). At these meetings, a standard agenda ensures a review of new FDA drug approvals or regimens, practice-changing therapies for customization and clinical trials that may need updated Epic Beacon regimens, and mapping from the ClinPath pathways to our Epic Beacon regimens.

A Pathways Committee was established before the initial go-live of VIA, now ClinPath, pathways in 2017. It has continued to meet monthly to review pathway use, on-pathway rates, off-pathway rates, and reasons for off-pathway choices by disease type and to review data reporting with Epic therapy orders. The group also oversees the recruitment of members to serve on-pathway disease committees and oversees improvements in Epic–ClinPath interfaces and clinical trial integrations.

COH clinicians or their team members document pathway navigation choices in one of two ways. A total of 30% of clinicians navigate to the pathway tool through our electronic health record when they plan to order a systemic medical oncology or hematology therapy (other than BMT, cellular therapies, or acute leukemias). By using an "order with pathways" link, the staging and biomarker data from the EPIC-staging forms are populated into the pathway tool. The clinician then only adds any additionally required information before being taken to the preferred treatment options, which start with our clinical trials. If a pathway choice is made, the clinician is then taken back to the mapped Epic Beacon treatment orders, which include the NCCN-compliant antiemetics and any other disease lead team-determined guidance. The second option for clinician navigation is to order their preferred therapy in Epic Beacon, and directly after which we ask that, asynchronously, they or their staff enter the regimen's pathway navigation information into the ClinPath tool. Elsevier provides a monthly report of pathway navigations by site, doctor, and disease, reporting choices for On-pathway, On-pathway-off treatment, Clinical Trial, and Other Trial, of which all 4 are considered On-pathway. Reporting also includes the other navigation option: Off-Pathway. Direct data feeds from ClinPath to our EDW populate Tableau reports. These reports show the 5 choices as well as when a provider enters the Not a Pathway diagnosis, No Pathway, and Off treatment choices. Clinicians are prompted to enter a reason for Off-Pathway choices. Tableau dashboard reports are sent to clinicians weekly to show any missing navigations to encourage completion. Other reports are sent to leadership showing pathway compliance by disease, site, region, and physician. Institutional or departmental incentive programs have encouraged at least 80% of ordered therapies for covered diseases to be navigated in the pathway tool. Payer metrics are incentivizing the enterprise to more fully capture available pathway navigations as well.

Formal and Informal Faculty–Network Clinician Consultations (FCC): Communication among faculty members is essential for supporting busy oncologists who see multiple types of cancers each day to optimize patient care. An opportunity for the growing networks of academic and network oncology clinicians is the establishment of true respect and collegiality. This has been a key goal of COH's enterprise and chair leadership. In medical oncology, regular symposia where academic and network clinicians co-present on cutting-edge topics have brought collegiality, respect, and awareness of specialty expertise and clinical challenges among the faculty. The option of formal consultation that can be provided to services both from academic faculty to colleagues on the academic campus and by network clinicians to academic campus clinicians when deemed necessary for any patient's care is available to all faculty. Additionally, network clinicians feel very comfortable reaching out to expert faculty on clinical issues when an informal 1:1 consultation can resolve a question. Academic faculty have also become comfortable sending their patients back to network sites for care to minimize travel time and facilitate more involvement of family and caregivers in local communities. While we can track formal consultations from medical oncologists to academic site faculty and academic faculty consultations to other academic colleagues, we currently have no formal method to collect the number of informal consultations that occur, the specific faculty, clinical issues raised, nor the impact on changes in the care plans.

Tumor Boards (TB): Over the three years studied, for selected cases where academic and network oncology clinicians desired additional decision support beyond EBP, COH has offered 13 weekly or biweekly multispecialty, disease-focused (11), molecular oncology/precision oncology (1), and multi-tumor (1) boards (TB). Traditionally, tumor boards were focused on specific organs, e.g., breast cancer. However, cross-cutting tumor boards, such as molecular tumor boards, are important for patient care and research [63].

The composition of disease-specific tumor boards includes the traditional disciplines of providers including a surgeon, medical oncologist, and radiation oncologist, as well as pathologist and radiologist, while the Precision Oncology tumor board has a higher number of geneticists and non-physicians with expertise in detection and discovery of the molecular composition of cancer. Total attendance at each TB meeting is recorded, though data on the subspecialties of all attendees were not specifically recorded.

Cases are submitted in advance to the tumor board. Submissions include relevant case information and any specific questions the submitting provider has for the multidisciplinary team. Cases are reviewed ahead of time by pathology and radiology specialists and additional materials such as digitized pathology slides are prepped for presentation. Similarly, images are loaded for review and discussion during the meeting. Tumor boards are also recognized as a forum for identifying the most appropriate place for patients to receive any elements of their care that require a highly specialized setting in coordination with care delivered at a network site selected by the patient or their health plan. For example, highly specialized surgeries are directed to the academic campus as are potential candidates for clinical trials when they are not open at a closer network site.

Until March 2020, TBs were held in person, which limited the ability of network oncology clinicians to attend. Subsequently, due to the COVID-19 pandemic, TBs were made virtual, leading to potentially improved accessibility [64]. Although TBs have been able to provide a critical additional level of decision support, we currently do not have a formal procedure across all TBs for reporting on submissions, the context of discussions, and specific recommendations. Nor do we have information as to whether the proposed care plan was approved, adopted, or revised [65], which limits the quantification of the overall impact of these TBs on patient management. Efforts to improve this deficit have come from this study and are the focus of a new institutional quality improvement project, which will help ensure TBs are both efficient and have reportable impacts on patient outcomes [11].

Our tumor boards, like most, vary in their content, disease focus, and membership composition [66]. Beyond the time commitment for faculty attendance, resources are committed to coordinating and preparing for the meetings. To understand the structure and process of these tumor boards, we collected data on the schedule of the tumor boards, membership and attendance, operating procedures for coordination of the tumor boards, data management for tumor boards, and potential patient impacts. Additionally, we used the alteration in the operating procedures during COVID-19 as an opportunity to assess any changes in the attendance at tumor boards and the opportunity for modernizing the concept of tumor boards [67]. Available case data from each tumor board that occurred from January 2019 through December 2021 were collected from respective TB administrators. TB attendance was collected from the Continuing Medical Education department.

Complex Oncology Case Discussions (COCD): The experience of our oncology practices during the COVID-19 pandemic highlighted a gap in our decision support offerings with a need to streamline network practice consultations due to the limitations of in-person referrals, the geographic growth of network sites, and growing care complexities, as noted by others [11,68]. We thus developed an additional level of faculty decision support to network clinicians called Complex Oncology Case Discussions (COCD). These are led by regional network practice sites in collaboration with the academic center. COCDs are constitute a multi-faceted approach to sharing academic site expertise in a timely and practical fashion with network physicians (Figure 2).

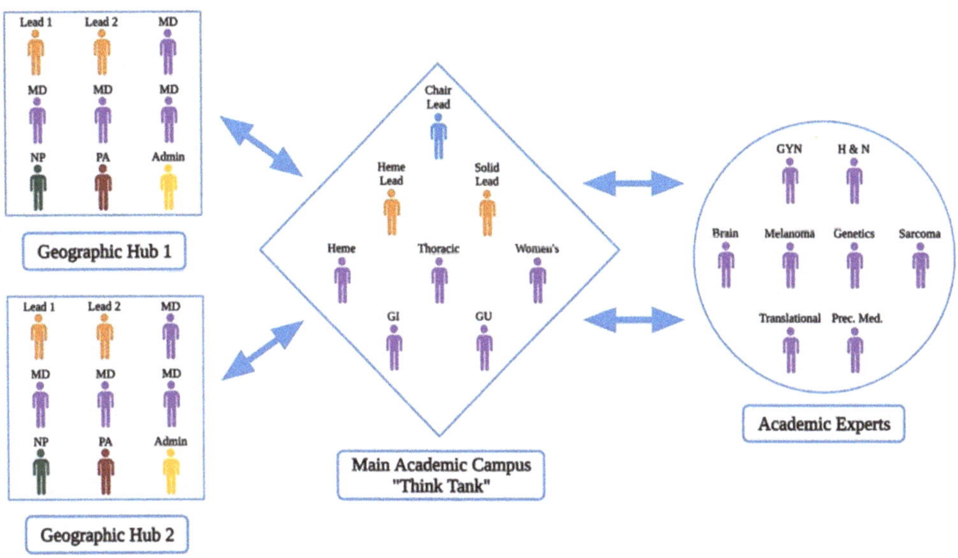

Figure 2. Complex Oncology Case Discussion: community practice hub and academic site 'think tank' integration of expertise.

The challenge in creating and establishing this model was in leveraging the availability of very busy individual team leaders and subspecialty experts with the very busy network practice physicians for regular weekly, bi-weekly, or monthly COCD. To overcome this challenge, our model was established so that the site leads of the individual sites assess their cases, determine the expertise that is required before the meeting, and only include individuals for the regional complex oncology case discussion as necessary based on their expertise. Every network practice site is assigned two site leads who supervise and organize the needed faculty experts for regional discussions with the network physicians. Our original academic site in Duarte includes a thinktank of disease team experts from oncology and hematology with access to chair leads of the various disease teams. This collaborative conference between the network and the academic site allows the oncologists to receive expertise from sub-specialized oncology experts, which would otherwise be absent and can be missing in pathways for common disease presentations as well as traditional tumor board models [69]. Evidence has shown that integration of such a collaborative model increases survival and outcomes in community practice and allows for expert intervention in situations where the network practice cannot provide the care or expertise required based on individual complexity [70]. Therefore, our model enhances the paradigm of precision medicine through the inclusion of individual disease team experts including rare diseases such as sarcoma and head and neck cancer, as well as the implementation of the expertise of a genomics expert and genetic counseling in the network practices.

In practice, this has been implemented efficiently and with minimal disruption of both the academic site and the community practice physician's schedules. We now have 3 regional COCDs that meet monthly and as needed whenever urgent COCDs are needed. We believe that the fluidity and structure of our multi-faceted team of experts have the potential to transform network practice interactions and traditional consultations into necessary functions for all our network site practices to provide as-needed, seamless, state-of-the-art academic inputs for precision care plans for patients. By only gathering the needed experts for each COCD, we optimize the demands on faculty time and have focused meetings to meet the specific needs identified by the network clinicians [71].

Process-wise, two leads at the regional network site along with the site administrator gather the individual cases and any specific questions from the regional network practitioners before convening the COCD. The leads assess the individual cases and determine the expertise that is required to attend the COCD and notify these experts 2–5 days before the meeting. The network practice physicians and administrators work with the academic site thinktank supervised by the disease team leaders and the chairs of the departments to invite the experts requested. While 4–6 leads are permanently assigned to the multidisciplinary team, other leads and experts of rare diseases such as brain cancer, sarcoma, melanoma, and head and neck cancers are invited as needed. The COCD is then convened virtually, and cases are assessed to answer any questions and determine the plan of care based on the consensus between the network practice and the academic site physicians (Figure 3).

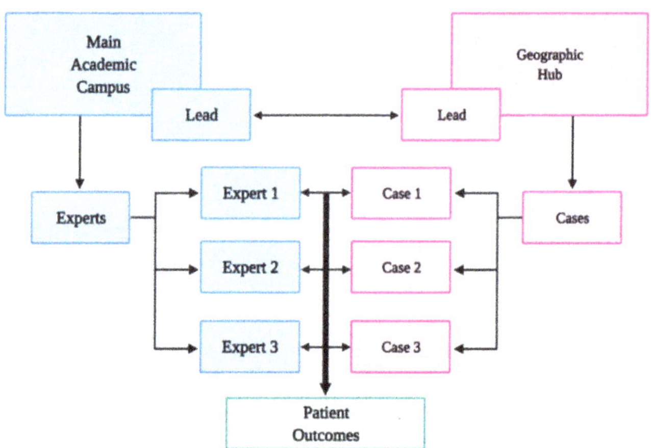

Figure 3. Complex Oncology Case Discussion algorithm for expert collaboration between academic and geographic network hubs.

If a plan of care cannot be decided due to a lack of further information such as a pending or recommended genomic-testing procedure or additional biopsy, then the network practice physicians can follow up with the academic site experts on the individual cases. Further consultation and intervention are available both virtually and in person for cases that require genetic counseling, radiation oncology intervention, specialty surgery, and clinical trial consultation. The integration of network practice sites into the academic site model also allows for clinical trial screening and trial onboarding at a few larger designated network sites—with the option for the patient to receive the trial drug treatment at the network site or the academic site. If the patient eventually relapses or undergoes progression, the network site leaders can alert the thinktank experts to convene another COCD or refer them directly for an academic consultation. This model allows for the fluidity of care and provides the patient with a consistent primary oncologist while maintaining state-of-art care that is associated with academic centers. We hope that the implementation of this model will result in greater survival outcomes as compared nationally and as have been seen at our academic site with the potential to transform networked practice care nationwide [23–27].

3. Results

The results of COH's Pyramidal Decision Support Framework will be shown for each component. The overall survival analytics have shown superior survival data for analytic patients for all stages and cancer types seen at City of Hope's Duarte academic campus compared to regional and national SEER data and published on our site for breast cancer, lung cancer, colorectal cancer, prostate cancer, and myeloma [23–27]. City of Hope began adding regional cancer care network sites in 2010, which grew to 5 sites in 2011, 17 by 2014, and 30 by the end of 2018. During the 3-year period, 2019–2021, City of Hope expanded from 30 to 38 network sites, which are included in this analysis. These sites were served by 29 academics, Duarte campus medical oncology faculty, and 50 network medical oncology faculty. The network medical oncologists also saw hematology patients. These clinicians saw 8479 new, 3752 consults, and 38,263 unique patients when 25,429 follow-up visits were included in 2019. They saw 6637 new, 2836 consults, and 43,329 unique patients when both 10,711 telehealth and 22,722 follow-up visits were counted in 2020. For 2021, the group saw 8807 new, 2915 consults, and 56,143 unique patients when 13,229 telehealth and 30,775 follow-up visits were counted.

A value framework has been built to track patient information, therapies, pathway choices, and survival for analytic and non-analytic patients seen across the enterprise since the implementation of our EPIC system in December 2017. Data are being tracked for 5- and 10-year survival outcomes but are not yet mature. This framework will provide clinical outcomes for patients whose treatments were guided by these four components of decision support. We present the initial use data on the most recent 3-year period: January 2019 through December 2021.

Evidence-Based Pathways (EBP): The data sent from ClinPath to our enterprise data warehouse (EDW) weekly are presented in Tableau reports according to the clinician, practice, region, network, and academic site, and according to the navigation choice for enterprise, Duarte academic center, and the network as well as its individual sites. Disease navigations can be reported for quarterly, annual, and time-bounded periods.

We have built Tableau charts so that clinicians and administrators can review all decisions by pathway disease. The On-pathway rates can vary significantly by disease and time period. The original goal of the pathway system was to cover 80% of therapy choices. Given the rising importance of understanding why a therapy was prescribed for a specific patient and their disease, some pathway committees have expanded guidance and flow sheet options to cover more episodes of care to provide a national group of cancer experts recommendations for the best options when appropriate. Not all committees have adoptedthis approach, so the on-pathway compliance rates will vary by how fast a national group adopts practice-changing information from presentations to publications and on to FDA approval and health plan coverage as well as by the depth of evidence-based recommendations the committee feels warranted. At City of Hope, we have no prespecified on-pathway compliance expectations but want the pathway system to provide the best standard care recommendations for the increasingly more complex range of therapy choices, especially as patients with some cancers benefit from more than three to four lines of therapy and since the therapies they had previously in any setting impact the currently recommended best option along with evolving molecular genomic and other diagnostic tests.

The evaluation of pathway navigations for the 22 solid tumors with pathway choice data from our Tableau system for the 3 years, January 2019 through December 2021, reveals over 35,000 pathway decisions across all seven decision categories: No Pathway/Not a pathway diagnosis, On-Pathway, On-Pathway–Off treatment, Clinical Trial, Other Trial, Off-Pathway, and Off Treatment (data available but not shown). Almost 15,000 were performed in network sites and over 20,000 were performed at Duarte's academic site with up to 28% of network and 8% of academic sites noting a non-pathway diagnosis. Of these total decision types, the overall on-pathway choices corresponded to 35–45% at Duarte and 38–60% in the network sites when there was a disease pathway, which is due to the percentage of other-than-On/Off-Pathway choices. Some diseases were noted to have low on-pathway rates for some quarters, which suddenly changed in a subsequent quarter. Investigating these changes showed the pathways had been updated through regularly set meetings that included key updates that our academic leads had already implemented. What had been considered 'off pathway' can change to 'on pathway'. This time delay from the early adoption of practice-changing reports and research can influence the timing of what is considered off- vs on-pathway and remains a challenge to harmonize.

The data from ClinPath on navigation choices for patients with diagnoses that have a pathway in the system are the largest subset of patients who receive systemic therapies at City of Hope. Figure 4 shows how patient data flows from all patients to those who get systemic therapies and from that subset, which diseases and their subtypes have a pathway in ClinPath for navigation and which do not. For diseases without a pathway, such as cervical cancer, hepatobiliary disease, myeloproliferative disease, ALL, AML, cellular therapies, and other rare, advanced line or very rare mutation-related diseases, a no pathway available category is available to enter in the pathway tool; however, most doctors who treat those diseases order therapies for non-pathway diseases and subtypes directly in the EPIC Beacon EHR and they may not be available from the navigation data. We are in the process of building new databases in our EDW to study the disease, stage, biomarker, line, type, ECOG, and therapy ordered for every patient seen. We can then divide those into diseases and settings with standard pathways available to pair with pathway choice navigations and track those without pathway tool pathways available. The addition of sarcoma pathways in 2019 was highly advocated and supported by City of Hope faculty given the numbers we see and the expertise we felt would be beneficial to have in our formal pathway system.

Reviews of the more commonly reported Off-Pathway vs. On-pathway (On-pathway, On-pathway-off treatment, Clinical Trial, and Other Trial) choices for patients who were navigated through ClinPath for the 29 covered diseases (22 solid tumors and 7 hematologic) from our Tableau system for the 3 years (January 2019 through December of 2021) are shown by quarter in Figure 5 for the academic center and network sites. These results are summarized in Table 1 for On-pathway vs. Off-Pathway results from the enterprise, academic, and network sites. The data show that there were 20,583 total On/Off-Pathway Decisions for the enterprise over the 3 years, of which 79% were On-pathway. A total of 8856 decisions were made at Network sites, of which 7324 were On-pathway for an 83% rate, while 11,727 total decisions were made in Duarte with 8901 being On-pathway for a 76% on-pathway rate.

Data by quarter can also be presented for the 19 solid tumor types in the 10 categories (breast, GU, GI, GUN, Head and Neck, Skin, Neuro, Neuro-endocrine, Lung, and Other) and for the eight commonly seen hematologic diseases with pathways in ClinPath (CML, CLL/Lymphoma (B Cell, T Cell, and common histology Hodgkin's), multiple myeloma, MDS, and ITP). Table 2 shows the on-pathway rate, the total number of on-pathway decisions, and the total decisions for 18 of the most common pathway diseases, where we had at least 200 decisions or an on-pathway rate of $\geq$80%. Upon review, breast, pancreatic, gastroesophageal, melanoma, neuro, bladder, and testis cancers all met the goal of ≥80% on-pathway therapies. Renal and colorectal cancers were on-pathway 78% of the time. An internal review of these, as well as the non-small cell's 64% on-pathway rate and small cell's 67% rate, reflect the rapid changes in therapy recommendations during these 3 years such that at the time of decision making, COH's disease lead choices were ordered, which was only later reflected in the pathway updates. There is no current report to compare off-pathway decisions that would, in a subsequent quarter or time period, be considered on-pathway. Such a report could show the early adoption of practice-changing therapies before their incorporation into pathway tools. This remains a challenge for the reporting of pathway data. Of note, each of the most common hematology diagnoses, multiple myeloma, and CLL/lymphomas had over 80% on-pathway choices made for their therapies, for which the reports are almost solely from network sites, as our academic colleagues plan to expand their use of the hematology pathway tool in the future.

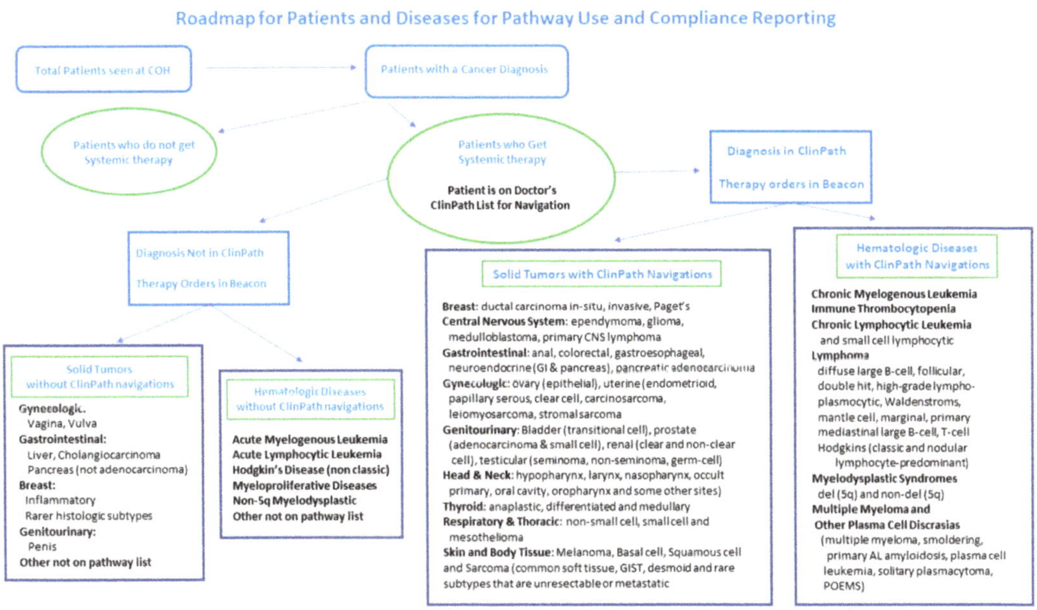

Figure 4. Roadmap of patients seen at City of Hope to understand those with diseases eligible for pathway use and reporting from ClinPath system. Both patients with and without a diagnosis in the ClinPath system will have Beacon therapy orders in the EHR. GI (gastrointestinal), CNS (central nervous system), GIST (gastro intestinal stromal tumor).

Table 1. Therapy Decisions from ClinPath Tool for Solid Tumors and Hematology by Enterprise, Duarte, and Network Sites. Pathway Compliance for Enterprise (E), Duarte Academic Campus (D), and Network (N) Clinician ClinPath Navigations over 3 years, 2019–2021. Total Number of Decisions On and Off-Pathway; total number of On-pathway Decisions for Enterprise, Duarte, and Network Sites; and total percent of On-pathway Navigations for Enterprise, Duarte, and Network sites. From ClinPath Reports.

Total # On/Off-Pathway Decisions			# On-Pathway			% On-Pathway		
E	D	N	E	D	N	E	D	N
20,583	11,727	8856	16,229	8901	7324	79%	76%	83%

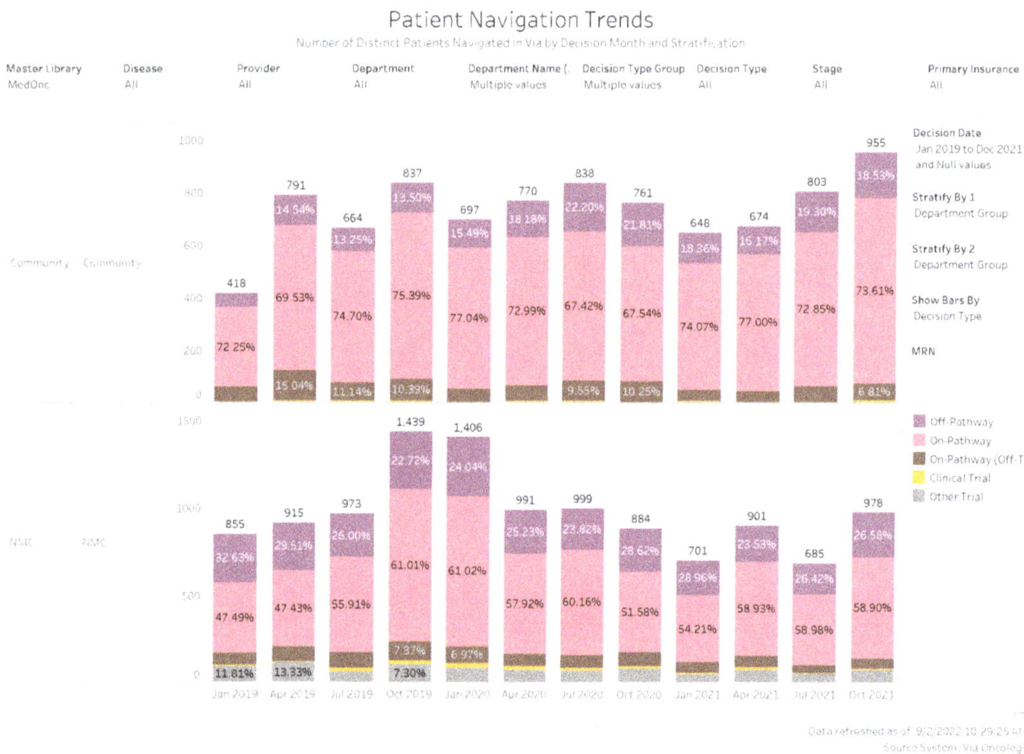

Figure 5. ClinPath Tableau Report of On-pathway vs Off-Pathway choices for covered solid tumors and hematology diagnoses by quarter in Duarte and Network sites for 1 January 2019–31 December 2021: Off-Pathway and On-pathway, (On-pathway, On-Pathway-Off Therapy, Clinical Trial, and Other Trials) navigation choices.

We also collect data on navigations for the eight commonly seen hematologic diseases through the pathways in ClinPath (chronic myelogenous leukemia (CML), chronic lymphocytic leukemia (CLL)/Lymphomas (B Cell, T Cell, and common histology Hodgkin's), multiple myeloma, myelodysplastic syndromes (MDS), and idiopathic thrombocytopenic purpura (ITP) and can generate Tableau reports. Network clinicians have valued and navigate care through the hematology pathways while academic center clinicians have only recently started using hematology pathways. The 3-year data, from 1 January 2019 to 31 December 2021, for network physicians' use of the pathways shows that 968 patients from network sites had therapies ordered for hematologic diseases with a range of 70–95% of decisions were On-pathway choices. Of note, ITP, a benign disease, is included in our pathways, as there are multiple very costly but effective options for therapies for this commonly seen disease in our network sites. Starting in a cost-effective sequence ensures that patients obtain the therapy with the best chance at efficacy and with the lowest cost or toxicity. Severely refractory patients often receive a long sequence of therapies, so choosing the most cost-effective approach at each step can lower their overall cost and improve quality of life by using oral therapies and those requiring fewer clinic visits early on and hoping that most will not need every type of therapy available.

Table 2. Enterprise 3-Year ClinPath Navigation Data: Percent (%) of On-pathway Decisions, total number (#) on-pathway decisions, and total number (#) of decisions by 18 tumor types with >200 decisions or >80% On-pathway rates. Tumors are ranked by the number of decisions from the highest number for breast cancers down to the lowest for testicular tumors. A blue highlight indicates On-pathway rates > 80%. Tumor type CLL is chronic lymphocytic leukemia, Neuro refers to brain tumors.

3-Year Enterprise On-Pathway Data			
Jan 2019 through Dec 2021			
Tumor Type	**% On Path**	**# On Path Decisions**	**Total # Decisions**
Breast	85%	5946	6954
Colorectal	77%	1834	2384
Non-Small Cell Lung	64%	1486	2317
Pancreatic	91%	1026	1124
Ovarian	70%	653	937
Gastroesophageal	81%	678	838
Prostate	91%	738	814
Head and Neck	69%	418	603
Lymphoma and CLL	83%	491	593
Uterine	67%	361	538
Melanoma + Skin	74%	366	494
Neuro	96%	355	369
Multiple Myeloma	88%	308	351
Renal	75%	242	325
Sarcoma	74%	226	304
Bladder	84%	251	300
Small Cell Lung	72%	153	214
Testicular	87%	77	89

As we engaged our academic disease leads for hematology and medical oncology, they identified many regimens that had not been built in our Epic Beacon system. Of approximately 900 regimens in our Epic Beacon system, 638 are regimens in the pathway system. Working with our disease leads in 2021 and 2022, we identified 300 therapy regimens that needed to be built, modified, merged, or mapped from the pathway tool into Epic to fully integrate our pathway and ordering system. 220 have been completed with 50 more due to be completed in December 2022 and the remaining 30 by February 2023. Completing standardized regimen builds in the EHR provides clinicians with a robust and efficient pathway ordering process from the pathway decision prompt. Standardized therapy orders support include the therapy agents, dosing and schedule as well as partnered antiemetic regimen by emetogenic risk level, laboratory, nursing and education visit orders to efficiently facilitate pre-authorizations, patient education, care delivery, payment metric reports and internal analytics. Two updates in progress will have the ClinPath team placing new clinical trials weekly into our pathways while an OnCore integration will provide real time status updates for clinical trials in the pathway tool. Clinical trials appear first in the pathway options and will be shown as pending, open, on hold or closed. These upgrades are expected to improve trial considerations and accruals.

Faculty Clinician Consultations: Formal consultations between the academic medical oncology clinicians to other academic faculty across disciplines totaled 4083 in 2019, decreased to 3978 consults during the first pandemic year (2020) and rebounded to 5635 consults in 2021 as the pandemic was mitigated in our region. The main specialties consulted over the 3 years from the academic medical oncology faculty were the surgical oncology specialties with 6119 consultations followed by radiation oncology with 3401 consultations and hematology with 703 consultations.

From the network of medical oncologists, there were 6203 consultations in 2019 referred to campus specialists in medical oncology (4883, 81%), surgical oncology specialties (744, 12%), hematology (393, 6%), and radiation oncology (91, 1%). By 2020, there were 5623 consultations in similar ratios for these main oncology specialties. This reflects that most network sites have City of Hope surgical and radiation oncology specialists locally who provide specialty oncology care. Thus, most consultations provided by medical oncologists to the academic center are for medical oncology to collaborate on complex patient presentations or clinical trials not offered in the community. Of note, there was only a 10% drop in overall consultations in 2020, which included the time after the global and regional COVID-19 pandemic was announced in March of 2020. This shows that despite the pandemic, cancer services could still be provided with the comprehensive safety measures instituted at the campus and network sites to protect staff, patients, and their families. By 2021, the third year of our study, similar tableau reporting on consultations from network medical oncologists provide to campus faculty showed a significant drop to 1494 consults overall, which is a 76% drop from 2019 and a 73% drop from 2020. These major decreases occurred in medical oncology consultations for gastrointestinal, breast, genitourinary, gynecologic, and thoracic subspecialties. The exact reasons for this drop have not been studied but are postulated to be from the expanded access to 1:1 informal faculty consultations, tumor boards, and complex oncology case discussions that that is providing high-quality input between the academic and network medical oncologists to meet patients' care-planning needs.

Data were not collected for the many informal consultations that occur between network and faculty clinicians nor for the questions raised, any changes in workup or care plans recommended, nor the potential impact on those discussions. Even though formal consultations result in consultation notes and communications, we do not currently track the impacts of recommendations from those consults. Most of the 29-member academic medical oncology faculty report at least one call per week from network clinicians and colleagues regarding oncology care planning for an oncology patient. Over 52 weeks, this would represent 1508 informal consultations or over 4500 consultations over 3 years. Our study identified this as an opportunity for improvement and to define and collect data on numbers, issues, and likely impacts, as well as the time invested for both formal and informal consults, in order to better understand the academic specialty faculty workloads and return on invested time.

Tumor Boards: A review of the available TB data demonstrated significant variability in the discrete data points collected and how the data were stored. A total of 4653 cases were presented across the eleven TBs. All cases contained the submitting provider's name and at least a general case description and question for the TB, though there was significant variation in the structure and extent of these data points. Case recommendations were recorded in six of the TBs, corresponding to 63% of all cases. However, whether these recommendations constituted a change from the initially proposed course of management was not recorded. The composite TB data that could be queried were only available for the Musculoskeletal Sarcoma TB, representing 8% of the total cases. This TB was led by only one disease expert since its inception, who established the methodology and oversaw the data collection. The data for other TBs were stored in the form of text documents or PDFs, which limited quantitative analysis (Table 3).

To characterize the impact of the transition from in-person only TBs to virtual TBs, the average TB attendance was compared between the 1 January 2019–30 June 2020 time period and the 1 July 2020–31 December 2021 time period (Figure 6). Absolute changes in the attendance of between −1 to 1 participant were considered stable. Of the eleven TBs, the attendance at four TBs increased while it remained stable in five TBs and decreased in two TBs. The largest increase in attendance occurred in the breast TB, with an average increase of 4.7 attendees ($p < 0.001$), while the largest decrease was −4.6 attendees in the neuroendocrine TB ($p < 0.001$). On average, across the eleven TBs, attendance increased by 0.6 attendees ($p < 0.001$).

Table 3. Tumor board data collected for 11 Multidisciplinary Tumor Boards Over the 3 years: 1 January 2019 to 31 December 2021.

Conference	Total Cases	Submitting Provider	Case Description	Specific Question	Case Recommendations	Data Centralized/Readily Accessible
Breast	585	Yes	Yes	Yes	Yes	No
Chest	600	Yes	Yes	Yes	No	No
Colorectal	326	Yes	Yes	Yes	No	No
Genitourinary	303	Yes	Yes	Yes	Yes	No
Gynecologic	521	Yes	Yes	Yes	Yes	No
Head and Neck	786	Yes	Yes	Yes	Yes	No
Melanoma	262	Yes	Yes	Yes	No	No
Musculoskeletal Sarcoma	375	Yes	Yes	Yes	Yes	Yes
Neuro-Endocrine	199	Yes	Yes	Yes	No	No
Neuro-Oncology	324	Yes	Yes	Yes	No	No
Upper Gastrointestinal	372	Yes	Yes	Yes	Yes	No
Total (%)	4653 (100%)	4653 (100%)	4653 (100%)	4653 (100%)	2942 (63%)	375 (8%)

There is no over-arching standard for the data collection or reporting of TB presentations nor a mechanism for tracking adherence to tumor board recommendations. The presenting faculty takes responsibility for executing the plan discussed in the tumor board. Another challenging issue is if more information is needed to render a final plan of care proposal, the re-presentation of the patient is not universally pursued, which could limit the potential for additional multidisciplinary decision making for this subset of patients.

Complex Oncology Case Discussions: We currently have three different Complex Oncology Case Discussions across the network of City of Hope. These meetings are organized regionally and have been held monthly with an option for urgent COCDs if needed. Table 4 shows the COCD data. COCDs' started at the Newport site, which now involves two other regions of practices that joined the network in 2020. As word spread on the value of COCDs to clinicians and patients, a second region started regular meetings in August 2020 and a third region started in March 2021. All are now held regularly with 14–21 attendees of which 4–10 are from the academic faculty. When answering questions for this study, the COCD leads uniformly described the COCDs as of high value to target complex questions very efficiently and completely. They also reported high patient satisfaction and peace of mind knowing their complex cancer diagnosis had been reviewed by specific experts to determine either the need for any additional workup, the option for a clinical trial, or the development or confirmation of a personalized cancer care plan. The Newport group highly values the involvement of their radiation oncologist as well as their onsite radiologist in the reviewing of films as needed. Other regional COCD directors hope to add such expertise as needed over time. COCDs' attendance is less diverse than that of tumor boards. These meetings are not supported by the pathology services, partly due to access to source materials. All patients receiving care at City of Hope, however, are required to have their pathology reviewed by COH pathologists. The case discussions are primarily focused on medical oncology interventions and transitions of care across lines of treatment as well as candidacy for clinical trials. The meetings are currently coordinated by the network physicians without using administrative staff. The format enables in-depth discussions, the engagement of the providers in the network practices, and the optimization of care via knowledge transfer. With these new meetings, we identified that having a standard intake-reporting form and meeting summary report with standardized categories of discussion and recommendations with possible likely impacts would be of value to the regional leads and the COH leadership. Discussions of a standardized format for such reporting are underway.

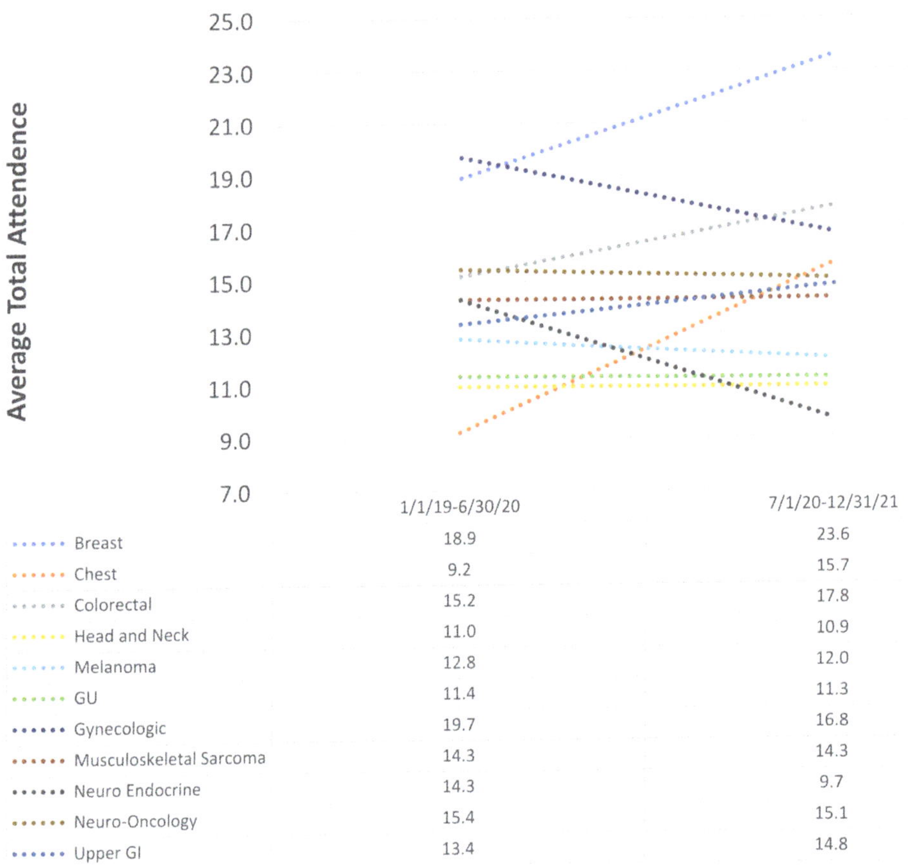

Figure 6. Tumor Board Attendance Over 3 years: 1 January 2019 to 31 December 2021.

Table 4. Details of comprehensive case discussion conferences.

	Orange County	Inland Empire	North Valley
Start	June 2020	August 2020	March 2021
Frequency	Monthly	Monthly	Monthly
Format	Virtual	Virtual	Virtual
Community Sites	Newport Beach, South Bay, Irvine	Upland, Corona, Arrowhead	Antelope Valley, Santa Clarita, Mission Hills, Thousand Oaks, Simi Valley
Attendees	14 (4 Duarte campus)	21 (10 Duarte campus)	16–20 (6–8 Duarte campus)
Cases Presented Per Meeting	4.5	4	5

4. Discussion

City of Hope's enterprise commitment to democratizing precision cancer care includes the provision of multispecialty, cutting-edge cancer knowledge to the chairside of every network clinician. With the rapid expansion of the City of Hope network regionally and now nationally, this goal is being realized through our pyramid of decision support. This pyramid expands on traditional evidence-based pathways, formal and informal faculty consultations, and tumor boards via its added Complex Oncology Case Discussions that bring rapidly evolving knowledge to clinicians so they can enhance the provision of customized cancer care plans to patients whose cancer diagnoses do not have standardized therapy approaches. The study of this framework—which has evolved over the last 4 years as we continue to expand our care delivery network regionally, nationally, and internationally—has identified strengths and opportunities to fill further gaps in understanding the efficacy and impact of these tools.

The pyramidal decision support project falls within our evidence-based care pillar in our comprehensive value-based care framework. These initiatives have been discussed previously [5]. They are supported by a comprehensive digital data strategy so as to have all discrete data within our enterprise data warehouse so that informative analytics can be made available to meet our expanding geography of care-delivery sites, expand access to clinical trials and state-of-the-art cancer care, achieve identified quality-of-care goals, and support our growing oncology-focused medical-home-type payor contracts.

The pathway improvements have come from the establishment of a formal program to capture disease leads directing new drugs and therapy builds in our Epic EHR, overseeing the addition of clinical trials, validating ClinPath recommendations, and identifying any customization to achieve the best outcomes for patients. New operational initiatives to incentivize real-time pathway navigation when ordering systemic medical oncology and hematologic therapies have been identified, as is performed by our radiation oncologists. This will improve the capturing of pathway choices for each line of therapy ordered, which can improve prior-authorization turnaround times, enhance analytics to support growing medical-home-type payor contracts, and inform our quality reporting and the disease leads of therapies being given throughout the enterprise for specific cancer subtypes. The improved discrete data capture of all entered elements from the pathway decision tool to our enterprise data warehouse (EDW) is underway to improve the validation of Epic and ClinPath data, which will support expanded value-based analytics.

Network and academic clinicians have long had the option to reach out to colleagues with subspecialized clinical and research expertise for the 1:1 discussion of patient issues and to order formal consultations when the standard of care is not applicable or optimal. We do not currently collect any data on these informal but very helpful consultations. However, we have data on formal requests for consultation. The consideration of a simple report of informal consultations with clinicians, patient issues, recommendation categories, and likely outcome impact via an efficient EHR tool or phone app might further capture valuable work performed by subspecialty cancer faculty for which they currently do not receive recognition, time, or compensation.

Disease-specific and precision oncology tumor boards carry out essential work by bringing multidisciplinary teams together to ensure that the care plans of presented patients are optimized. Given the substantial resources required to provide these tumor boards, we identified a need to understand the full impacts of this resource more formally. Our study led to the development and launch of a quality improvement study to capture (1) structured data about the issues raised for the patients presented, (2) attendees noting academic and network clinicians by specialty, and (3) the capture of structured decision impacts. Enhancing standardized data collection can better inform the enterprise about the impact of the many clinician and staff hours invested to improve patient care planning for patients' best health outcomes. The questionnaire for the study is shown in Supplemental Table S1.

Our novel Community Oncology Complex Discussions have been a welcome addition to our pyramid of decision support offerings. These discussions meet the requirement for providing the urgent expertise of academic site specialists who can be flexibly distributed to network practices without interrupting either the academic or network sites' operations. This integration is becoming more important as our enterprise grows to serve more diverse and geographically distant practices where general oncologists and their patients welcome input from subspecialty faculty who can rapidly share the newest therapeutic options, provide expert evaluation and recommendations for germline mutational-testing and results, advise on a clinical trial of a new agent, an agent available for compassionate use, or one not yet approved for an expanded indication. The impact of these complex decision-making discussions may provide patients with new therapeutic options from targeted therapies, to immune, or cellular therapies that may not be available or known to network clinicians but may significantly improve patients' cancer outcomes.

Our COCD model, however, could be further enhanced through cross-institutional collaborations for orphan diseases and rare tumor sub-types such as NUT carcinoma ('nuclear protein in testis' carcinomas, which can be found anywhere in the body but are often midline), which require national and international expertise to arrive at proper clinical trial options for this vulnerable population of patients. At the same time, there is potential to extend this model beyond clinical operations and integrate network practice physicians and patient data and specimens into research operations where specimens can be collected into a network-wide tumor bank that evaluates patients for potential new clinical trials based on the molecular and research-derived testing results [72]. Our institution has currently made strides in implementing this strategy of an institution-wide tumor bank for tissue samples and genetic results, including germline-testing results, but further efforts are required to ensure that all patients are captured in this model. The enhancement of the decision support pyramid with the COCD component has the potential to enhance and transform the enterprise practice of cancer care across the nation and allow for seamless, transformative precision medicine care for patients without the need for the traditional consultation model.

5. Conclusions

Decision support that can be efficiently and effectively provided in real-time or near real-time to every clinician before finalizing a patient's care plan for each episode of their cancer care has the best chance of optimizing patient outcomes across a spectrum of measures. Comprehensive decision support with integrated tools to share current and cutting-edge knowledge can improve diagnostic testing, identify the most effective therapies, and reduce toxicities and avoidable emergency room and hospital admissions, which can improve patient and providers' satisfaction and goal-concordant-end-of-life care. As payers move to more accountable, metric-based incentive contracts, having tools that incorporate reportable metrics and bring subspecialty faculty expertise to every network clinician can be informed by our Pyramidal Decision Support framework, which filled an unmet need with the addition of the COCD component. As enterprises such as City of Hope grow to expand access to high-quality cancer care, pyramidal decision support tools serve as critical components to democratize cancer care efficiently and with measurable outcomes.

Supplementary Materials: The following supporting information can be downloaded at: https://www.mdpi.com/article/10.3390/jcm11226738/s1, Figure S1: Tumor Types by Disease Categories in City of Hope's ClinPath Pathways. Table S1: Tumor Board Quality Improvement Questionnaire.

Author Contributions: Conceptualization, L.D.B., I.M., C.L., A.B., W.S., S.R., A.M., G.H., D.K., V.T., R.S.; methodology, L.D.B., I.M., C.L., A.B., D.J., D.M., W.S., S.R., A.M., G.H., D.K., J.F., V.T., R.S.; software, C.L., D.M., D.D.; validation, L.D.B., C.L., A.B., D.J., D.M., D.D.; formal analysis, L.D.B., I.M., C.L., A.B., D.J., D.D., D.M., W.S., S.R., A.M., G.H., D.K.; investigation, L.D.B., C.L., A.B., W.S., S.R., A.M., G.H., D.K., resources, L.D.B., V.T., R.S.; data curation, L.D.B., C.L., D.J., D.M., D.D.; writing—original draft preparation, L.D.B., I.M., C.L., A.B.; writing—review and editing, L.D.B., I.M., C.L., A.B., D.J., D.M., D.D., W.S., S.R., A.M., G.H., D.K., J.F., V.T., R.S.; visualization, L.D.B., I.M., C.L., D.J., D.M., D.D., J.F.; supervision, L.D.B., I.M., V.T., R.S.; project administration, L.D.B., D.M.; funding acquisition, R.S. All authors have read and agreed to the published version of the manuscript.

Funding: This research was funded by NIH, grant number P30CA033572.

Institutional Review Board Statement: Not applicable.

Informed Consent Statement: Not applicable.

Data Availability Statement: Not applicable.

Acknowledgments: The authors thank our dedicated City of Hope colleagues who enter data and navigate care with ClinPath Pathways during and after clinical care, which informs analytics that expands our understanding of our patients' diagnoses along with the care provided and their outcomes. We also thank our patients who trust us to partner with them to deliver the most up-to-date and effective care to improve their health outcomes. We acknowledge the many other informatics, nursing, administration, electronic health records, quality, pathways, supportive care, and genomics colleagues who made this study possible.

Conflicts of Interest: The authors declare no conflict of interest.

References

1. American Society of Clinical Oncology. The State of Cancer Care in America, 2016: A Report by the American Society of Clinical Oncology. *J. Oncol. Pract.* **2016**, *12*, 339–383. [CrossRef] [PubMed]
2. Kirkwood, M.K.; Hanley, A.; Bruinooge, S.S.; Garrett-Mayer, E.; Levit, L.A.; Schenkel, C.; Seid, J.E.; Polite, B.N.; Schilsky, R.L. The State of Oncology Practice in America, 2018: Results of the ASCO Practice Census Survey. *J. Oncol. Pract.* **2018**, *14*, e412–e420. [CrossRef] [PubMed]
3. Bosserman, L.D. Benefits and Challenges of Growing Oncology Networks in the United States. *J. Oncol. Pract.* **2018**, *14*, 761–762. [CrossRef] [PubMed]
4. Rajurkar, S.; Mambetsariev, I.; Pharaon, R.; Leach, B.; Tan, T.; Kulkarni, P.; Salgia, R. Non-Small Cell Lung Cancer from Genomics to Therapeutics: A Framework for Community Practice Integration to Arrive at Personalized Therapy Strategies. *J. Clin. Med.* **2020**, *9*, 1870. [CrossRef]
5. Bosserman, L.D.; Cianfrocca, M.; Yuh, B.; Yeon, C.; Chen, H.; Sentovich, S.; Polverini, A.; Zachariah, F.; Deaville, D.; Lee, A.B.; et al. Integrating Academic and Community Cancer Care and Research through Multidisciplinary Oncology Pathways for Value-Based Care: A Review and the City of Hope Experience. *J. Clin. Med.* **2021**, *10*, 188. [CrossRef]
6. Boland, G.M.; Chang, G.J.; Haynes, A.B.; Chiang, Y.J.; Chagpar, R.; Xing, Y.; Hu, C.Y.; Feig, B.W.; You, Y.N.; Cormier, J.N. Association between adherence to National Comprehensive Cancer Network treatment guidelines and improved survival in patients with colon cancer. *Cancer* **2013**, *119*, 1593–1601. [CrossRef]
7. Visser, B.C.; Ma, Y.; Zak, Y.; Poultsides, G.A.; Norton, J.A.; Rhoads, K.F. Failure to comply with NCCN guidelines for the management of pancreatic cancer compromises outcomes. *HPB* **2012**, *14*, 539–547. [CrossRef]
8. Kline, R.; Adelson, K.; Kirshner, J.J.; Strawbridge, L.M.; Devita, M.; Sinanis, N.; Conway, P.H.; Basch, E. The Oncology Care Model: Perspectives From the Centers for Medicare & Medicaid Services and Participating Oncology Practices in Academia and the Community. *Am. Soc. Clin. Oncol. Educ. Book* **2017**, *37*, 460–466. [CrossRef]
9. Seidman, J.; Masi, D.; Gomez-Rexrode, A.E. Personalizing Value in Cancer Care: The Case for Incorporating Patient Preferences Into Routine Clinical Decision Making. *J. Particip. Med.* **2019**, *11*, e13800. [CrossRef]
10. Pfohl, U.; Pflaume, A.; Regenbrecht, M.; Finkler, S.; Graf Adelmann, Q.; Reinhard, C.; Regenbrecht, C.R.A.; Wedeken, L. Precision Oncology Beyond Genomics: The Future Is Here-It Is Just Not Evenly Distributed. *Cells* **2021**, *10*, 928. [CrossRef]
11. El Saghir, N.S.; Keating, N.L.; Carlson, R.W.; Khoury, K.E.; Fallowfield, L. Tumor boards: Optimizing the structure and improving efficiency of multidisciplinary management of patients with cancer worldwide. *Am. Soc. Clin. Oncol. Educ. Book* **2014**, *34*, e461–e466. [CrossRef] [PubMed]
12. Knepper, T.C.; Bell, G.C.; Hicks, J.K.; Padron, E.; Teer, J.K.; Vo, T.T.; Gillis, N.K.; Mason, N.T.; McLeod, H.L.; Walko, C.M. Key Lessons Learned from Moffitt's Molecular Tumor Board: The Clinical Genomics Action Committee Experience. *Oncologist* **2017**, *22*, 144–151. [CrossRef] [PubMed]

13. Zon, R.T.; Frame, J.N.; Neuss, M.N.; Page, R.D.; Wollins, D.S.; Stranne, S.; Bosserman, L.D. American Society of Clinical Oncology Policy Statement on Clinical Pathways in Oncology. *J. Oncol. Pract.* **2016**, *12*, 261–266. [CrossRef] [PubMed]
14. Zon, R.T.; Edge, S.B.; Page, R.D.; Frame, J.N.; Lyman, G.H.; Omel, J.L.; Wollins, D.S.; Green, S.R.; Bosserman, L.D. American Society of Clinical Oncology Criteria for High-Quality Clinical Pathways in Oncology. *J. Oncol. Pract.* **2017**, *13*, 207–210. [CrossRef]
15. Chiang, A.C.; Lake, J.; Sinanis, N.; Brandt, D.; Kanowitz, J.; Kidwai, W.; Kortmansky, J.; LaSala, J.; Orell, J.; Sabbath, K.; et al. Measuring the Impact of Academic Cancer Network Development on Clinical Integration, Quality of Care, and Patient Satisfaction. *J. Oncol. Pract.* **2018**, *14*, e823–e833. [CrossRef]
16. Neubauer, M.A.; Hoverman, J.R.; Kolodziej, M.; Reisman, L.; Gruschkus, S.K.; Hoang, S.; Alva, A.A.; McArthur, M.; Forsyth, M.; Rothermel, T.; et al. Cost effectiveness of evidence-based treatment guidelines for the treatment of non-small-cell lung cancer in the community setting. *J. Oncol. Pract.* **2010**, *6*, 12–18. [CrossRef]
17. Hoverman, J.R.; Cartwright, T.H.; Patt, D.A.; Espirito, J.L.; Clayton, M.P.; Garey, J.S.; Kopp, T.J.; Kolodziej, M.; Neubauer, M.A.; Fitch, K.; et al. Pathways, outcomes, and costs in colon cancer: Retrospective evaluations in two distinct databases. *J. Oncol. Pract.* **2011**, *7*, 52s–59s. [CrossRef]
18. Weese, J.L.; Shamah, C.J.; Sanchez, F.A.; Moreno, A.C.P.; Mitchell, D.; Sessa, T.; Gutantes, J.; Huibregtse, C.; Clement, K.; Barry-Weers, A.M.; et al. Use of treatment pathways reduce cost and decrease ED utilization and unplanned hospital admissions in patients (pts) with stage II breast cancer. *J. Clin. Oncol.* **2019**, *37*, e12012. [CrossRef]
19. Jackman, D.M.; Zhang, Y.; Dalby, C.; Nguyen, T.; Nagle, J.; Lydon, C.A.; Rabin, M.S.; McNiff, K.K.; Fraile, B.; Jacobson, J.O. Cost and Survival Analysis Before and After Implementation of Dana-Farber Clinical Pathways for Patients With Stage IV Non-Small-Cell Lung Cancer. *J. Oncol. Pract.* **2017**, *13*, e346–e352. [CrossRef]
20. Newcomer, L.N.; Gould, B.; Page, R.D.; Donelan, S.A.; Perkins, M. Changing physician incentives for affordable, quality cancer care: Results of an episode payment model. *J. Oncol. Pract.* **2014**, *10*, 322–326. [CrossRef]
21. Doyle, C. Anthem's Clinical Pathways Demonstrate Value: The Payer Perspective. *Am. Health Drug Benefits* **2015**, *8*, 28. [PubMed]
22. Scott, J.A.; Milligan, S.; Wong, W.; Winn, D.; Cooper, J.; Schneider, N.; Parkes, S.; Feinberg, B.A. Validation of observed savings from an oncology clinical pathways program. *J. Clin. Oncol.* **2013**, *31*, 6553. [CrossRef]
23. City of Hope. How Innovations in Multiple Myeloma Care Result in Superior National Survival Outcomes. Available online: https://www.cityofhope.org/physician-news/myeloma-innovations-result-in-superior-survival-outcomes (accessed on 13 September 2022).
24. City of Hope. Lung Cancer Survival Rates Exceed National Averages*. Available online: https://www.cityofhope.org/patients/outcomes/best-lung-cancer-survival-results-in-southern-california (accessed on 13 September 2022).
25. City of Hope. Breast Cancer Survival Rates at City of Hope Exceed National Averages. Available online: https://www.cityofhope.org/physician-news/coh-breast-cancer-survival-rates-exceed-national-averages (accessed on 13 September 2022).
26. City of Hope. City of Hope Prostate Cancer Outcome Summary. Available online: https://www.cityofhope.org/sites/www/files/2022-05/prostate-outcomes-report-feb2022.pdf (accessed on 13 September 2022).
27. City of Hope. City of Hope Colorectal Cancer Outcome Summary. 2022. Available online: https://www.cityofhope.org/sites/www/files/2022-05/colorectal-outcome-summary.pdf (accessed on 13 September 2022).
28. Flaherty, K.T.; Gray, R.; Chen, A.; Li, S.; Patton, D.; Hamilton, S.R.; Williams, P.M.; Mitchell, E.P.; Iafrate, A.J.; Sklar, J.; et al. The Molecular Analysis for Therapy Choice (NCI-MATCH) Trial: Lessons for Genomic Trial Design. *J. Natl. Cancer Inst.* **2020**, *112*, 1021–1029. [CrossRef] [PubMed]
29. Inal, C.; Yilmaz, E.; Cheng, H.; Zhu, C.; Pullman, J.; Gucalp, R.A.; Keller, S.M.; Perez-Soler, R.; Piperdi, B. Effect of reflex testing by pathologists on molecular testing rates in lung cancer patients: Experience from a community-based academic center. *J. Clin. Oncol.* **2014**, *32*, 8098. [CrossRef]
30. Gutierrez, M.E.; Choi, K.; Lanman, R.B.; Licitra, E.J.; Skrzypczak, S.M.; Pe Benito, R.; Wu, T.; Arunajadai, S.; Kaur, S.; Harper, H.; et al. Genomic Profiling of Advanced Non-Small Cell Lung Cancer in Community Settings: Gaps and Opportunities. *Clin. Lung Cancer* **2017**, *18*, 651–659. [CrossRef]
31. Gierman, H.J.; Goldfarb, S.; Labrador, M.; Weipert, C.M.; Getty, B.; Skrzypczak, S.M.; Catasus, C.; Carbral, S.; Singaraju, M.; Singleton, N.; et al. Genomic testing and treatment landscape in patients with advanced non-small cell lung cancer (aNSCLC) using real-world data from community oncology practices. *J. Clin. Oncol.* **2019**, *37*, 1585. [CrossRef]
32. Presley, C.J.; Tang, D.; Soulos, P.R.; Chiang, A.C.; Longtine, J.A.; Adelson, K.B.; Herbst, R.S.; Zhu, W.; Nussbaum, N.C.; Sorg, R.A.; et al. Association of Broad-Based Genomic Sequencing With Survival Among Patients With Advanced Non-Small Cell Lung Cancer in the Community Oncology Setting. *JAMA* **2018**, *320*, 469–477. [CrossRef]
33. Illei, P.B.; Wong, W.; Wu, N.; Chu, L.; Gupta, R.; Schulze, K.; Gubens, M.A. ALK Testing Trends and Patterns Among Community Practices in the United States. *JCO Precis. Oncol.* **2018**, *2*, 1–11. [CrossRef]
34. Hussein, M.; Richards, D.A.; Ulrich, B.; Korytowsky, B.; Pandya, D.; Cogswell, J.; Batenchuk, C.; Burns, V. ORAL01.02: Biopsies in Initial Diagnosis of Non–Small Cell Lung Cancer in US Community Oncology Practices: Implications for First-Line Immunotherapy: Topic: Medical Oncology. *J. Thorac. Oncol.* **2016**, *11*, S249–S250. [CrossRef]
35. Mason, C.; Ellis, P.G.; Lokay, K.; Barry, A.; Dickson, N.; Page, R.; Polite, B.; Salgia, R.; Savin, M.; Shamah, C.; et al. Patterns of Biomarker Testing Rates and Appropriate Use of Targeted Therapy in the First-Line, Metastatic Non-Small Cell Lung Cancer Treatment Setting. *J. Clin. Pathw.* **2018**, *4*, 49–54. [CrossRef]

36. Khozin, S.; Abernethy, A.P.; Nussbaum, N.C.; Zhi, J.; Curtis, M.D.; Tucker, M.; Lee, S.E.; Light, D.E.; Gossai, A.; Sorg, R.A.; et al. Characteristics of Real-World Metastatic Non-Small Cell Lung Cancer Patients Treated with Nivolumab and Pembrolizumab During the Year Following Approval. *Oncologist* **2018**, *23*, 328–336. [CrossRef] [PubMed]
37. Nadler, E.; Espirito, J.L.; Pavilack, M.; Boyd, M.; Vergara-Silva, A.; Fernandes, A. Treatment Patterns and Clinical Outcomes Among Metastatic Non–Small-Cell Lung Cancer Patients Treated in the Community Practice Setting. *Clin. Lung Cancer* **2018**, *19*, 360–370. [CrossRef] [PubMed]
38. Nadler, E.; Pavilack, M.; Clark, J.; Espirito, J.; Fernandes, A. Biomarker Testing Rates in Patients with Advanced Non-Small Cell Lung Cancer Treated in the Community. *J. Cancer Ther.* **2019**, *10*, 971–984. [CrossRef]
39. Boehmer, L.; Shivakumar, L.; Weldon, C.B.; Trosman, J.R.; Cohen, S.A.; Nixon, D.; Catts, Z.A.-K.; Miesfeldt, S.; Wiryaman, S.; Tafra, L.; et al. Genetic counseling and testing rates among community cancer programs for patients with breast cancer following site-directed quality improvement. *J. Clin. Oncol.* **2021**, *39*, 10529. [CrossRef]
40. Kurian, A.W.; Ward, K.C.; Howlader, N.; Deapen, D.; Hamilton, A.S.; Mariotto, A.; Miller, D.; Penberthy, L.S.; Katz, S.J. Genetic Testing and Results in a Population-Based Cohort of Breast Cancer Patients and Ovarian Cancer Patients. *J. Clin. Oncol.* **2019**, *37*, 1305–1315. [CrossRef]
41. Chandler, Y.; Schechter, C.B.; Jayasekera, J.; Near, A.; O'Neill, S.C.; Isaacs, C.; Phelps, C.E.; Ray, G.T.; Lieu, T.A.; Ramsey, S.; et al. Cost Effectiveness of Gene Expression Profile Testing in Community Practice. *J. Clin. Oncol.* **2018**, *36*, 554–562. [CrossRef]
42. Levit, L.A.; Kim, E.S.; McAneny, B.L.; Nadauld, L.D.; Levit, K.; Schenkel, C.; Schilsky, R.L. Implementing Precision Medicine in Community-Based Oncology Programs: Three Models. *J. Oncol. Pract.* **2019**, *15*, 325–329. [CrossRef]
43. Patel, K.; Clinton, N.; Mukhi, H.; Patel, M. The Role of In-Office Next Generation Sequencing to Advance Precision Medicine in Community Oncology. Available online: https://www.targetedonc.com/view/the-role-of-in-office-next-generation-sequencing-to-advance-precision-medicine-in-community-oncology (accessed on 7 September 2022).
44. Norton, W.E.; McCaskill-Stevens, W.; Chambers, D.A.; Stella, P.J.; Brawley, O.W.; Kramer, B.S. DeImplementing Ineffective and Low-Value Clinical Practices: Research and Practice Opportunities in Community Oncology Settings. *JNCI Cancer Spectrum.* **2021**, *5*, pkab020. [CrossRef]
45. Shivakumar, L.; Weldon, C.B.; Lucas, L.; Perloff, T. Identifying obstacles to optimal integration of cancer immunotherapies in the community setting. *J. Oncol. Pract.* **2019**, *37*, 87. [CrossRef]
46. Concepcion, R.S. Germline testing for prostate cancer: Community urology perspective. *Can. J. Urol.* **2019**, *26*, 50–51.
47. Lincoln, S.E.; Nussbaum, R.L.; Kurian, A.W.; Nielsen, S.M.; Das, K.; Michalski, S.; Yang, S.; Ngo, N.; Blanco, A.; Esplin, E.D. Yield and Utility of Germline Testing Following Tumor Sequencing in Patients With Cancer. *JAMA Netw. Open* **2020**, *3*, e2019452. [CrossRef] [PubMed]
48. Weise, N.; Shaya, J.; Javier-Desloges, J.; Cheng, H.H.; Madlensky, L.; McKay, R.R. Disparities in germline testing among racial minorities with prostate cancer. *Prostate Cancer Prostatic Dis.* **2022**, *25*, 403–410. [CrossRef] [PubMed]
49. Ashing, K.T.; Jones, V.; Bedell, F.; Phillips, T.; Erhunmwunsee, L. Calling Attention to the Role of Race-Driven Societal Determinants of Health on Aggressive Tumor Biology: A Focus on Black Americans. *JCO Oncol. Pract.* **2022**, *18*, 15–22. [CrossRef] [PubMed]
50. Bruno, D.S.; Hess, L.M.; Li, X.; Su, E.W.; Zhu, Y.E.; Patel, M. Racial disparities in biomarker testing and clinical trial enrollment in non-small cell lung cancer (NSCLC). *J. Clin. Oncol.* **2021**, *39*, 9005. [CrossRef]
51. Frosch, Z.A.K.; Illenberger, N.; Mitra, N.; Boffa, D.J.; Facktor, M.A.; Nelson, H.; Palis, B.E.; Bekelman, J.E.; Shulman, L.N.; Takvorian, S.U. Trends in Patient Volume by Hospital Type and the Association of These Trends With Time to Cancer Treatment Initiation. *JAMA Netw. Open* **2021**, *4*, e2115675. [CrossRef] [PubMed]
52. Fingrut, W.; Beck, L.A.; Lo, D. Oncology communities of practice: Insights from a qualitative analysis. *Curr. Oncol.* **2018**, *25*, 378–383. [CrossRef] [PubMed]
53. Zon, R.T.; Bruinooge, S.S.; Lyss, A.P. The Changing Face of Research in Community Practice. *J. Oncol. Pract.* **2014**, *10*, 155–160. [CrossRef]
54. Freytag, M.; Herrlinger, U.; Hauser, S.; Bauernfeind, F.G.; Gonzalez-Carmona, M.A.; Landsberg, J.; Buermann, J.; Vatter, H.; Holderried, T.; Send, T.; et al. Higher number of multidisciplinary tumor board meetings per case leads to improved clinical outcome. *BMC Cancer* **2020**, *20*, 355. [CrossRef]
55. Hillner, B.E.; Smith, T.J.; Desch, C.E. Hospital and physician volume or specialization and outcomes in cancer treatment: Importance in quality of cancer care. *J. Clin. Oncol.* **2000**, *18*, 2327–2340. [CrossRef]
56. Onega, T.; Duell, E.J.; Shi, X.; Demidenko, E.; Gottlieb, D.; Goodman, D.C. Influence of NCI cancer center attendance on mortality in lung, breast, colorectal, and prostate cancer patients. *Med. Care Res. Rev.* **2009**, *66*, 542–560. [CrossRef]
57. Mambetsariev, I.; Pharaon, R.; Nam, A.; Knopf, K.; Djulbegovic, B.; Villaflor, V.M.; Vokes, E.E.; Salgia, R. Heuristic value-based framework for lung cancer decision-making. *Oncotarget* **2018**, *9*, 29877–29891. [CrossRef] [PubMed]
58. Zubiri, L.; Rosovsky, R.P.; Mooradian, M.J.; Piper-Vallillo, A.J.; Gainor, J.F.; Sullivan, R.J.; Marte, D.; Boland, G.M.; Gao, X.; Hochberg, E.P.; et al. Temporal Trends in Inpatient Oncology Census Before and During the COVID-19 Pandemic and Rates of Nosocomial COVID-19 Among Patients with Cancer at a Large Academic Center. *Oncologist* **2021**, *26*, e1427–e1433. [CrossRef] [PubMed]
59. Prigoff, J.; Hillyer, G.; Bell, F.; Accordino, M.K. Effects of COVID-19 on an academic breast oncology center in New York City. *J. Clin. Oncol.* **2020**, *38*, 51. [CrossRef]

60. Gong, J.; Pan, K.; Fakih, M.; Pal, S.; Salgia, R. Value-based genomics. *Oncotarget* **2018**, *9*, 15792–15815. [CrossRef]
61. Mullangi, S.; Schleicher, S.M.; Parikh, R.B. The Oncology Care Model at 5 Years-Value-Based Payment in the Precision Medicine Era. *JAMA Oncol.* **2021**, *7*, 1283–1284. [CrossRef]
62. Olver, I.; Carey, M.; Bryant, J.; Boyes, A.; Evans, T.; Sanson-Fisher, R. Second opinions in medical oncology. *BMC Palliat. Care* **2020**, *19*, 112. [CrossRef]
63. Pishvaian, M.J.; Blais, E.M.; Bender, R.J.; Rao, S.; Boca, S.M.; Chung, V.; Hendifar, A.E.; Mikhail, S.; Sohal, D.P.S.; Pohlmann, P.R.; et al. A virtual molecular tumor board to improve efficiency and scalability of delivering precision oncology to physicians and their patients. *JAMIA Open* **2019**, *2*, 505–515. [CrossRef]
64. Stevenson, M.M.; Irwin, T.; Lowry, T.; Ahmed, M.Z.; Walden, T.L.; Watson, M.; Sutton, L. Development of a virtual multidisciplinary lung cancer tumor board in a community setting. *J. Oncol. Pract.* **2013**, *9*, e77–e80. [CrossRef]
65. Newman, E.A.; Guest, A.B.; Helvie, M.A.; Roubidoux, M.A.; Chang, A.E.; Kleer, C.G.; Diehl, K.M.; Cimmino, V.M.; Pierce, L.; Hayes, D.; et al. Changes in surgical management resulting from case review at a breast cancer multidisciplinary tumor board. *Cancer* **2006**, *107*, 2346–2351. [CrossRef]
66. Keating, N.L.; Landrum, M.B.; Lamont, E.B.; Bozeman, S.R.; Shulman, L.N.; McNeil, B.J. Tumor boards and the quality of cancer care. *J. Natl. Cancer Inst.* **2013**, *105*, 113–121. [CrossRef]
67. Hall, P.; Weaver, L. Interdisciplinary education and teamwork: A long and winding road. *Med. Educ.* **2001**, *35*, 867–875. [CrossRef] [PubMed]
68. Wright, F.C.; De Vito, C.; Langer, B.; Hunter, A. Multidisciplinary cancer conferences: A systematic review and development of practice standards. *Eur. J. Cancer* **2007**, *43*, 1002–1010. [CrossRef] [PubMed]
69. Lesslie, M.; Parikh, J.R. Implementing a Multidisciplinary Tumor Board in the Community Practice Setting. *Diagnostics* **2017**, *7*, 55. [CrossRef] [PubMed]
70. Lesslie, M.D.; Parikh, J.R. Multidisciplinary Tumor Boards: An Opportunity for Radiologists to Demonstrate Value. *Acad. Radiol.* **2017**, *24*, 107–110. [CrossRef] [PubMed]
71. Gebbia, V.; Guarini, A.; Piazza, D.; Bertani, A.; Spada, M.; Verderame, F.; Sergi, C.; Potenza, E.; Fazio, I.; Blasi, L.; et al. Virtual Multidisciplinary Tumor Boards: A Narrative Review Focused on Lung Cancer. *Pulm. Ther.* **2021**, *7*, 295–308. [CrossRef]
72. Geiger, A.M.; O'Mara, A.M.; McCaskill-Stevens, W.J.; Adjei, B.; Tuovenin, P.; Castro, K.M. Evolution of Cancer Care Delivery Research in the NCI Community Oncology Research Program. *J. Natl. Cancer Inst.* **2020**, *112*, 557–561. [CrossRef]

Journal of
Clinical Medicine

MDPI

Review

Clinical Network Systems Biology: Traversing the Cancer Multiverse

Isa Mambetsariev [1,†], Jeremy Fricke [1,†], Stephen B. Gruber [1], Tingting Tan [1], Razmig Babikian [1], Pauline Kim [2], Priya Vishnubhotla [1,3], Jianjun Chen [4], Prakash Kulkarni [1,4] and Ravi Salgia [1,*]

[1] Department of Medical Oncology and Therapeutic Research, City of Hope National Medical Center, Duarte, CA 91010, USA
[2] Department of Pharmacy, City of Hope National Medical Center, Duarte, CA 91010, USA
[3] Department of Medical Oncology, City of Hope Atlanta, Newnan, GA 30265, USA
[4] Department of Systems Biology, City of Hope National Medical Center, Duarte, CA 91010, USA
* Correspondence: rsalgia@coh.org
† These authors contributed equally to this work and should be considered co-first authors.

Abstract: In recent decades, cancer biology and medicine have ushered in a new age of precision medicine through high-throughput approaches that led to the development of novel targeted therapies and immunotherapies for different cancers. The availability of multifaceted high-throughput omics data has revealed that cancer, beyond its genomic heterogeneity, is a complex system of microenvironments, sub-clonal tumor populations, and a variety of other cell types that impinge on the genetic and non-genetic mechanisms underlying the disease. Thus, a systems approach to cancer biology has become instrumental in identifying the key components of tumor initiation, progression, and the eventual emergence of drug resistance. Through the union of clinical medicine and basic sciences, there has been a revolution in the development and approval of cancer therapeutic drug options including tyrosine kinase inhibitors, antibody–drug conjugates, and immunotherapy. This 'Team Medicine' approach within the cancer systems biology framework can be further improved upon through the development of high-throughput clinical trial models that utilize machine learning models, rapid sample processing to grow patient tumor cell cultures, test multiple therapeutic options and assign appropriate therapy to individual patients quickly and efficiently. The integration of systems biology into the clinical network would allow for rapid advances in personalized medicine that are often hindered by a lack of drug development and drug testing.

Keywords: team medicine; precision medicine; cancer systems biology; clinical network systems biology

Citation: Mambetsariev, I.; Fricke, J.; Gruber, S.B.; Tan, T.; Babikian, R.; Kim, P.; Vishnubhotla, P.; Chen, J.; Kulkarni, P.; Salgia, R. Clinical Network Systems Biology: Traversing the Cancer Multiverse. *J. Clin. Med.* **2023**, *12*, 4535. https://doi.org/10.3390/jcm12134535

Academic Editor: Enrico Capobianco

Received: 12 May 2023
Revised: 29 June 2023
Accepted: 1 July 2023
Published: 7 July 2023

1. Introduction

Cancer is a complex disease that is caused by a dysfunction of normal cell biology through genetic and non-genetic changes including epigenetic changes that corrode the cell's ability to promote cell death, resulting in a process of dysregulated growth and proliferation. Every year, approximately over 1.9 million people are diagnosed with cancer and 609,820 die from cancer in the United States alone [1]. The discovery of new diagnostic tools, immunotherapy, and novel therapies has helped to reduce the cancer death rate by 33% since 1991, but despite this positive milestone, the improvement in outcomes has not been uniform across all tumor types [1]. This is largely in part due to the heterogeneity of cancer as a multi-modal disease that is driven by a collection of genetic and non-genetic mechanisms, which means tumors from a single tissue type do not respond to the same therapies despite similar histological profiles [2–4]. Therefore, considerable effort has been invested over the last 20 years to understand the biology of cancer and more importantly cancer within individual patients to decipher the heterogeneity of cancer types [3,5–7]. The revolution in next-generation sequencing, liquid biopsy, single-cell sequencing, proteomics, and other novel diagnostic techniques has generated large libraries of whole genome, transcriptome, epigenetic, proteomic, and metabolomic data [3,5–8]. However, the relationship between the individual gene and protein discoveries is not intrinsic in affecting tumor pathology, and often times, intricate cascade effects in transcriptional, translational and post-translational modification limit therapeutic efficacy [9,10]. In essence, effective cancer therapeutics cannot be achieved through understanding a cancer's individual parts but requires a systems biology approach where large cross-collaborations of multi-modal scientists, clinicians, and experts collaborate to understand the entirety of the oncogenic network.

Systems biology at its foundation is comprehending that the whole is greater than the sum of its individual parts and is a heuristic process of collaboration, prediction, and discovery that has yielded several scientific discoveries in the last century [11,12]. Within a biological system, key processes are necessary for system-level insight and understanding including system structures, systems dynamics, the control method, and the design method as initially described by Kitano et al. [13]. Within the cancer systems biology paradigm, the system structures can be separated into five networks including the gene regulatory network, molecular network, cellular network, organ network, and clinical and research network. The systems dynamics process aims to understand how cancer as a complex system of abnormal cell growth behaves and changes over time from an initial set of conditions [14]. The cancer control method of systems biology relies on modulating the state of the cell to limit cancer growth or induce apoptosis to validate potential therapeutic options [15,16]. The highest level of cancer systems biology is a design method or design principles where multi-dimensional models, from in silico mathematical models to cell cultures to organoids to mouse PDXs, are constructed to mimic and mirror the oncogenic properties of individual patients or a cohort of patients so that therapies can be tested and applied based on the definitive initial conditions of the tumor [16–18]. Due to this multi-scale and multi-modal persistence of cancer, we propose a novel highly adaptive approach of clinical network cancer systems biology that integrates basic science expertise and novel methodology with physician-level expertise and patient access to achieve the dream of personalized medicine. With the advent of modern technology, especially machine learning and artificial intelligence (AI), it is noticeably clear that cancer systems biology ought to take on an integrated approach where preclinical biology, patient translational specimens, and clinical care are all merged under a singular umbrella.

2. Systems Biology in Cancer

One of the primary challenges in cancer is that it cannot be understood through a simplistic lens due to the nonlinear nature of the disease process and its subsequent evolution. At the organ level, cancers exhibit differential patterns, and more evidence has shown that cancer metastasis may have a deterministic pattern to its chaotic process where certain genotypes show a preference toward target organs [19]. Furthermore, the tumor tissue and its tumor microenvironment (TME) vary by cancer type, and recent evidence shows that the TME may have an active role in the proliferation, migration, invasion, survival, angiogenesis, and EMT within the cancer cell network [20]. This is further complicated by protein signaling networks and biochemical signaling pathways involved in cancer progression that are difficult to predict and overcome therapeutically due to distinct perturbations in genotypes and phenotypes that drive their formation and interaction [21,22]. At the lower magnification, genomic instability in DNA repair and maintenance mechanisms as well as the disruption of epigenetic regulators has led to the discovery of several genomic alterations and chromatin modifications. This has unfortunately led to a high failure rate with only 6.7% of therapies reaching the phase II trial phase with regulatory approval between 2009 and 2018 [23]. Ultimately, the issue of cancer drug discovery is two-fold in that while with the help of next-generation sequencing, large cohorts of patients have been identified with novel targeted therapeutic options such as NSCLC EGFR-mutated patients or BRCA-2 positive breast cancer, there were also numerous cohorts of patients discovered with genomic alterations that have no clinically proven drug options such as TP53, ARID1A or PIK3CA [24].

The discovery of novel therapeutics based on recent preclinical biological discoveries is an iterative process within cancer systems biology that can be represented as a life cycle of research that combines wet-lab and dry-lab efforts to arrive at validated therapeutics (Figure 1). While traditionally systems biology begins its life cycle at preclinical basic research, this is different in cancer in that there is a wide berth of data that is publicly available from large cancer databases such as TCGA and publicly available results from individual large cohort studies. This makes the life cycle of cancer systems biology more fluid in that initial discoveries or drug targets can be made prior to any wet-lab experiments through bioinformatics analyses and in silico modeling. Nevertheless, wet-lab analytical modeling involving cell lines, 3D spheroids, tumoroids, and in vivo experiments is a required stepping stone toward verifying an underlying therapeutic hypothesis regardless of whether the foundation of that hypothesis was based on previous preclinical or clinical knowledge. Subsequently, predictive modeling and translational research go hand in hand in validating the clinical efficacy and viability of any therapeutic approach. This is then followed by biomarker discovery and computational modeling where potential therapeutics attempt to find the "best-fit niche" for their mechanism of action. However, it is important to underscore that the cancer systems biology life cycle is nonlinear, and each step may flow back into the previous step where further analytical modeling and predictive modeling work is required based on the computational and biomarker findings, which in turn may require new hypotheses to be made. This has further importance in clinical trials and personalized medicine where initial findings of the therapeutic in a clinical population such as toxicity or various omics profiling may yield results that require further drug optimization or drug repurposing.

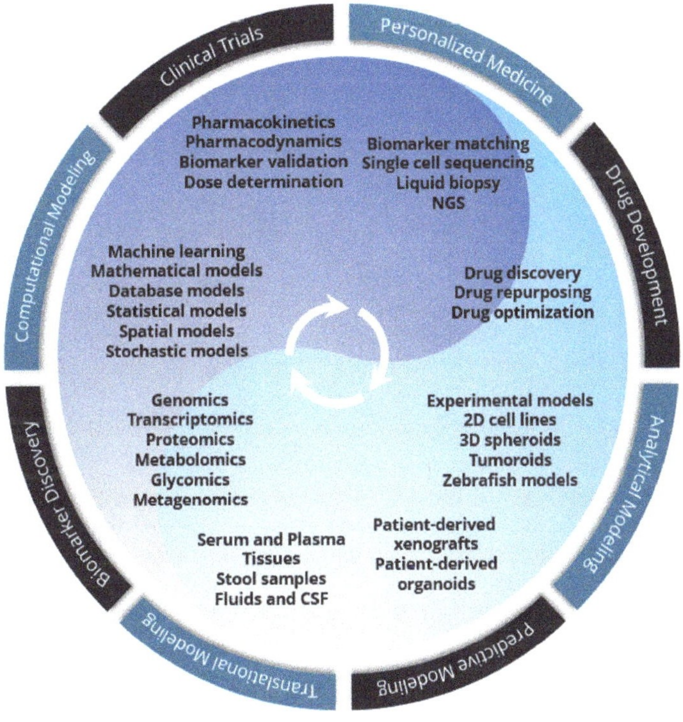

Figure 1. The life cycle of cancer systems biology drug discovery.

The arrival of next-generation sequencing in the clinical setting has allowed for the further stratification of individual cancer types beyond their histology or tumor locale. However, as mentioned previously, cancer systems biology is complicated by the fact that individual components of data do not represent the entire network of the cancer system. While genomic data has been valuable in developing targeted therapies and stratifying patients by biomarkers, it is not uniform with actionable mutation rates in patients varying from 10.8% to 90.6% depending on cancer type [25]. This leaves large cohorts of patients without viable therapeutic options. A recent example is EGFR to SCLC transformation following osimertinib therapy, which underscores the importance of non-genetic mechanisms at play in cancer resistance [26]. The underlying challenge for this beyond identifying the possible drug candidates and novel therapeutic approaches is clinical trial cost and a lack of clinical trial integration into the oncology standard of care, which in turn further increases clinical trial costs [27–29]. This is in part due to the traditional clinical trial model where cohorts of patients at different sites especially in the community network are screened for individual trials separately to identify an individual with a biomarker that is possibly present in less than 1% of that cancer population [27–29]. The implementation of large umbrella trials such as the Lung-MAP, ALCHEMIST, or NCI-MATCH trials that aim to screen patients' biomarkers and match the patients to appropriate therapies have been successful in the academic setting [30]. However, it has been reported that 40% of patients were more than 60 min away from a clinical trial location, which is a central issue in increasing NCI-MATCH trial recruitment in the community network setting [31,32]. This is further complicated by the lack of access to the community practice patients from the trial and drug development perspective in that often, the complex community network patients do not have access to trials that address their biomarker [33].

We believe the solution to these issues is a novel approach that integrates cancer systems biology with a concept that we previously identified called "Team Medicine" [34,35]. Team Medicine is a cross-collaborative effort to integrate basic scientists with clinicians to drive forward rapid-pace translational research. The merging of Team Medicine and cancer systems biology would result in a new paradigm called Clinical Network Systems Biology where the academic site, the clinical community network, and basic scientists at a research center would integrate under one umbrella to discover, develop, and test novel therapeutics at a rapid pace to achieve more personalized medicine (Figure 2). The framework embodies the four biological networks involved in cancer including the organ network, cellular network, molecular network, and gene regulatory network, and it combines it with the clinical and research network that encompasses the primary academic site and community practice network.

In the subsequent sections, we will delve deeper into the two components that comprise this framework by looking at the individual parts of the biological network that drive the patients' cancer and the various strategies that can be utilized in the clinical network to enhance the basics of systems biology toward precision medicine.

2.1. Biological Network in Cancer Systems Biology

Clinical Network Systems Biology is analogous to the Matryoshka nesting dolls: a set of wooden dolls of increasing size placed one inside another. Thus, Matryoshka serves as a great metaphor for a complex system. Analytically speaking, the metaphor is especially well suited since it is likened to thinking in systems. Relatively speaking, a system may be defined as an interconnected set of components that are organized toward a specific function or purpose. Complex systems are systems within a system. Indeed, a Clinical Network may be thought of as a complex system itself. Here, the biological network may be perceived as comprising the inner (smallest) doll representing a single cell with its gene network, i.e., the gene regulatory network (GRN) together with the non-genetic, protein interaction network (PIN), which is followed by the next (bigger) doll representing the cellular networks to form tissues that comprise the individual organs and, finally, a bigger doll representing a network of organs that constitute an individual. Thus, it follows that a Clinical Network is a complex system comprising many systems which may interact with each other with dependencies, competitions, relationships, or other types of interactions such as feedback loops between their parts or between the system and its environment. These interactive systems are traditionally called complex adaptive systems (CAS) where the biological behavior of one component does not predict the behavior of the other components. CASs are capable of self-organization that adapts to their environmental stimuli, which increases their chances of survival. Therefore, due to the unpredictable and temporal nature of these systems, they cannot be studied with traditional tools and require analysis using nonlinear dynamical models that can accurately predict emergent behaviors, cellular plasticity, and heterogeneous cells.

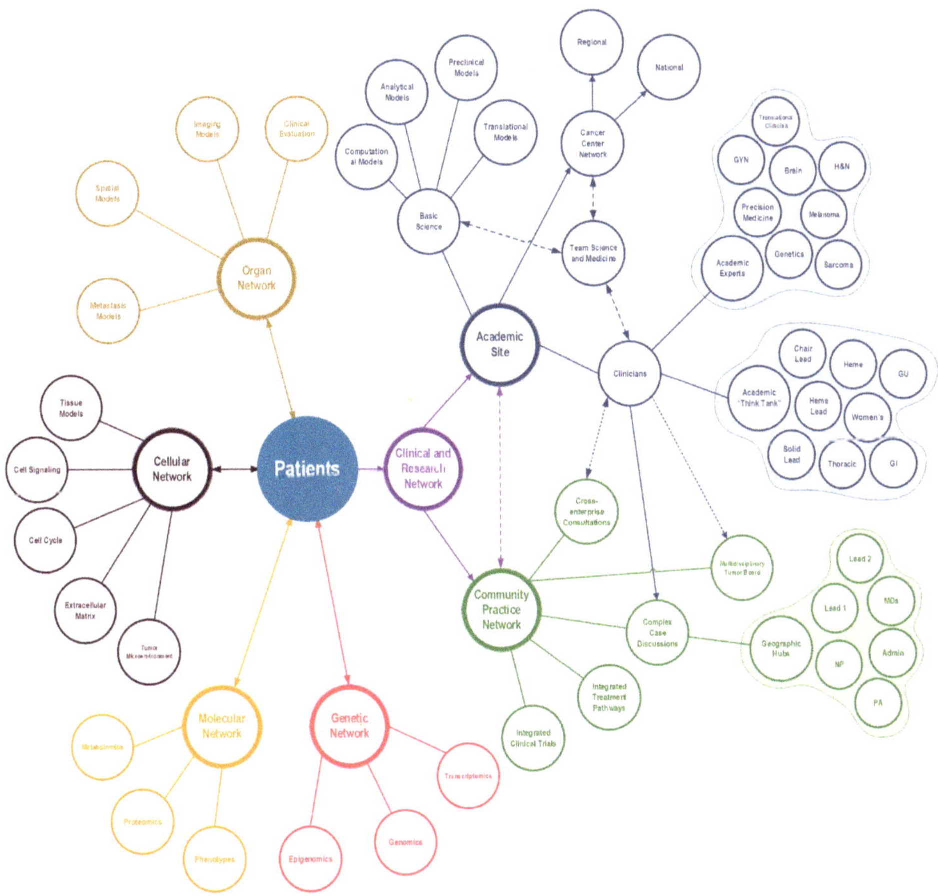

Figure 2. Clinical Network Systems Biology framework that integrates the biological networks with the clinical and research networks.

GRN (Gene Regulatory Network): At the principal level, a GRN is a group of genes that are characterized by gene expression and linked to one another through target gene nodes that regulate a specific cell function. Such interactions are genetically "wired" to ensure transgenerational transfer with high fidelity. Regulators of gene expression include transcription factors (TFs) that typically bind specific DNA sequence motifs and transcriptional regulators that typically interact with the basic transcriptional machinery and specific transcription factors. Both TFs and regulators can act as either activators of gene expression or as repressors that repress gene expression. Other molecules that may also play important roles in regulating gene expression include RNA-binding proteins and regulatory RNAs. Elucidating the intricate regulatory relationships between TFs, transcriptional regulators and their targets is essential to understand cellular functions such as cell growth and division, differentiation, and development. They can also help shed light on evolution, especially in the past half a billion years or so [36]. Furthermore, identifying GRNs can also aid in understanding how the dysregulation of gene expression contributes to complex heritable diseases as well as diseases such as cancer that have both genetic and non-genetic underpinnings [37,38].

PIN (Protein Interacting Network): The proteins that result from differential gene expression regulated by the GRNs interact with their cognate partners to form cellular PINs. While it was initially believed that PIN configurations occur randomly, Barabási and colleagues showed that PINs have a "scale-free" architecture in which the degree distribution P(k) expresses a power-law behavior as a function of the degree k [39,40]. A major advantage of scale-free networks is that they are largely resistant to random node failure, but they are vulnerable to critical hub failures [39].

Intrinsically disordered proteins (IDPs) are proteins that lack unique 3D structures and constitute a significant fraction of the proteome [41,42]. Because IDPs exist as conformational ensembles (are highly malleable), they can interact with multiple partners [43]. Consistent with their unique ability to interact with multiple partners, IDPs occupy hub positions in the scale-free network and play critical biological roles including transcriptional regulation [44–46]. Furthermore, they also regulate several key processes such as cell cycle regulation and facilitate phenotypic plasticity [47–50]. Nevertheless, IDP dysregulation of expression can often bring about non-specific interactions and generate phenotypic plasticity due to PIN modulation. This heuristic can often discover dormant pathways in the network and result in phenotypic variability. When the environmental stressors are removed, the IDPs are capable of reconfiguring the PIN to its original state, which suggests a non-genetic mechanism in phenotypic reversal. However, when the stressors persist, they can result in chronic network frustration through the acquisition of DNA mutations and other genetic alterations, which can result in permanent phenotypic alterations. This pinpoints the genetic/non-genetic duality in nature such as the evolution of drug resistance in tumor cells. This duality helps us understand how non-genetic mechanisms are involved in acquired resistance through irreversible genetic alterations at the single cell level.

Cellular Network: Individual cells, both in normal healthy tissue as well as in diseased tissue such as cancer for example, do not exist as individuals: they live in communities with other cells be it in their natural tissue environment or the tumor microenvironment. Therefore, they exhibit group behavior which can significantly influence their fitness. Thus, it is imperative to gain a systems perspective to fully understand their group behavior, leading to the expected physiological output or how cancer cells exploit group behavior to evade the toxic effects of a drug to eventually develop drug resistance. Nonetheless, previous studies have not investigated drug resistance from such a systems-level perspective. Most studies employ a reductionist approach focusing on a gene target, its mutated version(s), a pathway, or a small molecule. Alternatively, they endeavor to develop "intermittent/adaptive" therapy by studying group behavior at the population level but do not consider the role of individual molecules or the associated pathways.

Organ network: The human body is a complex interconnected organ system where individual organs have their own morphology and functional diversity, which leads to temporary, shifting, nonlinear output biological changes. This process is interlinked in that one organ in the system has a direct effect on the behavior of the other systems. The multi-component organ systems regularly interface with one another through continuous feedback mechanisms and throughout varying scales of space and time to arrive at a precise physiological output. The lack of such coordinated interactions and communications can lead to the malfunction of individual systems or the entire organism [51]. Thus, it follows that a systems perspective rather than a reductionist approach is required to gain an in-depth understanding of the integrated physiologic function, which is an emergent phenomenon resulting from interactions between the diverse organ systems. Indeed, in recent years, a new field called network physiology has emerged [52,53]. The goal of network physiology is to horizontally integrate physiological systems where individual structures and regulation mechanisms lead to biological behavior and unique physiological functions. There is a necessity to develop innovative analytical instruments and theoretical structures to address dynamical networks observed in physiological systems, which has further underscored the need for a highly interdisciplinary 'Team Medicine' approach to the problem.

2.2. Clinical Network in Cancer Systems Biology

A Clinical Network may be likened to a complex system comprising individual physicians and physician–scientists at both academic and community practice sites who enhance cancer systems biology through biomarker discovery, translational research, and clinical trial enrollment with a focus on cross-collaborative precision medicine. Precision medicine is the tool that drives cancer systems biology, where the technologies of precision medicine are utilized in tandem with the clinical network to study the distinct biological and environmental factors of each patient toward the development of new therapeutics [54]. Precision medicine has revolutionized the field of cancer over the last two decades from identifying new cancer biomarkers, genetic alterations, and treatments to improving patient outcomes [55–57]. Despite all the successes, there are several shortcomings of current precision medicine that need to be addressed such as its incorporation into the clinical cancer network, more consistent serial specimen collection, and increased collaboration between the researchers and clinicians to harness the research network in real time [58,59]. Here, we introduce the clinical network as a part of cancer systems biology and build upon our approach by proposing a model for an AI-driven drug-matching algorithm.

One crucial issue that needs to be resolved for the further widespread adaptation of precision oncology is the consistent use of biomarker platforms at the community and independent oncology clinic level. The availability of biomarker testing among practicing oncologists differs based on their geographical location and practice type with reported rates varying from 0.1% to 100% in actionable biomarkers in community practices, indicating the need for further policies that ensure all cancer patients have access to precision oncology [32,60]. Limited resources at both the clinics and in the community are a few of the multiple factors that contribute to this disparity [61]. Furthermore, the utilization of multiplex biomarker tests in clinical practice varied significantly among oncologists, and since many reported mixed confidences in interpreting these results, evidence-based guidelines and deploying pathways with the combination of physician education efforts may combat this issue [62]. The implementation of large panel omics testing across the clinical network would improve biomarker discovery in cancer systems biology. A multifaceted approach is needed to encompass as many solutions as possible comprising a wide array of parameters to include infrastructure changes such as the expansion of academic centers to incorporate community clinics or geographical sites into one large oncology network, the use of clinical pathways, and the development of molecular tumor boards within those networks and at the patient level such as community engagement, education, and empowerment [61,63–65].

Our previous work highlighted the importance of a strong integrated clinical and research network at both academic and community practice sites [29,32–34,66,67]. Oncology pathways that guide physicians have been implemented across the City of Hope network, and applying such a strategy can ensure that patients are assigned appropriate therapies based on their biomarker profile both in the academic and community practice settings [66,67]. Most cancer patients start their cancer journey with a community oncologist, and the main reason they are referred to an academic site is to enroll in a clinical trial; nevertheless, cooperation and communication between sites needs to increase [68]. The complete incorporation and cross-collaboration of clinical trial systems from the lowest levels (e.g., community sites) to the highest levels (e.g., national networks) is critical in expanding access to clinical trials, which are specifically biomarker-driven [30,67]. The decentralization of clinical trials conducted in the clinical network would address disparities of care, access to care, and raise trial accrual rates that will accelerate the cancer systems biology drug discovery pipeline [69]. We have previously designed a pyramidal decision support framework that leverages this cross-collaboration through four distinct levels including a clinical pathway program, network and academic clinician consultations, disease team tumor boards, and complex oncology case discussions [33]. This would allow for a better examination of rarer cancer-type populations such as Nuclear protein of the Testis (NUT) carcinomas or narrow targets for traditionally hard-to-treat cancers such as pancreatic cancer [30,70–72].

Additionally, multidisciplinary cancer teams' collaboration and sub-specialties are a vital component in the clinical network systems biology where knowledge and expertise need to be diversified beyond individual cancer specialists such as the involvement of pathologists, radiologists, and others to improve patient outcomes, particularly in complex cases and through the utilization of Precision Oncology Tumor Boards [33]. Baseline and serial sample collections need to be improved across the network. The use of technologies such as liquid biopsies and single-cell sequencing can help determine the early signs of possible recurrence of early-stage cancers, monitor treatment response, and follow the evolutionary heterogeneity diversity between cancer clones [73–75]. Yet, despite the prevalence and importance of biobanking protocols at institutions, many fail to capture the necessary specimens and data to accelerate its adaptation networkwide [76–78].

Previously mentioned stopgaps to precision medicine and more recently personalized medicine have largely been due to the high cost of various sequencing techniques as well as the cost of drug development or repurposing, and they have been limited by a lack of high-throughput drug screening. With the advent of liquid biopsies, it is now possible to study circulating tumor cells and detect protein expression from standard blood as well as cerebrospinal fluid (CSF) in patients with leptomeningeal metastases [74,79]. Advances in microbiome analysis have resulted in the identification of temporal changes in microbiome composition as a potential marker for immunotherapy response [80]. Microbiome discoveries have resulted in novel techniques of fecal microbiota transplants and have been shown in advanced melanoma to help immunotherapy-resistant patients overcome anti-PD-1 resistance [81]. Novel biopsy analysis techniques to detect and study circulating cancer cells, epigenetic modifications, point mutations, translocations, amplifications, deletions, chromosomal abnormalities, protein expression, and phosphorylation are now more readily used for liquid and tissue samples. Alongside this, the development of rapid 3D cell cultures and tumor organoids allows for high-throughput drug screening [82–84]. The recent developments in artificial intelligence, specifically machine learning, can further enhance personalized drug screening and match patients quickly with appropriate therapies and discover new therapeutics or candidates for drug repurposing [82,85,86]. Taken altogether, harnessing the clinical data and specimens and the research network, we have designed and proposed a novel real-time AI-driven drug-matching algorithm that could be utilized to enhance future personalized medicine (Figure 3). Additionally, the hope is that this technology ultimately assists in drug discovery and the development of novel therapies by taking advantage of retrospective samples leading to clinical trials.

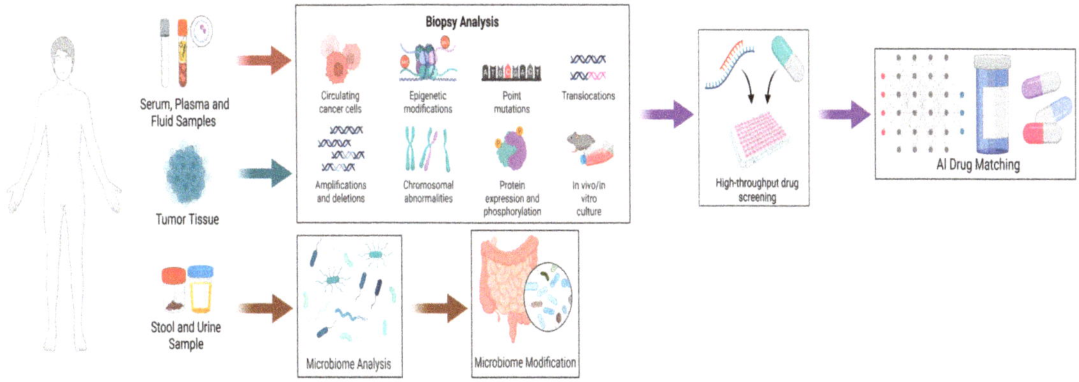

Figure 3. AI-driven drug-matching algorithm for future personalized medicine (created with BioRender.com, accessed on 1 May 2023).

3. Conclusions

Cancer systems biology has been instrumental in the recent discoveries of precision medicine. Furthermore, the integration of traditional basic science and clinical cancer researchers with a multidisciplinary team of scientists from other fields has allowed for the study of cancer at multiple scales with a deeper understanding of its biology and evolution. While previously, sequencing cost remained a barrier for clinical research, novel technologies have made it possible to quantitate tumor samples beyond genomic sequencing toward understanding protein expression and phosphorylation, epigenetic, chromosomal abnormalities, and other non-genetic mechanisms in real-world clinical samples. Furthermore, adaptive therapy (also known as intermittent therapy) based on the principles of ecology and evolution may help address the issue of drug resistance, which is almost inevitable [87,88]. This has allowed for the study of cancer biology at multiple scales enhanced by the traditional experimental and computational models. However, further cross-collaboration and integration between individual academic sites, national cancer networks, and community practices is required to achieve truly personalized medicine. The implementation of these ideas powered by recent advances in artificial intelligence and machine learning would in the future allow for personalized high-throughput drug screenings that would yield faster drug discoveries and approved therapeutics.

Author Contributions: Conceptualization, I.M., J.F., P.K. (Prakash Kulkarni) and R.S.; visualization—I.M., J.F., P.K. (Prakash Kulkarni) and R.S.; writing—original draft preparation, I.M., J.F., P.K. (Prakash Kulkarni) and R.S.; review and editing—I.M., J.F., S.B.G., T.T., R.B., P.K. (Pauline Kim), P.V., J.C., P.K. (Prakash Kulkarni) and R.S.; supervision—R.S. All authors have read and agreed to the published version of the manuscript.

Funding: The work was supported by the National Cancer Institute of the National Institutes of Health under award number P30CA033572.

Data Availability Statement: No new data were created or analyzed in this study. Data sharing is not applicable to this article.

Conflicts of Interest: The authors declare no conflict of interest.

References

1. Siegel, R.L.; Miller, K.D.; Wagle, N.S.; Jemal, A. Cancer statistics, 2023. *CA Cancer J. Clin.* **2023**, *73*, 17–48. [CrossRef]
2. Alexandrov, L.B.; Kim, J.; Haradhvala, N.J.; Huang, M.N.; Tian Ng, A.W.; Wu, Y.; Boot, A.; Covington, K.R.; Gordenin, D.A.; Bergstrom, E.N.; et al. The repertoire of mutational signatures in human cancer. *Nature* **2020**, *578*, 94–101. [CrossRef] [PubMed]
3. The ICGC/TCGA Pan-Cancer Analysis of Whole Genomes Consortium. Pan-cancer analysis of whole genomes. *Nature* **2020**, *578*, 82–93. [CrossRef] [PubMed]

4. Hoadley, K.A.; Yau, C.; Hinoue, T.; Wolf, D.M.; Lazar, A.J.; Drill, E.; Shen, R.; Taylor, A.M.; Cherniack, A.D.; Thorsson, V.; et al. Cell-of-Origin Patterns Dominate the Molecular Classification of 10,000 Tumors from 33 Types of Cancer. *Cell* **2018**, *173*, 291–304 e296. [CrossRef] [PubMed]
5. Fox, E.J.; Salk, J.J.; Loeb, L.A. Cancer genome sequencing—An interim analysis. *Cancer Res.* **2009**, *69*, 4948–4950. [CrossRef]
6. Loeb, L.A. Human cancers express mutator phenotypes: Origin, consequences and targeting. *Nat. Rev. Cancer* **2011**, *11*, 450–457. [CrossRef]
7. Roosan, M.R.; Mambetsariev, I.; Pharaon, R.; Fricke, J.; Baroz, A.R.; Chao, J.; Chen, C.; Nasser, M.W.; Chirravuri-Venkata, R.; Jain, M.; et al. Evaluation of Somatic Mutations in Solid Metastatic Pan-Cancer Patients. *Cancers* **2021**, *13*, 2776. [CrossRef]
8. Rohatgi, N.; Ghoshdastider, U.; Baruah, P.; Kulshrestha, T.; Skanderup, A.J. A pan-cancer metabolic atlas of the tumor microenvironment. *Cell Rep.* **2022**, *39*, 110800. [CrossRef]
9. Wong, D.J.; Nuyten, D.S.; Regev, A.; Lin, M.; Adler, A.S.; Segal, E.; van de Vijver, M.J.; Chang, H.Y. Revealing targeted therapy for human cancer by gene module maps. *Cancer Res.* **2008**, *68*, 369–378. [CrossRef]
10. Bild, A.H.; Potti, A.; Nevins, J.R. Linking oncogenic pathways with therapeutic opportunities. *Nat. Rev. Cancer* **2006**, *6*, 735–741. [CrossRef]
11. Butcher, E.C.; Berg, E.L.; Kunkel, E.J. Systems biology in drug discovery. *Nat. Biotechnol.* **2004**, *22*, 1253–1259. [CrossRef] [PubMed]
12. Trewavas, A. A brief history of systems biology. "Every object that biology studies is a system of systems." Francois Jacob (1974). *Plant Cell* **2006**, *18*, 2420–2430. [CrossRef] [PubMed]
13. Kitano, H. Systems biology: A brief overview. *Science* **2002**, *295*, 1662–1664. [CrossRef] [PubMed]
14. Saeed, K.; Ryder, E.F.; Manning, A.L. Cancer as a System Dysfunction. *Systems* **2021**, *9*, 14. [CrossRef]
15. Viktorsson, K.; Lewensohn, R.; Zhivotovsky, B. Systems biology approaches to develop innovative strategies for lung cancer therapy. *Cell Death Dis.* **2014**, *5*, e1260. [CrossRef]
16. Filipp, F.V. Precision medicine driven by cancer systems biology. *Cancer Metastasis Rev.* **2017**, *36*, 91–108. [CrossRef]
17. Rocca, A.; Kholodenko, B.N. Can Systems Biology Advance Clinical Precision Oncology? *Cancers* **2021**, *13*, 6312. [CrossRef]
18. Joo, J.I.; Choi, M.; Jang, S.-H.; Choi, S.; Park, S.-M.; Shin, D.; Cho, K.-H. Realizing Cancer Precision Medicine by Integrating Systems Biology and Nanomaterial Engineering. *Adv. Mater.* **2020**, *32*, 1906783. [CrossRef] [PubMed]
19. Nguyen, D.X.; Bos, P.D.; Massague, J. Metastasis: From dissemination to organ-specific colonization. *Nat. Rev. Cancer* **2009**, *9*, 274–284. [CrossRef]
20. Truffi, M.; Sorrentino, L.; Corsi, F. Fibroblasts in the Tumor Microenvironment. *Adv. Exp. Med. Biol.* **2020**, *1234*, 15–29. [CrossRef]
21. Sahni, N.; Yi, S.; Taipale, M.; Fuxman Bass, J.I.; Coulombe-Huntington, J.; Yang, F.; Peng, J.; Weile, J.; Karras, G.I.; Wang, Y.; et al. Widespread macromolecular interaction perturbations in human genetic disorders. *Cell* **2015**, *161*, 647–660. [CrossRef] [PubMed]
22. Zhao, W.; Li, J.; Chen, M.M.; Luo, Y.; Ju, Z.; Nesser, N.K.; Johnson-Camacho, K.; Boniface, C.T.; Lawrence, Y.; Pande, N.T.; et al. Large-Scale Characterization of Drug Responses of Clinically Relevant Proteins in Cancer Cell Lines. *Cancer Cell* **2020**, *38*, 829–843 e824. [CrossRef] [PubMed]
23. Wouters, O.J.; McKee, M.; Luyten, J. Estimated Research and Development Investment Needed to Bring a New Medicine to Market, 2009-2018. *JAMA* **2020**, *323*, 844–853. [CrossRef]
24. Mendiratta, G.; Ke, E.; Aziz, M.; Liarakos, D.; Tong, M.; Stites, E.C. Cancer gene mutation frequencies for the U.S. population. *Nat. Commun.* **2021**, *12*, 5961. [CrossRef] [PubMed]
25. Toomey, S.; Carr, A.; Mezynski, M.J.; Elamin, Y.; Rafee, S.; Cremona, M.; Morgan, C.; Madden, S.; Abdul-Jalil, K.I.; Gately, K.; et al. Identification and clinical impact of potentially actionable somatic oncogenic mutations in solid tumor samples. *J. Transl. Med.* **2020**, *18*, 99. [CrossRef] [PubMed]
26. Mambetsariev, I.; Arvanitis, L.; Fricke, J.; Pharaon, R.; Baroz, A.R.; Afkhami, M.; Koczywas, M.; Massarelli, E.; Salgia, R. Small Cell Lung Cancer Transformation following Treatment in EGFR-Mutated Non-Small Cell Lung Cancer. *J. Clin. Med.* **2022**, *11*, 1429. [CrossRef]
27. Lu, C.Y.; Terry, V.; Thomas, D.M. Precision medicine: Affording the successes of science. *NPJ Precis. Oncol.* **2023**, *7*, 3. [CrossRef]
28. Cutler, D.M. Early Returns from the Era of Precision Medicine. *JAMA* **2020**, *323*, 109–110. [CrossRef]
29. Brooks, G.A.; Bosserman, L.D.; Mambetsariev, I.; Salgia, R. Value-Based Medicine and Integration of Tumor Biology. *Am. Soc. Clin. Oncol. Educ. Book* **2017**, *37*, 833–840. [CrossRef]
30. Fountzilas, E.; Tsimberidou, A.M.; Vo, H.H.; Kurzrock, R. Clinical trial design in the era of precision medicine. *Genome Med.* **2022**, *14*, 101. [CrossRef]
31. Galsky, M.D.; Stensland, K.D.; McBride, R.B.; Latif, A.; Moshier, E.; Oh, W.K.; Wisnivesky, J. Geographic accessibility to clinical trials for advanced cancer in the United States. *JAMA Intern. Med.* **2015**, *175*, 293–295. [CrossRef] [PubMed]
32. Rajurkar, S.; Mambetsariev, I.; Pharaon, R.; Leach, B.; Tan, T.; Kulkarni, P.; Salgia, R. Non-Small Cell Lung Cancer from Genomics to Therapeutics: A Framework for Community Practice Integration to Arrive at Personalized Therapy Strategies. *J. Clin. Med.* **2020**, *9*, 1870. [CrossRef] [PubMed]
33. Bosserman, L.D.; Mambetsariev, I.; Ladbury, C.; Barzi, A.; Johnson, D.; Morse, D.; Deaville, D.; Smith, W.; Rajurkar, S.; Merla, A.; et al. Pyramidal Decision Support Framework Leverages Subspecialty Expertise across Enterprise to Achieve Superior Cancer Outcomes and Personalized, Precision Care Plans. *J. Clin. Med.* **2022**, *11*, 6738. [CrossRef] [PubMed]
34. Salgia, R.; Kulkarni, P. Integrating Clinical and Translational Research Networks-Building Team Medicine. *J. Clin. Med.* **2020**, *9*, 2975. [CrossRef]

35. Kulkarni, P.; Mohanty, A.; Bhattacharya, S.; Singhal, S.; Guo, L.; Ramisetty, S.; Mirzapoiazova, T.; Mambetsariev, B.; Mittan, S.; Malhotra, J.; et al. Addressing Drug Resistance in Cancer: A Team Medicine Approach. *J. Clin. Med.* **2022**, *11*, 5701. [CrossRef]
36. Davidson, E.H. 1—Introduction: The Variable Gene Activity Theory of Cell Differentiation. In *Gene Activity in Early Development*, 2nd ed.; Davidson, E.H., Ed.; Academic Press: Cambridge, MA, USA, 1976; pp. 1–26.
37. Barabasi, A.L.; Gulbahce, N.; Loscalzo, J. Network medicine: A network-based approach to human disease. *Nat. Rev. Genet.* **2011**, *12*, 56–68. [CrossRef]
38. Hu, J.X.; Thomas, C.E.; Brunak, S. Network biology concepts in complex disease comorbidities. *Nat. Rev. Genet.* **2016**, *17*, 615–629. [CrossRef]
39. Barabasi, A.L.; Albert, R. Emergence of scaling in random networks. *Science* **1999**, *286*, 509–512. [CrossRef]
40. Barabasi, A.L. Scale-free networks: A decade and beyond. *Science* **2009**, *325*, 412–413. [CrossRef]
41. Xue, B.; Dunker, A.K.; Uversky, V.N. Orderly order in protein intrinsic disorder distribution: Disorder in 3500 proteomes from viruses and the three domains of life. *J. Biomol. Struct. Dyn.* **2012**, *30*, 137–149. [CrossRef]
42. Peng, Z.; Yan, J.; Fan, X.; Mizianty, M.J.; Xue, B.; Wang, K.; Hu, G.; Uversky, V.N.; Kurgan, L. Exceptionally abundant exceptions: Comprehensive characterization of intrinsic disorder in all domains of life. *Cell Mol. Life Sci.* **2015**, *72*, 137–151. [CrossRef] [PubMed]
43. Kulkarni, P.; Solomon, T.L.; He, Y.; Chen, Y.; Bryan, P.N.; Orban, J. Structural metamorphism and polymorphism in proteins on the brink of thermodynamic stability. *Protein Sci.* **2018**, *27*, 1557–1567. [CrossRef] [PubMed]
44. Hu, G.; Wu, Z.; Uversky, V.N.; Kurgan, L. Functional Analysis of Human Hub Proteins and Their Interactors Involved in the Intrinsic Disorder-Enriched Interactions. *Int. J. Mol. Sci.* **2017**, *18*, 2761. [CrossRef]
45. Bondos, S.E.; Dunker, A.K.; Uversky, V.N. Intrinsically disordered proteins play diverse roles in cell signaling. *Cell Commun. Signal* **2022**, *20*, 20. [CrossRef] [PubMed]
46. Wright, P.E.; Dyson, H.J. Intrinsically disordered proteins in cellular signalling and regulation. *Nat. Rev. Mol. Cell Biol.* **2015**, *16*, 18–29. [CrossRef] [PubMed]
47. Mitrea, D.M.; Yoon, M.K.; Ou, L.; Kriwacki, R.W. Disorder-function relationships for the cell cycle regulatory proteins p21 and p27. *Biol. Chem.* **2012**, *393*, 259–274. [CrossRef]
48. Krasinska, L.; Fisher, D. A Mechanistic Model for Cell Cycle Control in Which CDKs Act as Switches of Disordered Protein Phase Separation. *Cells* **2022**, *11*, 2189. [CrossRef]
49. Camponeschi, I.; Damasco, A.; Uversky, V.N.; Giuliani, A.; Bianchi, M.M. Phenotypic suppression caused by resonance with light-dark cycles indicates the presence of a 24-hours oscillator in yeast and suggests a new role of intrinsically disordered protein regions as internal mediators. *J. Biomol. Struct. Dyn.* **2021**, *39*, 2490–2501. [CrossRef]
50. Kulkarni, P.; Achuthan, S.; Bhattacharya, S.; Jolly, M.K.; Kotnala, S.; Leite, V.B.P.; Mohanty, A.; Orban, J.; Roy, S.; Rangarajan, G.; et al. Protein conformational dynamics and phenotypic switching. *Biophys. Rev.* **2021**, *13*, 1127–1138. [CrossRef]
51. Bartsch, R.P.; Liu, K.K.; Bashan, A.; Ivanov, P. Network Physiology: How Organ Systems Dynamically Interact. *PLoS ONE* **2015**, *10*, e0142143. [CrossRef]
52. Bashan, A.; Bartsch, R.P.; Kantelhardt, J.W.; Havlin, S.; Ivanov, P. Network physiology reveals relations between network topology and physiological function. *Nat. Commun.* **2012**, *3*, 702. [CrossRef] [PubMed]
53. Ivanov, P.C. The New Field of Network Physiology: Building the Human Physiolome. *Front. Netw. Physiol.* **2021**, *1*, 711778. [CrossRef]
54. Elemento, O. The future of precision medicine: Towards a more predictive personalized medicine. *Emerg. Top Life Sci.* **2020**, *4*, 175–177. [CrossRef]
55. Krzyszczyk, P.; Acevedo, A.; Davidoff, E.J.; Timmins, L.M.; Marrero-Berrios, I.; Patel, M.; White, C.; Lowe, C.; Sherba, J.J.; Hartmanshenn, C.; et al. The growing role of precision and personalized medicine for cancer treatment. *Technol. (Singapore World Sci.)* **2018**, *6*, 79–100. [CrossRef] [PubMed]
56. Hoeben, A.; Joosten, E.A.J.; van den Beuken-van Everdingen, M.H.J. Personalized Medicine: Recent Progress in Cancer Therapy. *Cancers* **2021**, *13*, 242. [CrossRef] [PubMed]
57. Mambetsariev, I.; Wang, Y.; Chen, C.; Nadaf, S.; Pharaon, R.; Fricke, J.; Amanam, I.; Amini, A.; Bild, A.; Chu, P.; et al. Precision medicine and actionable alterations in lung cancer: A single institution experience. *PLoS ONE* **2020**, *15*, e0228188. [CrossRef] [PubMed]
58. Sorich, M.J.; McKinnon, R.A. Personalized medicine: Potential, barriers and contemporary issues. *Curr. Drug Metab.* **2012**, *13*, 1000–1006. [CrossRef]
59. Tannock, I.F.; Hickman, J.A. Limits to Personalized Cancer Medicine. *N. Engl. J. Med.* **2016**, *375*, 1289–1294. [CrossRef]
60. Gardner, B.; Doose, M.; Sanchez, J.I.; Freedman, A.N.; de Moor, J.S. Distribution of Genomic Testing Resources by Oncology Practice and Rurality: A Nationally Representative Study. *JCO Precis. Oncol* **2021**, *5*, 1060–1068. [CrossRef]
61. Melas, M.; Subbiah, S.; Saadat, S.; Rajurkar, S.; McDonnell, K.J. The Community Oncology and Academic Medical Center Alliance in the Age of Precision Medicine: Cancer Genetics and Genomics Considerations. *J. Clin. Med.* **2020**, *9*, 2125. [CrossRef]
62. Gray, S.W.; Hicks-Courant, K.; Cronin, A.; Rollins, B.J.; Weeks, J.C. Physicians' attitudes about multiplex tumor genomic testing. *J. Clin. Oncol.* **2014**, *32*, 1317–1323. [CrossRef] [PubMed]

63. Levit, L.A.; Kim, E.S.; McAneny, B.L.; Nadauld, L.D.; Levit, K.; Schenkel, C.; Schilsky, R.L. Implementing Precision Medicine in Community-Based Oncology Programs: Three Models. *J. Oncol. Pract.* **2019**, *15*, 325–329. [CrossRef] [PubMed]
64. Pritchard, D.E.; Moeckel, F.; Villa, M.S.; Housman, L.T.; McCarty, C.A.; McLeod, H.L. Strategies for integrating personalized medicine into healthcare practice. *Per. Med.* **2017**, *14*, 141–152. [CrossRef] [PubMed]
65. Fohner, A.E.; Volk, K.G.; Woodahl, E.L. Democratizing Precision Medicine Through Community Engagement. *Clin. Pharmacol. Ther.* **2019**, *106*, 488–490. [CrossRef]
66. Salgia, R.; Mambetsariev, I.; Tan, T.; Schwer, A.; Pearlstein, D.P.; Chehabi, H.; Baroz, A.; Fricke, J.; Pharaon, R.; Romo, H.; et al. Complex Oncological Decision-Making Utilizing Fast-and-Frugal Trees in a Community Setting-Role of Academic and Hybrid Modeling. *J. Clin. Med.* **2020**, *9*, 1884. [CrossRef]
67. Bosserman, L.D.; Cianfrocca, M.; Yuh, B.; Yeon, C.; Chen, H.; Sentovich, S.; Polverini, A.; Zachariah, F.; Deaville, D.; Lee, A.B.; et al. Integrating Academic and Community Cancer Care and Research through Multidisciplinary Oncology Pathways for Value-Based Care: A Review and the City of Hope Experience. *J. Clin. Med.* **2021**, *10*, 188. [CrossRef]
68. Salgia, N.J.; Chehrazi-Raffle, A.; Hsu, J.; Zengin, Z.; Salgia, S.; Chawla, N.S.; Meza, L.; Malhotra, J.; Dizman, N.; Muddasani, R.; et al. Characterizing the relationships between tertiary and community cancer providers: Results from a survey of medical oncologists in Southern California. *Cancer Med.* **2021**, *10*, 5671–5680. [CrossRef]
69. Goodson, N.; Wicks, P.; Morgan, J.; Hashem, L.; Callinan, S.; Reites, J. Opportunities and counterintuitive challenges for decentralized clinical trials to broaden participant inclusion. *NPJ Digit. Med.* **2022**, *5*, 58. [CrossRef]
70. Flaherty, K.T.; Gray, R.J.; Chen, A.P.; Li, S.; McShane, L.M.; Patton, D.; Hamilton, S.R.; Williams, P.M.; Iafrate, A.J.; Sklar, J.; et al. Molecular Landscape and Actionable Alterations in a Genomically Guided Cancer Clinical Trial: National Cancer Institute Molecular Analysis for Therapy Choice (NCI-MATCH). *J. Clin. Oncol.* **2020**, *38*, 3883–3894. [CrossRef]
71. Lauer, U.M.; Hinterleitner, M.; Horger, M.; Ohnesorge, P.V.; Zender, L. NUT Carcinoma-An Underdiagnosed Malignancy. *Front. Oncol.* **2022**, *12*, 914031. [CrossRef]
72. O'Kane, G.M.; Lowery, M.A. Moving the Needle on Precision Medicine in Pancreatic Cancer. *J. Clin. Oncol.* **2022**, *40*, 2693–2705. [CrossRef] [PubMed]
73. Zarinshenas, R.; Amini, A.; Mambetsariev, I.; Abuali, T.; Fricke, J.; Ladbury, C.; Salgia, R. Assessment of Barriers and Challenges to Screening, Diagnosis, and Biomarker Testing in Early-Stage Lung Cancer. *Cancers* **2023**, *15*, 1595. [CrossRef]
74. Roosan, M.R.; Mambetsariev, I.; Pharaon, R.; Fricke, J.; Husain, H.; Reckamp, K.L.; Koczywas, M.; Massarelli, E.; Bild, A.H.; Salgia, R. Usefulness of Circulating Tumor DNA in Identifying Somatic Mutations and Tracking Tumor Evolution in Patients with Non-small Cell Lung Cancer. *Chest* **2021**, *160*, 1095–1107. [CrossRef] [PubMed]
75. Nath, A.; Bild, A.H. Leveraging Single-Cell Approaches in Cancer Precision Medicine. *Trends Cancer* **2021**, *7*, 359–372. [CrossRef] [PubMed]
76. Kinkorova, J. Biobanks in the era of personalized medicine: Objectives, challenges, and innovation: Overview. *EPMA J.* **2015**, *7*, 4. [CrossRef] [PubMed]
77. Patil, S.; Majumdar, B.; Awan, K.H.; Sarode, G.S.; Sarode, S.C.; Gadbail, A.R.; Gondivkar, S. Cancer oriented biobanks: A comprehensive review. *Oncol. Rev.* **2018**, *12*, 357. [CrossRef]
78. Hung, R.J.; Khodayari Moez, E.; Kim, S.J.; Budhathoki, S.; Brooks, J.D. Considerations of biomarker application for cancer continuum in the era of precision medicine. *Curr. Epidemiol. Rep.* **2022**, *9*, 200–211. [CrossRef]
79. Wooster, M.; McGuinness, J.E.; Fenn, K.M.; Singh, V.M.; Franks, L.E.; Lee, S.; Cieremans, D.; Lassman, A.B.; Hershman, D.L.; Crew, K.D.; et al. Diagnosis of Leptomeningeal Metastasis in Women with Breast Cancer Through Identification of Tumor Cells in Cerebrospinal Fluid Using the CNSide Assay. *Clin. Breast Cancer* **2022**, *22*, e457–e462. [CrossRef]
80. Salgia, N.J.; Bergerot, P.G.; Maia, M.C.; Dizman, N.; Hsu, J.; Gillece, J.D.; Folkerts, M.; Reining, L.; Trent, J.; Highlander, S.K.; et al. Stool Microbiome Profiling of Patients with Metastatic Renal Cell Carcinoma Receiving Anti-PD-1 Immune Checkpoint Inhibitors. *Eur. Urol.* **2020**, *78*, 498–502. [CrossRef]
81. Davar, D.; Dzutsev, A.K.; McCulloch, J.A.; Rodrigues, R.R.; Chauvin, J.M.; Morrison, R.M.; Deblasio, R.N.; Menna, C.; Ding, Q.; Pagliano, O.; et al. Fecal microbiota transplant overcomes resistance to anti-PD-1 therapy in melanoma patients. *Science* **2021**, *371*, 595–602. [CrossRef]
82. Gorshkov, K.; Chen, C.Z.; Marshall, R.E.; Mihatov, N.; Choi, Y.; Nguyen, D.T.; Southall, N.; Chen, K.G.; Park, J.K.; Zheng, W. Advancing precision medicine with personalized drug screening. *Drug Discov. Today* **2019**, *24*, 272–278. [CrossRef] [PubMed]
83. Letai, A.; Bhola, P.; Welm, A.L. Functional precision oncology: Testing tumors with drugs to identify vulnerabilities and novel combinations. *Cancer Cell* **2022**, *40*, 26–35. [CrossRef] [PubMed]
84. Phan, N.; Hong, J.J.; Tofig, B.; Mapua, M.; Elashoff, D.; Moatamed, N.A.; Huang, J.; Memarzadeh, S.; Damoiseaux, R.; Soragni, A. A simple high-throughput approach identifies actionable drug sensitivities in patient-derived tumor organoids. *Commun. Biol.* **2019**, *2*, 78. [CrossRef]
85. Johnson, K.B.; Wei, W.Q.; Weeraratne, D.; Frisse, M.E.; Misulis, K.; Rhee, K.; Zhao, J.; Snowdon, J.L. Precision Medicine, AI, and the Future of Personalized Health Care. *Clin. Transl. Sci.* **2021**, *14*, 86–93. [CrossRef] [PubMed]
86. Heaven, W. AI is Dreaming up Drugs That No One Has Ever Seen. Now We've Got to See If They Work. Available online: https://www.technologyreview.com/2023/02/15/1067904/ai-automation-drug-development/ (accessed on 11 May 2023).

87. Gatenby, R.A.; Brown, J.S. Integrating evolutionary dynamics into cancer therapy. *Nat. Rev. Clin. Oncol.* **2020**, *17*, 675–686. [CrossRef] [PubMed]
88. Gatenby, R.A.; Brown, J.S. The Evolution and Ecology of Resistance in Cancer Therapy. *Cold Spring Harb. Perspect. Med.* **2020**, *10*, a040972. [CrossRef]

Journal of
Clinical Medicine

MDPI

Review

A Closer Look at EGFR Inhibitor Resistance in Non-Small Cell Lung Cancer through the Lens of Precision Medicine

Martin Sattler [1,2,*], Isa Mambetsariev [3], Jeremy Fricke [3], Tingting Tan [3], Sariah Liu [3], Nagarajan Vaidehi [4], Evan Pisick [5], Tamara Mirzapoiazova [3], Adam G. Rock [3], Amartej Merla [3], Sunil Sharma [6] and Ravi Salgia [3]

1 Department of Medical Oncology, Dana-Farber Cancer Institute, 450 Brookline Ave., Boston, MA 02215, USA
2 Department of Medicine, Harvard Medical School, Boston, MA 02115, USA
3 Department of Medical Oncology and Therapeutics Research, City of Hope, 1500 E Duarte Road, Duarte, CA 91010, USA
4 Department of Computational and Quantitative Medicine, City of Hope, 1500 E Duarte Road, Duarte, CA 91010, USA
5 City of Hope Chicago, 2520 Elisha Avenue, Zion, IL 60099, USA
6 Division of Applied Cancer Research and Drug Discovery, Translational Genomic Research Institute (Tgen), 445 N 5th St, Phoenix, AZ 85004, USA
* Correspondence: martin_sattler@dfci.harvard.edu

Abstract: The development of EGFR small-molecule inhibitors has provided significant benefit for the affected patient population. Unfortunately, current inhibitors are no curative therapy, and their development has been driven by on-target mutations that interfere with binding and thus inhibitory activity. Genomic studies have revealed that, in addition to these on-target mutations, there are also multiple off-target mechanisms of EGFR inhibitor resistance and novel therapeutics that can overcome these challenges are sought. Resistance to competitive 1st-generation and covalent 2nd- and 3rd-generation EGFR inhibitors is overall more complex than initially thought, and novel 4th-generation allosteric inhibitors are expected to suffer from a similar fate. Additional nongenetic mechanisms of resistance are significant and can include up to 50% of the escape pathways. These potential targets have gained recent interest and are usually not part of cancer panels that look for alterations in resistant patient specimen. We discuss the duality between genetic and nongenetic EGFR inhibitor drug resistance and summarize current team medicine approaches, wherein clinical developments, hand in hand with drug development research, drive potential opportunities for combination therapy.

Keywords: EGFR; non-small cell lung cancer; drug resistance; genetic/nongenetic; epigenetics

Citation: Sattler, M.; Mambetsariev, I.; Fricke, J.; Tan, T.; Liu, S.; Vaidehi, N.; Pisick, E.; Mirzapoiazova, T.; Rock, A.G.; Merla, A.; et al. A Closer Look at EGFR Inhibitor Resistance in Non-Small Cell Lung Cancer through the Lens of Precision Medicine. *J. Clin. Med.* **2023**, *12*, 1936. https://doi.org/10.3390/jcm12051936

Academic Editor: David Barnes

Received: 25 January 2023
Revised: 22 February 2023
Accepted: 26 February 2023
Published: 1 March 2023

1. EGFR Mutations in Cancer

The *epidermal growth factor receptor* (*EGFR*) gene encodes for a transmembrane tyrosine kinase, is expressed in many tissues at various levels and is normally activated by its ligand epidermal growth factor [1]. Alterations of EGFR are common in solid tumors, including amplifications and activating mutations, in particular for patients with glioblastoma and non-small cell lung cancer (NSCLC). There are striking differences in NSCLC patients, with a significantly higher frequency of EGFR mutations in patients of East Asian heritage versus Caucasian patients and somewhat higher incidence in women and never-smokers [2–4]. About 90% of the identified activating EGFR mutations in lung adenocarcinoma involve either the L858R substitution in exon 21 or in-frame deletions in exon 19, leading to malignant transformation with ligand-independent activation of growth and anti-apoptotic pathways. Additional rare EGFR alterations include point mutations, insertions or deletions in exon 18–21 [5,6]. The development of ATP-competitive small-molecule EGFR inhibitors demonstrated the significance of these EGFR mutations for cancer growth. Gefitinib was the first molecularly targeted EGFR inhibitor that showed remarkable efficacy in lung cancer patients with EGFR mutations [7,8] (Figure 1). Currently, different FDA-approved EGFR inhibitors are available, and it appears that common and uncommon oncogenic EGFR mutations are preferably inhibited by some of these drugs [5]. Unfortunately, as it is the case for most small-molecule inhibitors that target oncogenic tyrosine kinases, resistance eventually ensues. An initial major focus was identifying on-target mutations in the drug binding pocket of the inhibitor. Depending on the mechanisms of action of the drug, mutations can also occur in other parts of EGFR and it appears that resistant mutations are context-specific. It is also possible that transformation in resistant cells is driven by off-target genetic alterations that lead to the activation of other transforming proteins and circumvent EGFR dependency. However, a significant portion of EGFR inhibitor resistance is driven by mechanisms that have not yet been identified and it has become clear that nongenetic mechanisms of resistance may play a larger role than previously thought. In this review, we take a closer look at the various genetic and nongenetic mechanisms of resistance towards EGFR inhibitors, both of which can either be acquired or pre-existing. The overall mechanism of how clonal selection occurs may be similar between the drug resistance mechanism towards different classes of drugs and towards different drug targets in growth pathways. We have previously suggested an intermediate drug-tolerant state for acquired genetic as well as nongenetic KRAS inhibitor resistance that may also apply here for the occurrence of genetic and nongenetic mechanisms of EGFR inhibitor resistance [9,10]. Nevertheless, the molecular mechanisms that determine the resistance pathways are not known. It has become apparent that EGFR-targeted monotherapy, even with next-generation inhibitors, is prone to drug resistance. Initially, research was focused on catching up to emerging resistance mutations and now it appears that combination therapy may be a more viable approach to at least delay treatment failure and increase clinical benefit. Within this context, we will further summarize recent team medicine approaches directed towards overcoming EGFR inhibitor resistance.

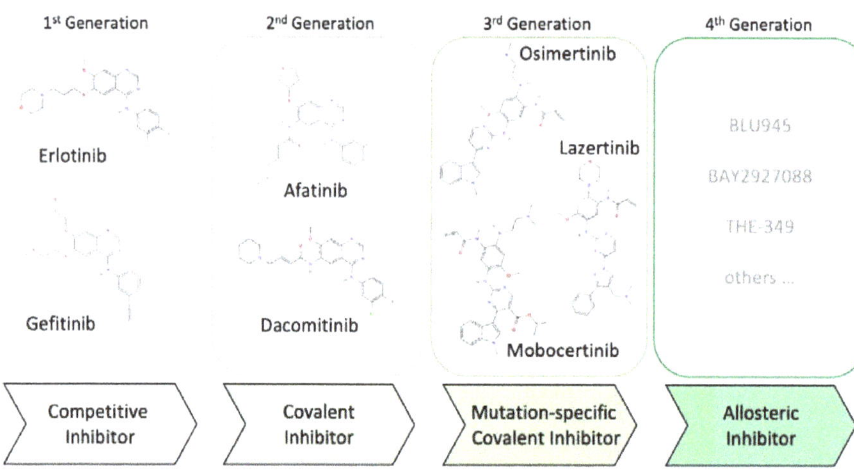

Figure 1. Evolution of EGFR tyrosine kinase inhibitors.

2. Evolution of EGFR Inhibitors

First-generation small-molecule EGFR inhibitors (erlotinib, gefitinib) were designed to interfere with EGFR tyrosine kinase activity by competing with binding of the adenine base of ATP to its binding pocket. The core interacting residues for ATP in EGFR include L718, V726, A743, M793 and L844 [11] and many competitive inhibitors commonly form interactions, in particular with M793 at the hinge region, including gefitinib and erlotinib [12]. About half the patients treated with these drugs acquire the T790M mutation at the highly conserved 'gatekeeper' residue and it has been suggested that future covalent inhibitors may circumvent this escape mechanism [13–15]. Indeed, second-generation inhibitors (afatinib, dacomitinib) were designed to be structurally related to these compounds but contained additional moieties to facilitate covalent binding to C797, in addition to the inhibition of the EGFR tyrosine kinase activity. Unfortunately, these drugs are also susceptible to the emergence of T790M mutations as they cannot distinguish between wild-type and mutant EGFR. This broad inhibition of EGFR is associated with dose-limiting side-effects, which do not allow for sufficient inhibition of the T790M mutation [16,17]. Mutant-selective third-generation inhibitors (osimertinib, lazertinib) were a true product of team medicine, where efforts from clinicians, clinical scientist, medicinal chemists, pathologists, biostatisticians and others led to the design of new EGFR inhibitors to circumvent these limitations and to specifically target EGFR. These contain activating mutations, including the T790M resistance mutation [18,19]. Even though these inhibitors do not target wild-type EGFR and have therefore little side-effects associated with this particular target, they are also susceptible to on-target drug resistance, such as C797S mutations or others.

In 2018, osimertinib became the first drug in this class to receive FDA-approval for first-line treatment of metastatic NSCLC with EGFR exon 19 deletions or exon 21 L858R mutations. The double-blind, phase III FLAURA trial with 556 patients established the efficacy of osimertinib, demonstrating prolonged progression-free survival [20]. When compared to standard EGFR inhibitor therapy, osimertinib increased median progression-free survival from 10.2 months to 18.9 months (hazard ratio (HR) 0.46; 95% CI: 0.37 to 0.57). Both, standard EGFR inhibitor therapy and osimertinib had similar objective response rates (76% vs. 80%), but the median duration of response with standard EGFR inhibitor therapy was 8.5 months (95% CI, 7.3 to 9.8) and osimertinib resulted in a 17.2-month response (95% CI, 13.8 to 22.0). Additionally, standard EGFR inhibitor therapy had a higher rate of grade 3/4 adverse reactions compared to osimertinib (45% vs. 34%). A further long-term follow up also demonstrated increased overall survival with osimertinib, compared to standard EGFR inhibitor therapy, in previously untreated patients [21]. These results further suggest that adverse reactions with osimertinib maybe somewhat higher (42% vs. 47% in the standard EGFR inhibitor therapy group) than previously reported. In general, osimertinib increased overall survival by almost 7 months to 38.6 months (95% CI: 34.5 to 41.8) compared to the standard therapy group result at 31.8 months (95% CI: 26.6 to 36.0). After 3 years, 28% (79/279; 20.7 months median exposure) of osimertinib treated patients were still on trial, versus 9% (26/277; 11.5 months median exposure) in the comparison group. A meta-analysis of 15 studies with 324 patients further supported a role for osimertinib in the control of intracranial metastatic disease, with complete intracranial response rates of 7% to 23% [22]. The objective response rate was calculated for 195 patients at 64% (95% CI: 53–76) and the disease control rate was 90% (95% CI: 85–93), calculated for 246 patients.

The efficacy of osimertinib in EGFR-mutated NSCLC patients harboring exon 20 insertion mutations (up to 12% of mutated EGFR) is limited and this population can benefit from mobocertinib, a C797 covalent EGFR inhibitor, which may be resistant to C797S mutations as well [23–25]. In 2021, mobocertinib was FDA-approved as the first drug for NSCLC patients with locally advanced or metastatic disease that had an EGFR exon 20 insertion mutations and progressed with platinum-based chemotherapy. The rate of adverse events was found to be similar to that of other EGFR inhibitors and in general manageable. In the phase I/II dose-escalation/expansion trial a response rate was achieved in 43% (12/28) of the patients (95% CI: 24 to 63) and the median progression-free survival was 7.3 months [24]. Further, in a larger cohort of 114 patients, the objective response rate was 28% (95% CI: 20 to 37), the median progression-free survival was 7.3 months (95% CI: 5.5 to 9.2) and median overall survival was 24.0 months (95% CI: 14.6 to 28.8) [26]. Additional data suggest that the intracranial activity of mobocertinib could be limited. Mobocertinib may provide better responses in patients without brain metastases, who benefited from longer treatment periods, and the intracranial anti-tumor activity appears to be insufficient [27].

Lazertinib was tested in a phase I/II clinical trial with 38 patients in the dose escalation group and 89 patients in the dose expansion group, where it was generally well tolerated. Treatment-related grade 3 or 4 adverse events occurred in only 3% (4/127) of patients, without any events that lead to death or treatment-related deaths. A total of 54% (69/127) of patients achieved an objective response (95% CI: 46 to 63) [19]. In another South Korean phase 1/2 study with 78 T790M-positive NSCLC patients, lazertinib caused a complete response in one patient and 53.9% (41/78) of the patients had partial responses, resulting in a similar objective response rate of 55.3% (95% CI: 44 to 66) [28]. Median progression-free survival was 11.1 months (95% CI: 5.5 to 16.4) and the median overall survival did not reach 22 months. As expected, loss of EGFR T790M was identified as a major resistance mechanism. Lazertinib was also active in the brain and suppressed intracranial tumor growth, with one patient showing a complete response and five patients showing partial responses, resulting in an intracranial objective response rate of 85.7% (95% CI: 60 to 100.0). Lazertinib received local approval in South Korea in 2021 for NSCLC patients with EGFR T790M mutations that had previously received treatment with EGFR inhibitors and that had locally advanced or metastatic disease. However, it has not yet gained FDA approval in the USA.

Fourth-generation allosteric mutant-selective EGFR inhibitors that have different binding sites are currently being tested. These mutations can be compounded and include the activating mutation (e.g., exon 19 deletion, L858R), the first-/second-generation inhibitor resistance mutation (e.g., T790M) and/or an osimertinib resistance mutations. Occasionally, mutations that cause osimertinib resistance may not necessarily emerge from clones that contain the T790M resistance mutations but can also originate from the original clone containing the oncogenic mutation, such as L858R with the M766Q exon 20 resistance mutation. Interestingly, this double-mutant can be sensitive towards the tyrosine kinase inhibitor neratinib, which was originally developed against EGFR family members [29]. The occurrence of multiple mutations in EGFR complicates the development of next generation inhibitors but allosteric mutant-selective fourth generation EGFR inhibitors are designed to show efficacy in this context. Whether they must be combined with other targeted therapies, standard chemotherapy or immunotherapy will have to be determined. Team medicine takes center stage in the development of new therapeutics that are essentially initiated by results from precision medicine approaches. The development of second-generation EGFR inhibitors has demonstrated that pre-clinical results may not easily be transferrable to clinical practice and that with a combined effort of various pre-clinical and clinical groups of scientist significant progress can be achieved.

3. Genetic Mechanisms of EGFR Inhibitor Resistance beyond On-Target Mutations

Different on-target mutations within EGFR have been reviewed previously (e.g., [30]). Additional genetic changes frequently target the signaling molecules that substitute for the functional activation of pathways, which are otherwise dysregulated by oncogenic EGFR (Figure 2). Mutated proteins within these pathways can present themselves as therapeutic targets or hint at potential targets for combination therapy.

There are at least two categories of signaling targets: (a) genetic alterations of receptors that substitute for EGFR signaling and (b) mutations within signaling pathways activating mechanisms downstream of EGFR (Figure 3). The first category includes amplification of the gene for the receptor tyrosine kinases (RTKs) MET, HER2 and FGFR1 or the activating fusion of ALK, FGFR3 and RET, essentially substituting for loss of EGFR kinase activity [31–37]. The second mechanism affects downstream effectors, including targets within the mitogen-activated protein kinase (MAPK) pathway (*BRAF* mutation and fusion [36,38], *KRAS* mutation [39,40]), *CRKL* amplification [41]), the phosphatidylinositol-3-kinase (PI3K) pathway (*PIK3CA* (phosphatidylinositol-3-kinase catalytic α subunit) mutation [42], *PTEN* deletion [43]), or cell cycle pathways (CCNE1, CCND1, CDK6 [44]) were observed. There is also a reduced expression of the tumor suppressor NF-1, a KRAS-specific GTPase-activating protein (GAP), but whether these changes are genetic and/or nongenetic has not been well established. However, concomitant exon 19 deletion and stop-gain mutation in NF1 can lead to poor clinical activity of gefitinib and osimertinib, suggesting that these mechanisms may be involved in EGFR inhibitor resistance [45,46]. Further, BIM deletion polymorphisms did not directly cause EGFR inhibitor resistance but resulted in significantly shorter progression-free survival and therefore affected the overall efficacy of the treatment [47]. Additional rare mutations have been found in various models of cell line-based EGFR inhibitor resistance.

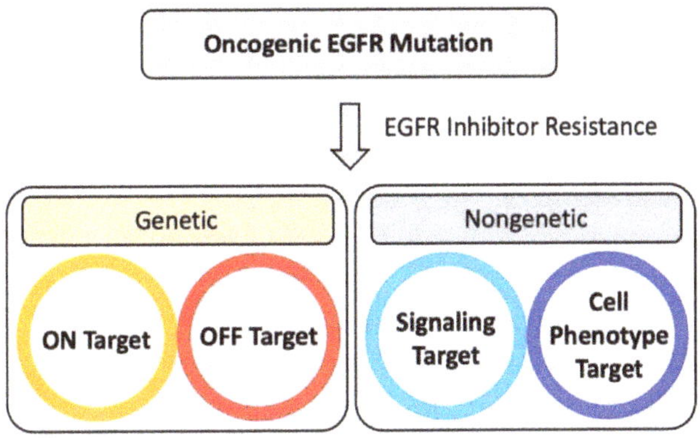

Figure 2. Major genetic and nongenetic escape mechanisms for acquired EGFR inhibitor resistance.

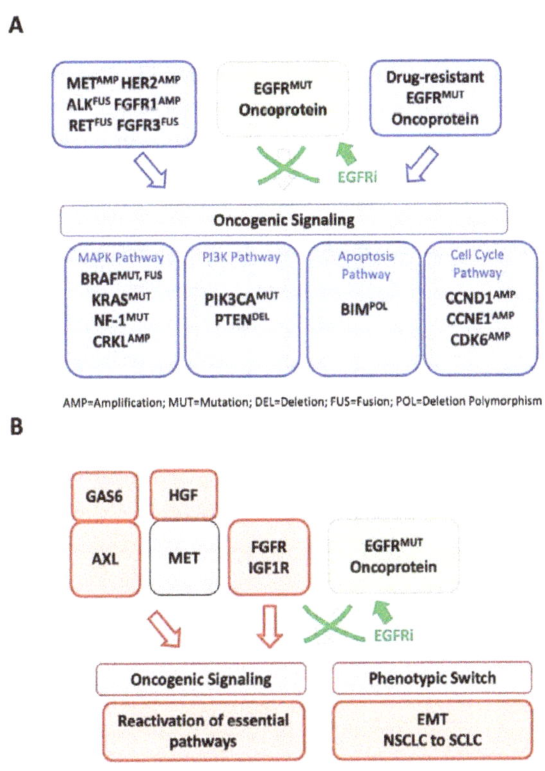

Figure 3. Model of genetic and nongenetic drug resistance mechanisms. Simplified model of possible (**A**) genetic and (**B**) nongenetic alteration identified in patients with therapy-related resistance to EGFR inhibitors.

4. The EGFR Inhibitor Genetic Resistance Gap

Not all mechanisms of resistance are based on genetic changes that can be attributed to mutations (Figure 2). There is no defined overall proportion of specific mechanisms and it also likely depends on multiple factors, such as patient selection, pretreatment or cotreatment, and the type and class of EGFR inhibitor used, to name a few variables. For example, in a cohort of 37 patients resistant to first-generation EGFR inhibitors, 44% showed a nongenetic mechanisms of drug resistance (including phenotypic changes) [42]. First-line osimertinib resistance in NSCLC can be caused by 53–69% and second-line osimertinib resistance can be caused by 30–60% of unknown mechanisms that are likely mostly nongenetic [48]. Moreover, there is little information about what causes the regulation of signaling molecules through nongenetic mechanisms of EGFR inhibitor resistance. It is likely that these mechanisms involve typical modifications that regulate gene expression, such as changes in DNA methylation or changes in histone modifications, ultimately changing DNA accessibility and allowing for changes in gene expression. These alterations are unlikely to affect single genes but may contain unique vulnerabilities that could be exploited.

A complex and poorly understood mechanism of EGFR inhibitor resistance involves transformation into new histologic subtypes, including epithelial-to-mesenchymal transition (EMT) [42,49] and small cell lung cancer (SCLC) transformation of NSCLC cells [42,50] (Figure 3). The altered tumor cells are potentially substituting EGFR-dependency with other mechanisms that also lead to phenotypic changes. These mechanisms are expected to involve transcription factors and their effectors, which could be opportune targets for drug development in this patient population. Nevertheless, EGFR inhibitor-resistant cells, when transformed into SCLC, are genetically diverse and acquired resistant mutations may play a larger role in them than initially thought. The current standard-of-care SCLC therapy is utilized in this population, but overall survival is significantly lower than that of the non-transformed EGFR population and more therapeutic options specific to this population are required [42,51–53].

Similar to the genetic mechanism of resistance, nongenetic mechanisms also involve activation of other RTKs and their ligands, including increased expression of hepatocyte growth factor and the ligand for MET [54,55], as well as the upregulation of fibroblast growth factor receptors (FGFR) [56,57] and insulin-like growth factor 1 receptor (IGF1R) [58] in cell line models or AXL and its ligand GAS6 [49] in EGFR inhibitor-resistant cells. Another interesting mechanism involves the function of Aurora kinase A (AURKA) in the development of EGFR inhibitor resistance. Both AURKA and the related AURKB can share oncogenic features but possess different substrates, and there is considerable interest in targeting Aurora kinase activity in cancers [59]. AURKA is thought to induce at least some level of drug tolerance towards third-generation EGFR inhibitors, which can be reverted by AURKA inhibitors [60]. AURKB may play a more prominent role in EMT-transformed EGFR inhibitor resistance, where it is thought that its co-inhibition with EGFR enhances BIM- and PUMA-mediated apoptosis [61]. Even though not defined in EGF inhibitor resistance models, the activation of signal transduction and activator of transcription 3 (STAT3) [62], the RIG-I-TBK1-IRF3 axis [63] or nuclear factor-κB (NF-κB) [64] may induce residual signaling during the inhibition of EGFR in dependent cells that could be sufficient for the evolution of resistant clones.

5. Clinical Strategies in the Treatment of EGFR Inhibitor Resistance

Clinical strategies for third-generation EGFR inhibitor resistance are mainly focused on combination therapies that inhibit emerging off-targets or on trying to inhibit EGFR with on-target mutations at C797, the binding site of covalent EGFR inhibitors in NSCLC and glioblastoma (Table 1). Combinations include the inhibition of oncogenic EGFR with osimertinib or lazertinib and the targeting of MET with tepotinib (NCT03940703, NCT05120960) and savolitinib (NCT03944772) or the FDA-approved bispecific EGFR-MET antibody amivantamab, respectively (NCT05299125, NCT02609776). MET could also be targeted outside of clinical trials with FDA-approved drugs that are active against this RTK, including crizotinib or capmatinib. Other RTKs that are targeted in combination with osimertinib are mainly in line with resistance mechanisms that are described above and include ALK with alectinib (NCT03944772), RET with selpercatinib (NCT03944772) or HER2 with trastuzumab (NCT04285671) or EGFR, HER2, or HER4 with dacomitinib (NCT03755102). Additional approaches involve combinations with traditional chemotherapy (pemetrexed plus platinum chemotherapy) (NCT03944772, NCT05153408, NCT05299125, NCT02609776) or targeting cancer dependency pathways that are known to be activated downstream of EGFR, including cell cycle (CDK4/CDK6) (NCT04545710), MAPK pathway (NCT03944772, NCT03944772), PI3K pathway (NCT05284994) and others, depending on the resistance mechanism. The development of 4th-generation EGFR inhibitors or inhibitors that are active in the presence of C797 mutations are exciting, including BLU-945 (Blueprint Medicines), WJ13405 (Suzhou Junjing BioSciences), BAY2927088 (Bayer), JIN-A02 (J Ints Bio), HS-10375 (Jiangsu Hansoh Pharmaceutical), QLH11811 (Qilu Pharmaceutical), BPI-361175 (Xcovery Holding Company), and BDTX-1535 (Black Diamond Therapeutics). None of these drugs have yet been approved and little is known about their efficacy, but preclinical information published for BLU-945 [65] or BDTX-1535 [66] is promising and there are additional drugs that will reach clinical stage soon, such as THE-349 (Theseus Pharmaceuticals). It will be important to see whether genetic and nongenetic mechanisms of resistance will apply for these drugs as well and whether there is a significant increase in overall survival. Nongenetic mechanism are difficult to discern, and better biomarker strategies are needed to identify therapeutic targets.

Table 1. Ongoing or planned registered clinical trials of patients with resistance to 3rd generation EGFR inhibitors (query date: 3 January 2023). Indicated are oncogenic EGFR-targeted combination therapies (top) and their targets and matching therapeutic as well as monotherapies (bottom) targeting EGFR with mutations at C797 (C797X).

Primary Target	Primary Therapeutic	Secondary Target	Secondary Therapeutic	ClinicalTrials.gov Identifier
EGFR	Osimertinib	CDK4/CDK6	Abemaciclib	NCT04545710
EGFR	Osimertinib	mTOR Aurora A	Sapanisertib Alisertib	NCT04479306
EGFR	Osimertinib	Anti-EGFR	Necitumumab	NCT02496663
EGFR	Osimertinib	MET	Tepotinib	NCT03940703
EGFR	Osimertinib	MET	Tepotinib	NCT05120960
EGFR	Osimertinib	COX1/COX2 (AKT/BIM)	Aspirin	NCT04184921
EGFR	Osimertinib	MET EGFR Anti-EGFR Antifolate + Anti-PD1 ALK RET Antifolate + Platinum Chemotherapy MEK1/MEK2 TROP2 ADC	Savolitinib Gefitinib Necitumumab Pemetrexed + Durvalumab Alectinib Selpercatinib Pemetrexed + Carboplatin or Cisplatin Selumetinib Datopotamab-deruxtecan	NCT03944772
-	-	Topoisomerase + Anti PD-L1 + Platinum Chemotherapy	Etoposide + Durvalumab + Carboplatin or Cisplatin	
EGFR	Osimertinib	BCL-2/BCL-xL	Pelcitoclax	NCT04001777
EGFR	Osimertinib	BCL-2/BCL-xL	Navitoclax	NCT02520778

Table 1. *Cont.*

Primary Target	Primary Therapeutic	Secondary Target	Secondary Therapeutic	ClinicalTrials.gov Identifier
EGFR	Osimertinib	SRC	Dasatinib	NCT02954523
EGFR	Osimertinib	α/δ Phosphatidylinositol 3-kinase	TQ-B3525	NCT05284994
EGFR	Osimertinib	EGFR HER2	Necitumumab + Trastuzumab	NCT04285671
EGFR, HER2, HER4	Dacomitinib	EGFR	Alone or + Osimertinib	NCT03755102
EGFR-MET bispecific antibody	Amivantamab	EGFR Antifolate Chemotherapy	Lazertinib or + Pemetrexed + Carboplatin	NCT05299125, NCT02609776, NCT04077463
EGFR-MET bispecific antibody	EMB-01			NCT03797391
Anti-HER3 ADC	Patritumab Deruxtecan	EGFR	Osimertinib	NCT04676477
EGFR	Nazartinib (EGF816)	MEK1/MEK2	Trametinib	NCT03516214
PARP	Olaparib	Anti-PD-L1	Durvalumab	NCT04538378
Antifolate + Chemotherapy	Pemetrexed + Platinum Chemotherapy	Anti-PD-1	Alone or + Pembrolizumab	NCT03515837
EGFR (C797X)	BLU-701	EGFR Antifolate Chemotherapy	Alone or + Osimertinib + Pemetrexed + Carboplatin	NCT05153408
EGFR (C797X)	BLU-945	EGFR	Alone or + Osimertinib	NCT04862780
EGFR (C797X)	WJ13405			NCT05662670
EGFR (C797X)	BAY2927088			NCT05099172
EGFR (C797X)	JIN-A02			NCT05394831
EGFR (C797X)	HS-10375			NCT05435248
EGFR (C797X)	QLH11811			NCT05555212
EGFR (C797X)	BPI-361175			NCT05393466
EGFR (C797X)	BDTX-1535			NCT05256290

Traditionally, immune checkpoint inhibitors did poorly as first-line therapeutic treatment in patients with oncogenic EGFR mutations. However, in certain contexts, some patients do benefit from reactivating the T-cell immune response [67]. Nevertheless, this class of therapeutics is also considered for the EGFR inhibitor resistance NSCLC population, including the anti-PD-1 antibody pembrolizumab (NCT03515837) and the anti PD-L1 antibody durvalumab (NCT03944772). Combinatory therapeutic options of EGFR inhibitors with immunotherapy in advanced NSCLC have been reported to result in an increase in the amount of grade 3 or higher toxicities, most notable pneumonitis, with no significant improvement in survival or response [68–70]. However, combination immunotherapy with chemotherapy plus antiangiogenics may be a more viable path towards the development of therapeutics. IMpower150 evaluated atezolizumab (anti PD-L1) with bevacizumab (anti VEGF-A) plus chemotherapy (carboplatin plus paclitaxel) in first-line nonsquamous NSCLC, and EGFR patients showed significantly improved progression-free survival and overall survival [71]. Even so, the majority of EGFR patients still receive TKI first-line therapy and evaluation of immunotherapy with chemotherapy plus antiangiogenic agents following the persistence of resistance is still ongoing. The ORIENT-31 evaluated sintilimab (anti PD-1) immunotherapy with bevacizumab plus chemotherapy (cisplatin plus pemetrexed) in EGFR patients following TKIs and results showed improved progression-free survival of 9.8 months and a response rate of 44% as compared to chemotherapy alone [72]. Overall, these results suggest immunotherapy, chemotherapy, and antiangiogenic therapy combinations may be the most promising therapeutic options. However, we must await more trial results to definitively determine the role of immunotherapy in the resistance setting.

6. Conclusions and Future Direction

The molecular mechanisms that cause genetic or nongenetic drug resistance are unknown, making it difficult to predict treatment strategies. There is some overlap between these two mechanisms, such as the activation of bypass RTK pathways, but there are also unique phenotypic changes induced by nongenetic mechanisms. There is currently no EGFR inhibitor that halts disease progression or even provides curative benefits due to drug resistance. A major goal of clinical strategies involves the identification of on-target and off-target mutations. These mutations allow for the maintenance of oncogenic signaling, either by blocking inhibitor binding or targeting bypass mechanisms. Liquid biopsies may help to provide useful insights into the type of mutations and will lead to the use of possible bypass pathway inhibitors. Additional tests that capture nongenetic changes should be considered for patients where no acquired driver mutations can be identified. Team medicine will lead the way and identify the best treatment strategies for novel or current therapeutics and evaluate their risk–benefit relationship, pinpoint novel escape mechanisms through precision medicine, evaluate and adjust treatments for differences in ethnicity, sex or age through clinical trials, provide the best possible care for patients while optimizing quality of life during treatment, and attempt to make affordable care available to all patients. Current common therapeutic approaches include combinations with drugs that target (a) bypass mechanism, such as MET inhibitors in cells with MET amplification, (b) common cancer pathways, such as apoptosis, cell cycle or MAPK pathways, (c) critical downstream effectors of EGFR, and (d) anti-folate and/or platinum chemotherapy. Alternatively, antibody drug conjugates (ADCs) and experimental vaccines can also be considered. The goal is not only to target cancer cells more efficaciously but possibly also to eliminate potential emerging drug resistant subclones to at least delay disease progression. A question that has been discussed for other oncogenes is that, in the presence of an inhibitor, the scaffold function of the mutated oncoprotein may contribute to some oncogenic signaling and therefore targeted degradation of EGFR should be considered [73]. Many inhibitors for bypass mechanisms are already available and there is hope that better biomarker strategies will help to identify the patient population that can benefit from these therapeutics, in particular for those with nongenetic resistance against EGFR inhibitors.

Author Contributions: Conceptualization, M.S. and R.S.; resources, M.S. and R.S.; original draft preparation, M.S. and R.S.; review and editing, M.S., I.M., J.F., T.T., S.L., N.V., E.P., T.M., A.G.R., A.M., S.S. and R.S. All authors have read and agreed to the published version of the manuscript.

Funding: This work was supported in part by Robert and Kathleen Henderson via the Robert and Kathleen Henderson Lung Cancer Research Accelerator Fund at the City of Hope and by William and Anna Tenenblatt via The William & Anna Tenenblatt Foundation.

Institutional Review Board Statement: Not applicable for studies not involving humans or animals.

Informed Consent Statement: Not applicable for studies not involving humans.

Data Availability Statement: Not applicable.

Conflicts of Interest: The authors declare no conflict of interest.

References

1. Ramani, S.; Samant, S.; Manohar, S.M. The story of EGFR: From signaling pathways to a potent anticancer target. *Future Med. Chem.* **2022**, *14*, 1267–1288. [CrossRef] [PubMed]
2. Kosaka, T.; Yatabe, Y.; Endoh, H.; Kuwano, H.; Takahashi, T.; Mitsudomi, T. Mutations of the epidermal growth factor receptor gene in lung cancer: Biological and clinical implications. *Cancer Res.* **2004**, *64*, 8919–8923. [CrossRef] [PubMed]
3. Midha, A.; Dearden, S.; McCormack, R. EGFR mutation incidence in non-small-cell lung cancer of adenocarcinoma histology: A systematic review and global map by ethnicity (mutMapII). *Am. J. Cancer Res.* **2015**, *5*, 2892–2911.
4. Shigematsu, H.; Lin, L.; Takahashi, T.; Nomura, M.; Suzuki, M.; Wistuba, I.I.; Fong, K.M.; Lee, H.; Toyooka, S.; Shimizu, N.; et al. Clinical and biological features associated with epidermal growth factor receptor gene mutations in lung cancers. *J. Natl. Cancer Inst.* **2005**, *97*, 339–346. [CrossRef] [PubMed]
5. Passaro, A.; Mok, T.; Peters, S.; Popat, S.; Ahn, M.J.; de Marinis, F. Recent Advances on the Role of EGFR Tyrosine Kinase Inhibitors in the Management of NSCLC With Uncommon, Non Exon 20 Insertions, EGFR Mutations. *J. Thorac. Oncol.* **2021**, *16*, 764–773. [CrossRef] [PubMed]
6. Harrison, P.T.; Vyse, S.; Huang, P.H. Rare epidermal growth factor receptor (EGFR) mutations in non-small cell lung cancer. *Semin. Cancer Biol.* **2020**, *61*, 167–179. [CrossRef] [PubMed]
7. Lynch, T.J.; Bell, D.W.; Sordella, R.; Gurubhagavatula, S.; Okimoto, R.A.; Brannigan, B.W.; Harris, P.L.; Haserlat, S.M.; Supko, J.G.; Haluska, F.G.; et al. Activating mutations in the epidermal growth factor receptor underlying responsiveness of non-small-cell lung cancer to gefitinib. *N. Engl. J. Med.* **2004**, *350*, 2129–2139. [CrossRef] [PubMed]
8. Paez, J.G.; Janne, P.A.; Lee, J.C.; Tracy, S.; Greulich, H.; Gabriel, S.; Herman, P.; Kaye, F.J.; Lindeman, N.; Boggon, T.J.; et al. EGFR mutations in lung cancer: Correlation with clinical response to gefitinib therapy. *Science* **2004**, *304*, 1497–1500. [CrossRef] [PubMed]
9. Sattler, M.; Mohanty, A.; Kulkarni, P.; Salgia, R. Precision oncology provides opportunities for targeting KRAS-inhibitor resistance. *Trends Cancer* **2023**, *9*, 42–54. [CrossRef]
10. Hata, A.N.; Niederst, M.J.; Archibald, H.L.; Gomez-Caraballo, M.; Siddiqui, F.M.; Mulvey, H.E.; Maruvka, Y.E.; Ji, F.; Bhang, H.E.; Krishnamurthy Radhakrishna, V.; et al. Tumor cells can follow distinct evolutionary paths to become resistant to epidermal growth factor receptor inhibition. *Nat. Med.* **2016**, *22*, 262–269. [CrossRef]
11. Zhao, Z.; Xie, L.; Bourne, P.E. Structural Insights into Characterizing Binding Sites in Epidermal Growth Factor Receptor Kinase Mutants. *J. Chem. Inf. Model.* **2019**, *59*, 453–462. [CrossRef] [PubMed]
12. Todsaporn, D.; Mahalapbutr, P.; Poo-Arporn, R.P.; Choowongkomon, K.; Rungrotmongkol, T. Structural dynamics and kinase inhibitory activity of three generations of tyrosine kinase inhibitors against wild-type, L858R/T790M, and L858R/T790M/C797S forms of EGFR. *Comput. Biol. Med.* **2022**, *147*, 105787. [CrossRef] [PubMed]
13. Kobayashi, S.; Boggon, T.J.; Dayaram, T.; Janne, P.A.; Kocher, O.; Meyerson, M.; Johnson, B.E.; Eck, M.J.; Tenen, D.G.; Halmos, B. EGFR mutation and resistance of non-small-cell lung cancer to gefitinib. *N. Engl. J. Med.* **2005**, *352*, 786–792. [CrossRef] [PubMed]
14. Pao, W.; Miller, V.A.; Politi, K.A.; Riely, G.J.; Somwar, R.; Zakowski, M.F.; Kris, M.G.; Varmus, H. Acquired resistance of lung adenocarcinomas to gefitinib or erlotinib is associated with a second mutation in the EGFR kinase domain. *PLoS Med.* **2005**, *2*, e73. [CrossRef]
15. Kwak, E.L.; Sordella, R.; Bell, D.W.; Godin-Heymann, N.; Okimoto, R.A.; Brannigan, B.W.; Harris, P.L.; Driscoll, D.R.; Fidias, P.; Lynch, T.J.; et al. Irreversible inhibitors of the EGF receptor may circumvent acquired resistance to gefitinib. *Proc. Natl. Acad. Sci. USA* **2005**, *102*, 7665–7670. [CrossRef]
16. Janne, P.A.; Boss, D.S.; Camidge, D.R.; Britten, C.D.; Engelman, J.A.; Garon, E.B.; Guo, F.; Wong, S.; Liang, J.; Letrent, S.; et al. Phase I dose-escalation study of the pan-HER inhibitor, PF299804, in patients with advanced malignant solid tumors. *Clin. Cancer Res.* **2011**, *17*, 1131–1139. [CrossRef]
17. Yap, T.A.; Vidal, L.; Adam, J.; Stephens, P.; Spicer, J.; Shaw, H.; Ang, J.; Temple, G.; Bell, S.; Shahidi, M.; et al. Phase I trial of the irreversible EGFR and HER2 kinase inhibitor BIBW 2992 in patients with advanced solid tumors. *J. Clin. Oncol.* **2010**, *28*, 3965–3972. [CrossRef]
18. Janne, P.A.; Yang, J.C.; Kim, D.W.; Planchard, D.; Ohe, Y.; Ramalingam, S.S.; Ahn, M.J.; Kim, S.W.; Su, W.C.; Horn, L.; et al. AZD9291 in EGFR inhibitor-resistant non-small-cell lung cancer. *N. Engl. J. Med.* **2015**, *372*, 1689–1699. [CrossRef]
19. Ahn, M.J.; Han, J.Y.; Lee, K.H.; Kim, S.W.; Kim, D.W.; Lee, Y.G.; Cho, E.K.; Kim, J.H.; Lee, G.W.; Lee, J.S.; et al. Lazertinib in patients with EGFR mutation-positive advanced non-small-cell lung cancer: Results from the dose escalation and dose expansion parts of a first-in-human, open-label, multicentre, phase 1–2 study. *Lancet Oncol.* **2019**, *20*, 1681–1690. [CrossRef]
20. Soria, J.C.; Ohe, Y.; Vansteenkiste, J.; Reungwetwattana, T.; Chewaskulyong, B.; Lee, K.H.; Dechaphunkul, A.; Imamura, F.; Nogami, N.; Kurata, T.; et al. Osimertinib in Untreated EGFR-Mutated Advanced Non-Small-Cell Lung Cancer. *N. Engl. J. Med.* **2018**, *378*, 113–125. [CrossRef]
21. Ramalingam, S.S.; Vansteenkiste, J.; Planchard, D.; Cho, B.C.; Gray, J.E.; Ohe, Y.; Zhou, C.; Reungwetwattana, T.; Cheng, Y.; Chewaskulyong, B.; et al. Overall Survival with Osimertinib in Untreated, EGFR-Mutated Advanced NSCLC. *N. Engl. J. Med.* **2020**, *382*, 41–50. [CrossRef]
22. Erickson, A.W.; Brastianos, P.K.; Das, S. Assessment of Effectiveness and Safety of Osimertinib for Patients With Intracranial Metastatic Disease: A Systematic Review and Meta-analysis. *JAMA Netw. Open* **2020**, *3*, e201617. [CrossRef]

23. Gonzalvez, F.; Vincent, S.; Baker, T.E.; Gould, A.E.; Li, S.; Wardwell, S.D.; Nadworny, S.; Ning, Y.; Zhang, S.; Huang, W.S.; et al. Mobocertinib (TAK-788): A Targeted Inhibitor of EGFR Exon 20 Insertion Mutants in Non-Small Cell Lung Cancer. *Cancer Discov.* **2021**, *11*, 1672–1687. [CrossRef]
24. Riely, G.J.; Neal, J.W.; Camidge, D.R.; Spira, A.I.; Piotrowska, Z.; Costa, D.B.; Tsao, A.S.; Patel, J.D.; Gadgeel, S.M.; Bazhenova, L.; et al. Activity and Safety of Mobocertinib (TAK-788) in Previously Treated Non-Small Cell Lung Cancer with EGFR Exon 20 Insertion Mutations from a Phase I/II Trial. *Cancer Discov.* **2021**, *11*, 1688–1699. [CrossRef] [PubMed]
25. Riess, J.W.; Gandara, D.R.; Frampton, G.M.; Madison, R.; Peled, N.; Bufill, J.A.; Dy, G.K.; Ou, S.I.; Stephens, P.J.; McPherson, J.D.; et al. Diverse EGFR Exon 20 Insertions and Co-Occurring Molecular Alterations Identified by Comprehensive Genomic Profiling of NSCLC. *J. Thorac. Oncol.* **2018**, *13*, 1560–1568. [CrossRef] [PubMed]
26. Zhou, C.; Ramalingam, S.S.; Kim, T.M.; Kim, S.W.; Yang, J.C.; Riely, G.J.; Mekhail, T.; Nguyen, D.; Garcia Campelo, M.R.; Felip, E.; et al. Treatment Outcomes and Safety of Mobocertinib in Platinum-Pretreated Patients With EGFR Exon 20 Insertion-Positive Metastatic Non-Small Cell Lung Cancer: A Phase 1/2 Open-label Nonrandomized Clinical Trial. *JAMA Oncol.* **2021**, *7*, e214761. [CrossRef]
27. Kian, W.; Christopoulos, P.; Remilah, A.A.; Levison, E.; Dudnik, E.; Shalata, W.; Krayim, B.; Marei, R.; Yakobson, A.; Faehling, M.; et al. Real-world efficacy and safety of mobocertinib in EGFR exon 20 insertion-mutated lung cancer. *Front. Oncol.* **2022**, *12*, 1010311. [CrossRef]
28. Cho, B.C.; Han, J.Y.; Kim, S.W.; Lee, K.H.; Cho, E.K.; Lee, Y.G.; Kim, D.W.; Kim, J.H.; Lee, G.W.; Lee, J.S.; et al. A Phase 1/2 Study of Lazertinib 240 mg in Patients With Advanced EGFR T790M-Positive NSCLC After Previous EGFR Tyrosine Kinase Inhibitors. *J. Thorac. Oncol.* **2022**, *17*, 558–567. [CrossRef]
29. Castellano, G.M.; Aisner, J.; Burley, S.K.; Vallat, B.; Yu, H.A.; Pine, S.R.; Ganesan, S. A Novel Acquired Exon 20 EGFR M766Q Mutation in Lung Adenocarcinoma Mediates Osimertinib Resistance but is Sensitive to Neratinib and Poziotinib. *J. Thorac. Oncol.* **2019**, *14*, 1982–1988. [CrossRef] [PubMed]
30. Tumbrink, H.L.; Heimsoeth, A.; Sos, M.L. The next tier of EGFR resistance mutations in lung cancer. *Oncogene* **2021**, *40*, 1–11. [CrossRef]
31. Engelman, J.A.; Zejnullahu, K.; Mitsudomi, T.; Song, Y.; Hyland, C.; Park, J.O.; Lindeman, N.; Gale, C.M.; Zhao, X.; Christensen, J.; et al. MET amplification leads to gefitinib resistance in lung cancer by activating ERBB3 signaling. *Science* **2007**, *316*, 1039–1043. [CrossRef] [PubMed]
32. Bean, J.; Brennan, C.; Shih, J.Y.; Riely, G.; Viale, A.; Wang, L.; Chitale, D.; Motoi, N.; Szoke, J.; Broderick, S.; et al. MET amplification occurs with or without T790M mutations in EGFR mutant lung tumors with acquired resistance to gefitinib or erlotinib. *Proc. Natl. Acad. Sci. USA* **2007**, *104*, 20932–20937. [CrossRef] [PubMed]
33. Takezawa, K.; Pirazzoli, V.; Arcila, M.E.; Nebhan, C.A.; Song, X.; de Stanchina, E.; Ohashi, K.; Janjigian, Y.Y.; Spitzler, P.J.; Melnick, M.A.; et al. HER2 amplification: A potential mechanism of acquired resistance to EGFR inhibition in EGFR-mutant lung cancers that lack the second-site EGFRT790M mutation. *Cancer Discov.* **2012**, *2*, 922–933. [CrossRef] [PubMed]
34. Kim, T.M.; Song, A.; Kim, D.W.; Kim, S.; Ahn, Y.O.; Keam, B.; Jeon, Y.K.; Lee, S.H.; Chung, D.H.; Heo, D.S. Mechanisms of Acquired Resistance to AZD9291: A Mutation-Selective, Irreversible EGFR Inhibitor. *J. Thorac. Oncol.* **2015**, *10*, 1736–1744. [CrossRef]
35. Ou, S.I.; Horn, L.; Cruz, M.; Vafai, D.; Lovly, C.M.; Spradlin, A.; Williamson, M.J.; Dagogo-Jack, I.; Johnson, A.; Miller, V.A.; et al. Emergence of FGFR3-TACC3 fusions as a potential by-pass resistance mechanism to EGFR tyrosine kinase inhibitors in EGFR mutated NSCLC patients. *Lung Cancer* **2017**, *111*, 61–64. [CrossRef]
36. Oxnard, G.R.; Hu, Y.; Mileham, K.F.; Husain, H.; Costa, D.B.; Tracy, P.; Feeney, N.; Sholl, L.M.; Dahlberg, S.E.; Redig, A.J.; et al. Assessment of Resistance Mechanisms and Clinical Implications in Patients With EGFR T790M-Positive Lung Cancer and Acquired Resistance to Osimertinib. *JAMA Oncol.* **2018**, *4*, 1527–1534. [CrossRef]
37. von Buttlar, X.; Reuss, J.E.; Liu, S.V.; Kim, C. EML4-ALK Rearrangement as a Mechanism of Resistance to Osimertinib in Metastatic Lung Adenocarcinoma: A Case Report. *JTO Clin. Res. Rep.* **2021**, *2*, 100179. [CrossRef]
38. Ohashi, K.; Sequist, L.V.; Arcila, M.E.; Moran, T.; Chmielecki, J.; Lin, Y.L.; Pan, Y.; Wang, L.; de Stanchina, E.; Shien, K.; et al. Lung cancers with acquired resistance to EGFR inhibitors occasionally harbor BRAF gene mutations but lack mutations in KRAS, NRAS, or MEK1. *Proc. Natl. Acad. Sci. USA* **2012**, *109*, E2127–E2133. [CrossRef]
39. Del Re, M.; Tiseo, M.; Bordi, P.; D'Incecco, A.; Camerini, A.; Petrini, I.; Lucchesi, M.; Inno, A.; Spada, D.; Vasile, E.; et al. Contribution of KRAS mutations and c.2369C > T (p.T790M) EGFR to acquired resistance to EGFR-TKIs in EGFR mutant NSCLC: A study on circulating tumor DNA. *Oncotarget* **2017**, *8*, 13611–13619. [CrossRef]
40. Chabon, J.J.; Simmons, A.D.; Lovejoy, A.F.; Esfahani, M.S.; Newman, A.M.; Haringsma, H.J.; Kurtz, D.M.; Stehr, H.; Scherer, F.; Karlovich, C.A.; et al. Circulating tumour DNA profiling reveals heterogeneity of EGFR inhibitor resistance mechanisms in lung cancer patients. *Nat. Commun.* **2016**, *7*, 11815. [CrossRef]
41. Cheung, H.W.; Du, J.; Boehm, J.S.; He, F.; Weir, B.A.; Wang, X.; Butaney, M.; Sequist, L.V.; Luo, B.; Engelman, J.A.; et al. Amplification of CRKL induces transformation and epidermal growth factor receptor inhibitor resistance in human non-small cell lung cancers. *Cancer Discov.* **2011**, *1*, 608–625. [CrossRef]
42. Sequist, L.V.; Waltman, B.A.; Dias-Santagata, D.; Digumarthy, S.; Turke, A.B.; Fidias, P.; Bergethon, K.; Shaw, A.T.; Gettinger, S.; Cosper, A.K.; et al. Genotypic and histological evolution of lung cancers acquiring resistance to EGFR inhibitors. *Sci. Transl. Med.* **2011**, *3*, 75ra26. [CrossRef]

43. Sos, M.L.; Koker, M.; Weir, B.A.; Heynck, S.; Rabinovsky, R.; Zander, T.; Seeger, J.M.; Weiss, J.; Fischer, F.; Frommolt, P.; et al. PTEN loss contributes to erlotinib resistance in EGFR-mutant lung cancer by activation of Akt and EGFR. *Cancer Res.* **2009**, *69*, 3256–3261. [CrossRef] [PubMed]
44. Lim, S.M.; Yang, S.D.; Lim, S.; Shim, H.S.; Cho, B.C. Brief Report: Heterogeneity of Acquired Resistance Mechanisms to Osimertinib and Savolitinib. *JTO Clin. Res. Rep.* **2021**, *2*, 100180. [CrossRef] [PubMed]
45. de Bruin, E.C.; Cowell, C.; Warne, P.H.; Jiang, M.; Saunders, R.E.; Melnick, M.A.; Gettinger, S.; Walther, Z.; Wurtz, A.; Heynen, G.J.; et al. Reduced NF1 expression confers resistance to EGFR inhibition in lung cancer. *Cancer Discov.* **2014**, *4*, 606–619. [CrossRef]
46. Pan, Y.; Yuan, C.; Cheng, C.; Zhang, Y.; Ma, Y.; Zheng, D.; Zheng, S.; Li, Y.; Jin, Y.; Sun, Y.; et al. Frequency and clinical significance of NF1 mutation in lung adenocarcinomas from East Asian patients. *Int. J. Cancer* **2019**, *144*, 290–296. [CrossRef] [PubMed]
47. Isobe, K.; Hata, Y.; Tochigi, N.; Kaburaki, K.; Kobayashi, H.; Makino, T.; Otsuka, H.; Sato, F.; Ishida, F.; Kikuchi, N.; et al. Clinical significance of BIM deletion polymorphism in non-small-cell lung cancer with epidermal growth factor receptor mutation. *J. Thorac. Oncol.* **2014**, *9*, 483–487. [CrossRef]
48. He, J.; Huang, Z.; Han, L.; Gong, Y.; Xie, C. Mechanisms and management of 3rd-generation EGFR-TKI resistance in advanced non-small cell lung cancer (Review). *Int. J. Oncol.* **2021**, *59*, 90. [CrossRef]
49. Zhang, Z.; Lee, J.C.; Lin, L.; Olivas, V.; Au, V.; LaFramboise, T.; Abdel-Rahman, M.; Wang, X.; Levine, A.D.; Rho, J.K.; et al. Activation of the AXL kinase causes resistance to EGFR-targeted therapy in lung cancer. *Nat. Genet.* **2012**, *44*, 852–860. [CrossRef]
50. Yu, H.A.; Arcila, M.E.; Rekhtman, N.; Sima, C.S.; Zakowski, M.F.; Pao, W.; Kris, M.G.; Miller, V.A.; Ladanyi, M.; Riely, G.J. Analysis of tumor specimens at the time of acquired resistance to EGFR-TKI therapy in 155 patients with EGFR-mutant lung cancers. *Clin. Cancer Res.* **2013**, *19*, 2240–2247. [CrossRef]
51. Niederst, M.J.; Sequist, L.V.; Poirier, J.T.; Mermel, C.H.; Lockerman, E.L.; Garcia, A.R.; Katayama, R.; Costa, C.; Ross, K.N.; Moran, T.; et al. RB loss in resistant EGFR mutant lung adenocarcinomas that transform to small-cell lung cancer. *Nat. Commun.* **2015**, *6*, 6377. [CrossRef]
52. Marcoux, N.; Gettinger, S.N.; O'Kane, G.; Arbour, K.C.; Neal, J.W.; Husain, H.; Evans, T.L.; Brahmer, J.R.; Muzikansky, A.; Bonomi, P.D.; et al. EGFR-Mutant Adenocarcinomas That Transform to Small-Cell Lung Cancer and Other Neuroendocrine Carcinomas: Clinical Outcomes. *J. Clin. Oncol.* **2019**, *37*, 278–285. [CrossRef] [PubMed]
53. Mambetsariev, I.; Arvanitis, L.; Fricke, J.; Pharaon, R.; Baroz, A.R.; Afkhami, M.; Koczywas, M.; Massarelli, E.; Salgia, R. Small Cell Lung Cancer Transformation following Treatment in EGFR-Mutated Non-Small Cell Lung Cancer. *J. Clin. Med.* **2022**, *11*, 1429. [CrossRef] [PubMed]
54. Yano, S.; Wang, W.; Li, Q.; Matsumoto, K.; Sakurama, H.; Nakamura, T.; Ogino, H.; Kakiuchi, S.; Hanibuchi, M.; Nishioka, Y.; et al. Hepatocyte growth factor induces gefitinib resistance of lung adenocarcinoma with epidermal growth factor receptor-activating mutations. *Cancer Res.* **2008**, *68*, 9479–9487. [CrossRef]
55. Yano, S.; Yamada, T.; Takeuchi, S.; Tachibana, K.; Minami, Y.; Yatabe, Y.; Mitsudomi, T.; Tanaka, H.; Kimura, T.; Kudoh, S.; et al. Hepatocyte growth factor expression in EGFR mutant lung cancer with intrinsic and acquired resistance to tyrosine kinase inhibitors in a Japanese cohort. *J. Thorac. Oncol.* **2011**, *6*, 2011–2017. [CrossRef] [PubMed]
56. Ware, K.E.; Hinz, T.K.; Kleczko, E.; Singleton, K.R.; Marek, L.A.; Helfrich, B.A.; Cummings, C.T.; Graham, D.K.; Astling, D.; Tan, A.C.; et al. A mechanism of resistance to gefitinib mediated by cellular reprogramming and the acquisition of an FGF2-FGFR1 autocrine growth loop. *Oncogenesis* **2013**, *2*, e39. [CrossRef]
57. Ware, K.E.; Marshall, M.E.; Heasley, L.R.; Marek, L.; Hinz, T.K.; Hercule, P.; Helfrich, B.A.; Doebele, R.C.; Heasley, L.E. Rapidly acquired resistance to EGFR tyrosine kinase inhibitors in NSCLC cell lines through de-repression of FGFR2 and FGFR3 expression. *PLoS ONE* **2010**, *5*, e14117. [CrossRef]
58. Cortot, A.B.; Repellin, C.E.; Shimamura, T.; Capelletti, M.; Zejnullahu, K.; Ercan, D.; Christensen, J.G.; Wong, K.K.; Gray, N.S.; Janne, P.A. Resistance to irreversible EGF receptor tyrosine kinase inhibitors through a multistep mechanism involving the IGF1R pathway. *Cancer Res.* **2013**, *73*, 834–843. [CrossRef]
59. Du, R.; Huang, C.; Liu, K.; Li, X.; Dong, Z. Targeting AURKA in Cancer: Molecular mechanisms and opportunities for Cancer therapy. *Mol. Cancer* **2021**, *20*, 15. [CrossRef]
60. Shah, K.N.; Bhatt, R.; Rotow, J.; Rohrberg, J.; Olivas, V.; Wang, V.E.; Hemmati, G.; Martins, M.M.; Maynard, A.; Kuhn, J.; et al. Aurora kinase A drives the evolution of resistance to third-generation EGFR inhibitors in lung cancer. *Nat. Med.* **2019**, *25*, 111–118. [CrossRef]
61. Tanaka, K.; Yu, H.A.; Yang, S.; Han, S.; Selcuklu, S.D.; Kim, K.; Ramani, S.; Ganesan, Y.T.; Moyer, A.; Sinha, S.; et al. Targeting Aurora B kinase prevents and overcomes resistance to EGFR inhibitors in lung cancer by enhancing BIM- and PUMA-mediated apoptosis. *Cancer Cell* **2021**, *39*, 1245–1261.e6. [CrossRef] [PubMed]
62. Lee, H.J.; Zhuang, G.; Cao, Y.; Du, P.; Kim, H.J.; Settleman, J. Drug resistance via feedback activation of Stat3 in oncogene-addicted cancer cells. *Cancer Cell* **2014**, *26*, 207–221. [CrossRef] [PubMed]
63. Gong, K.; Guo, G.; Panchani, N.; Bender, M.E.; Gerber, D.E.; Minna, J.D.; Fattah, F.; Gao, B.; Peyton, M.; Kernstine, K.; et al. EGFR inhibition triggers an adaptive response by co-opting antiviral signaling pathways in lung cancer. *Nat. Cancer* **2020**, *1*, 394–409. [CrossRef]

64. Blakely, C.M.; Pazarentzos, E.; Olivas, V.; Asthana, S.; Yan, J.J.; Tan, I.; Hrustanovic, G.; Chan, E.; Lin, L.; Neel, D.S.; et al. NF-kappaB-activating complex engaged in response to EGFR oncogene inhibition drives tumor cell survival and residual disease in lung cancer. *Cell Rep.* **2015**, *11*, 98–110. [CrossRef] [PubMed]
65. Eno, M.S.; Brubaker, J.D.; Campbell, J.E.; De Savi, C.; Guzi, T.J.; Williams, B.D.; Wilson, D.; Wilson, K.; Brooijmans, N.; Kim, J.; et al. Discovery of BLU-945, a Reversible, Potent, and Wild-Type-Sparing Next-Generation EGFR Mutant Inhibitor for Treatment-Resistant Non-Small-Cell Lung Cancer. *J. Med. Chem.* **2022**, *65*, 9662–9677. [CrossRef] [PubMed]
66. Poh, A. BDTX-1535 Goes after Osimertinib Resistance. *Cancer Discov.* **2021**, *11*, 2952–2953. [CrossRef]
67. Wiest, N.; Majeed, U.; Seegobin, K.; Zhao, Y.; Lou, Y.; Manochakian, R. Role of Immune Checkpoint Inhibitor Therapy in Advanced EGFR-Mutant Non-Small Cell Lung Cancer. *Front. Oncol.* **2021**, *11*, 751209. [CrossRef]
68. Gettinger, S.; Hellmann, M.D.; Chow, L.Q.M.; Borghaei, H.; Antonia, S.; Brahmer, J.R.; Goldman, J.W.; Gerber, D.E.; Juergens, R.A.; Shepherd, F.A.; et al. Nivolumab Plus Erlotinib in Patients With EGFR-Mutant Advanced NSCLC. *J. Thorac. Oncol.* **2018**, *13*, 1363–1372. [CrossRef]
69. Yang, J.C.; Gadgeel, S.M.; Sequist, L.V.; Wu, C.L.; Papadimitrakopoulou, V.A.; Su, W.C.; Fiore, J.; Saraf, S.; Raftopoulos, H.; Patnaik, A. Pembrolizumab in Combination With Erlotinib or Gefitinib as First-Line Therapy for Advanced NSCLC With Sensitizing EGFR Mutation. *J. Thorac. Oncol.* **2019**, *14*, 553–559. [CrossRef]
70. Ahn, M.J.; Cho, B.C.; Ou, X.; Walding, A.; Dymond, A.W.; Ren, S.; Cantarini, M.; Janne, P.A. Osimertinib Plus Durvalumab in Patients With EGFR-Mutated, Advanced NSCLC: A Phase 1b, Open-Label, Multicenter Trial. *J. Thorac. Oncol.* **2022**, *17*, 718–723. [CrossRef]
71. Socinski, M.A.; Jotte, R.M.; Cappuzzo, F.; Orlandi, F.; Stroyakovskiy, D.; Nogami, N.; Rodriguez-Abreu, D.; Moro-Sibilot, D.; Thomas, C.A.; Barlesi, F.; et al. Atezolizumab for First-Line Treatment of Metastatic Nonsquamous NSCLC. *N. Engl. J. Med.* **2018**, *378*, 2288–2301. [CrossRef] [PubMed]
72. Lu, S.; Wu, L.; Jian, H.; Chen, Y.; Wang, Q.; Fang, J.; Wang, Z.; Hu, Y.; Sun, M.; Han, L.; et al. Sintilimab plus bevacizumab biosimilar IBI305 and chemotherapy for patients with EGFR-mutated non-squamous non-small-cell lung cancer who progressed on EGFR tyrosine-kinase inhibitor therapy (ORIENT-31): First interim results from a randomised, double-blind, multicentre, phase 3 trial. *Lancet Oncol.* **2022**, *23*, 1167–1179. [CrossRef] [PubMed]
73. Hong, D.; Zhou, B.; Zhang, B.; Ren, H.; Zhu, L.; Zheng, G.; Ge, M.; Ge, J. Recent advances in the development of EGFR degraders: PROTACs and LYTACs. *Eur. J. Med. Chem.* **2022**, *239*, 114533. [CrossRef] [PubMed]

Journal of
Clinical Medicine

MDPI

Review

Talimogene Laherparepvec (T-VEC): A Review of the Recent Advances in Cancer Therapy

Tiantian Zhang [1], Tony Hong-Ting Jou [2], Jerline Hsin [3], Zhe Wang [4], Kelly Huang [5], Jian Ye [6], Holly Yin [4] and Yan Xing [5,*]

1 Toni Stephenson Lymphoma Center, Department of Hematology and Hematopoietic Stem Cell Transplantation, Beckman Research Institute, City of Hope, Duarte, CA 91010, USA
2 School of Medicine, National Yang Ming Chiao Tung University, Taipei 11217, Taiwan
3 Department of Pharmacy, City of Hope, Duarte, CA 91010, USA
4 High Throughput Screening Core, Department of Share Resources, Beckman Research Institute, City of Hope, Duarte, CA 91010, USA
5 Department of Medical Oncology and Therapeutics Research, City of Hope, Duarte, CA 91010, USA
6 Department of Immuno-Oncology, Beckman Research Institute, City of Hope, Duarte, CA 91010, USA
* Correspondence: yxing@coh.org

Abstract: The landscape of melanoma treatment has undergone a dramatic revolution in the past decade. The use of oncolytic viruses (OVs) represents a novel therapeutic approach that can selectively infect and lyse tumor cells and induce local and systemic antitumor immune responses. As the first OV approved by the Food and Drug Administration (FDA) for melanoma treatment, talimogene laherparepvec (T-VEC), a genetically modified herpes simplex virus (HSV), has shown promising therapeutic effects in the treatment of advanced melanoma, both as a monotherapy or in combination with other immunotherapies, such as the immune checkpoint inhibitors (ICIs). With proven efficacy, T-VEC has been evaluated against a variety of other cancer types in a clinical trial setting. In this article, we will provide a review on OVs and the application of T-VEC in melanoma monotherapy and combination therapy. In addition, we will review the recent progress of T-VEC application in other cutaneous cancer types. Moreover, we will briefly describe our experience of T-VEC therapy at City of Hope, aiming to provide more insight for expanding its future application.

Keywords: oncolytic virotherapy; T-VEC; immune checkpoint inhibitors; immunotherapy; targeted therapy; combinational therapy; melanoma; cutaneous cancers; clinical trials

Citation: Zhang, T.; Jou, T.H.-T.; Hsin, J.; Wang, Z.; Huang, K.; Ye, J.; Yin, H.; Xing, Y. Talimogene Laherparepvec (T-VEC): A Review of the Recent Advances in Cancer Therapy. *J. Clin. Med.* **2023**, *12*, 1098. https://doi.org/10.3390/jcm12031098

Academic Editor: Maria Lina Tornesello

Received: 23 December 2022
Revised: 13 January 2023
Accepted: 28 January 2023
Published: 31 January 2023

1. Introduction

The last decade has witnessed a dramatic transformation of the landscape of melanoma treatment. Based on the deeper understanding of the molecular features of melanoma and the tumor microenvironment, the current melanoma therapies have progressed to mainly include targeted therapy, immune checkpoint inhibitors (ICIs), and virotherapy. The elucidation of BRAF V600 mutations and the dysregulated RAS/RAF/MEK/ERK pathway in melanoma cells has led to the development of targeted therapies including BRAF inhibitors (BRAFi) and MEK inhibitors (MEKi), which have shown significant efficacy in melanoma eradication and been approved by the Food and Drug Administration (FDA). The discovery of ICIs mainly includes the antibodies against cytotoxic T lymphocyte antigen-4 (CTLA-4), programmed cell death protein-1 (PD-1), and programmed death-ligand 1 (PDL-1). ICIs have provided another approach by releasing the inhibitory brakes on the T cells and facilitating robust immune responses, rendering them effective in melanoma treatment [1,2].

Oncolytic viruses (OVs) represent a novel class of cancer therapy in which wild-type or genetically modified viruses are used. Historically, viruses have been explored as therapeutics in two ways—as viral vectors for gene therapy and tumor-lysing ("oncolytic")

viruses [3]. The key difference between these two categories lies in that the OVs are typically replication-competent, whereas the viral vectors for gene therapy are usually replication-defective viruses. Interestingly, modern OVs have often been engineered to express immunostimulatory proteins, which also fulfill the function as viral vectors. A variety of viruses, such as herpes simplex virus (HSV), vaccinia virus, adenovirus, and reovirus, have been evaluated for their oncolytic potency. While some of these viruses have completed different phases of clinical trials, talimogene laherparepvec (T-VEC), which is an engineered HSV-1 with the insertion of the granulocyte monocyte colony-stimulating factor (GM-CSF) gene and deletion of infected cell protein 34.5 (ICP34.5) and ICP47 genes, is the first OV approved by the FDA for melanoma treatment [4,5]. In this review, we will focus on T-VEC and its effects on melanoma and other cutaneous malignancies as a monotherapy and in combination with other cancer therapies (Table 1). We will also discuss ongoing trials involving T-VEC (Table 2). Moreover, we will look at how City of Hope Comprehensive Cancer Center provides T-VEC treatment to its patients, which will provide insight into the implementation of T-VEC in the real-world. In this review, the novelty lies in (1) providing an overview of the path the T-VEC took from initial testing to widespread use, (2) offering detailed information on the past and ongoing clinical trials involving the use of T-VEC as a monotherapy and in combination therapy, (3) and presenting a general description of the clinical experience with T-VEC at City of Hope.

Table 1. Efficacy and safety of T-VEC monotherapies and combination therapies in the treatment of skin cancers.

	Reference	Study Drugs/ Mechanisms of Action	Phase (n)	Disease	Treatment	Overall Response Rate	Progression-Free Survival (Month)	Overall Survival (Month)
I	**T-VEC monotherapy for melanoma**							
1	A Phase I Study of OncoVEXGM-CSF, a Second-Generation Oncolytic Herpes Simplex Virus Expressing Granulocyte Macrophage Colony-Stimulating Factor [6]	talimogene laherparepvec (TVEC)/oncolytic virus therapy (OVT)	Phase I (n = 30)	Refractory cutaneous and subcutaneous metastases from breast cancer, gastrointestinal adenocarcinoma, Malignant Melanoma, and Epithelial cancer of head and neck	TVEC	N/A	N/A	N/A
2	Phase II clinical trial of a granulocyte-macrophage colony-stimulating factor-encoding, second-generation oncolytic herpesvirus in patients with unresectable metastatic melanoma [7]	TVEC/OVT	Phase II (n = 50)	Stage IIIc unresectable metastatic melanoma	TVEC	26%	N/A	16
3	Final analyses of OPTiM: a randomized phase III trial of talimogene laherparepvec versus granulocyte-macrophage colony-stimulating factor in unresectable stage III–IV melanoma (NCT00769704) [8,9]	TVEC/OVT GM-CSF/bone marrow stimulation	Phase III (n = 436)	Stage IIIB to IV melanoma	A: TVEC B: GM-CSF	31.50% 6.40%	N/A	23.3 18.9 A: 73.7% at 1 year, 49.8% at 2 year, and 38.9% at 3 year B: 69.1% at 1 year, 40.3% at 2 year, and 30.4% at 3 year
II	**T-VEC combinational therapy for melanoma**							
1	Randomized, open-label phase II study evaluating the efficacy and safety of talimogene laherparepvec in combination with ipilimumab versus ipilimumab alone in patients with advanced, unresectable melanoma (NCT01740297) [10]	TVEC/OVT Ipilimumab/CTLA-4 inhibitor	Phase II (n = 198)	Melanoma	A: TVEC + ipilimumab B: ipilimumab	39% 18%	8.2 6.4	86.9% at 1 year, 76.6% at 2 year 81.4% at 1 year, 67.7% at 2 year
2	A phase 1/3 multicenter trial of talimogene laherparepvec in combination with pembrolizumab for unresected, stage IIIB-IV melanoma. MASTERKEY-265 (NCT02263508) [11]	TVEC/OVT Pembrolizumab/PD-1 inhibitor	Phase 1b (n = 21)	unresectable, stage IIIB-IVM1c melanoma	A: TVEC + Pembrolizumab B: Placebo + Pembrolizumab	N/A	25.6 25.5	N/A

Table 1. *Cont.*

	Reference	Study Drugs/ Mechanisms of Action	Phase (n)	Disease	Treatment	Overall Response Rate	Progression-Free Survival (Month)	Overall Survival (Month)
3	1037O MASTERKEY-265: A phase III, randomized, placebo (Pbo)-controlled study of talimogene laherparepvec (T) plus pembrolizumab (P) for unresectable stage IIIB–IVM1c melanoma (MEL). KENNOTE-034 (NCT02263508) [12]	TVEC/OVT Pembrolizumab/PD-1 inhibitor	Phase III (n = 692)	unresectable stage III-IVM1c melanoma	A: TVEC + Pembrolizumab B: Placebo + Pembrolizumab	48.60% 41.30%	14.3 8.5	66% at 2 year 49.2
4	PV-10 vs Chemotherapy or Oncolytic Viral Therapy for Treatment of Locally Advanced Cutaneous Melanoma (NCT02288897) [13]	PV-10 (10% rose Bengal disodium)/oncolytic immunotherapy	Phase III (n = 20)	Cutaneous Melanoma	A: PV-10 (10% rose Bengal disodium) B: Dacarbazine, temozolomide or TVEC	N/A Only has complete response rate (CRR)	6.1 (1.5 to 28.9) 8.6 (1.8 to 14.4)	N/A N/A
III	**T-VEC treatment in other cutaneous cancer types**							
1	Talimogene laherparepvec induces durable response of regionally advanced Merkel cell carcinoma in 4 consecutive patients [14]	TVEC/OVT	(n = 4)	Regionally advanced Merkel cell carcinoma	TVEC	100%	16 +	18.5 +
2	Pretreated anti-PD-1 refractory Merkel cell carcinoma successfully treated with the combination of PD-1/PD-L1 axis inhibitors and TVEC: a report of two cases [15]	TVEC/OVT	(n = 2)	Anti-PD-1 refractory Merkel cell Carcinoma	T-VEC and a PD-1/PD-L1 inhibitor	100%	N/A	N/A
3	Immunotherapy for Nonmelanoma skin cancer: Facts and Hopes (NCT02819843) [16]	TVEC/OVT	Phase II (n = 19)	Cutaneous Melanoma Merkel Cell Carcinoma Other Solid Tumors	TVEC + Radiotherapy	Study completion June 2023	Study completion June 2023	Study completion June 2023

Table 2. Ongoing clinical trials of T-VEC in skin cancers.

	Reference	Study Drugs/ Mechanisms of Action	Stage (n)	Disease	Treatment	Key Outcomes
1	Talimogene Laherparepvec and Pembrolizumab in Treating Patients With Stage III-IV Melanoma (NCT02965716)	TVEC/OVT Pembrolizumab/ PD-1 inhibitor	Phase II (n = 47)	Advanced Melanoma Refractory Melanoma	Pembrolizumab and TVEC combination	Objective response rate, median progression-free survival, median overall survival
2	T-VEC in Non-melanoma Skin Cancer (NCT03458117)	TVEC/OVT	Phase I (n = 26)	Non-melanoma Skin Cancer Basal Cell Carcinoma Squamous Cell Carcinoma Cutaneous Lymphoma Merkel Cell Carcinoma	TVEC	Local immune response, systemic immune response
3	Talimogene Laherparepvec and Nivolumab in Treating Patients with Refractory Lymphomas or Advanced or Refractory Non-melanoma Skin Cancers (NCT02978625)	TVEC/OVT Nivolumab/PD-1 inhibitor	Phase II (n = 68)	Refractory T cell Lymphoma Refractory NK cell lymphoma Cutaneous Squamous Cell Carcinoma Merkel Cell Carcinoma Other Rare Skin Tumors	TVEC followed by nivolumab and TVEC combination	Response rate, best overall response rate, progression-free survival, overall survival
4	Study of TVEC in Patients With Cutaneous Squamous Cell Cancer (NCT03714828)	TVEC/OVT	Phase II (n = 11)	Cutaneous Squamous Cell Cancer	TVEC	Overall response rate (ultrasound, targeted lesions, non-injected lesions)
5	Talimogene Laherparepvec and Panitumumab for the Treatment of Locally Advanced or Metastatic Squamous Cell Carcinoma of the Skin (NCT04163952)	TVEC/OVT Panitumumab/Anti-EGFT monoclonal antibodies	Phase I (n = 5)	Advanced Squamous Cell Cancer	Panitumumab and TVEC combination	Response rate, best overall response rate, progression-free survival, overall survival

2. Overview of Oncolytic Virus and T-VEC

OVs have emerged as a novel class of immunotherapies with remarkable efficacy through possessing two closely related properties: the capability to kill cancer cells and the potential to enhance anti-tumor immune responses [17]. The viruses, either native or modified, are able to infect and replicate within tumor cells, causing cell lysis and the release of viral progenies that will proceed to infect neighboring cells. Moreover, virus infection is able to trigger an apoptosis cascade in the surrounding cancer cells, which limits the viral replication and tumor cell proliferation. Meanwhile, the rupture of the tumor cells releases tumor-derived antigens that are new to the immune system, thereby facilitating the development of systemic tumor-specific immune responses [17].

In comparison to normal cells, which possess intact antiviral mechanisms, tumor cells have been found to have abnormally regulated pathways that can be manipulated to facilitate OV infection and replication. For instance, melanoma cells have been shown to harbor Ras overexpression and defective interferon (IFN)-signaling pathways, which can be readily targeted by the oncolytic vesicular stomatitis virus (VSV) and reovirus [18]. Additionally, while tumor cells often overexpress tyrosinase and survivin, the genetic modification of the viral genome to incorporate the promoters of tyrosinase or survivin genes has been found to increase the oncospecificity of oncolytic viruses. Moreover, to stimulate tumor-specific immune reactions, OVs have been genetically engineered to express an array of immunomodulatory or immunostimulatory proteins, such as interleukin (IL)-2, IFNγ, and GM-CSF [17].

In the past two decades, a wide variety of viruses, including adenovirus, HSV, and poxvirus, have been studied for their potency as oncolytic viruses [19–21]. T-VEC, an attenuated HSV expressing GM-CSF, became the first oncolytic agent that achieved regulatory approval in the United States, Europe, and Australia. As a JS1 strain of HSV-1, the preferential tumor infection and replication of T-VEC is enhanced via the deletion of the ICP34.5 gene, which also attenuates the natural neurovirulence of the virus and improves the safety [22]. The insertion of two copies of human GM-CSF gene in the genome of T-VEC leads to local expression, which enhances the recruitment of antigen-presenting cells (APCs). The activation of APCs facilitates the tumor antigen presentation to tumor-specific T cells, which further elevates the antitumor immunity [23]. Another key modification is the deletion of the ICP47 gene. While ICP47 normally reduces antigen presentation by binding to the transport-associated protein to prevent the antigen loading of MHC-I molecules, the deletion of the ICP47 gene enhances tumor antigen presentation. Additionally, the deletion of ICP47 permits the earlier and increased expression of the herpes unique short 11 (US 11) gene, leading to increased selectivity for tumor cells [24].

3. T-VEC Treatment for Melanoma

3.1. T-VEC Monotherapy for Melanoma and Path to FDA Approval

T-VEC was first tested in a phase I clinical trial published by Hu et al. in 2006, in which T-VEC was administered via intratumoral injection in patients with a wide diversity of tumor types, including refractory breast, head and neck, and gastrointestinal cancers and malignant melanoma. In total, thirty patients were segregated into either a single-dose group, where doses of 10^6, 10^7, and 10^8 plaque-forming units (pfu)/mL were tested, or into a multidose group, which tested a number of dose regimens. While 26 of the enrolled 30 patients were evaluable, 19 of the 26 posttreatment biopsies showed residual tumors, of which 14 exhibited extensive necrosis and apoptosis, and all demonstrated strong staining for HSV in the necrotic areas. A mild toxicity profile was reported, which mainly comprised low-grade fever, chills, myalgia, and local reactions. The dose regimen that consisted of an initial dose of 10^6 pfu/mL followed by 2 doses of 10^8 pfu/mL every two to three weeks was reported to be the most effective approach in both seropositive and seronegative patients [6].

In the following phase II clinical trial published by Senzer at al. in 2009, T-VEC (4 mL of 10^6 pfu/mL followed by 4 mL of 10^8 pfu/mL every 2 to 3 weeks for up to 24 treatments) was tested in fifty patients with stage IIIc unresectable metastatic melanomas. A mild toxicity profile, including transient flu-like symptoms, was reported. The overall response rate (ORR) per the Response Evaluation Criteria in Solid Tumors (RECIST) was 26%; the complete response (CR) rate was 16% and the partial response (PR) rate was 10%. The regression of both injected and distant lesions was observed, with 92% of the responses being maintained for nearly three years. The overall survival (OS) rates were 58% at 1 year and 52% at 2 years [7].

In the subsequent phase III OPTIM study, intralesional T-VEC was compared with subcutaneous GM-CSF when treating 436 patients with unresected stage IIIB to IV melanomas. While the primary end point was a durable response rate (DRR), which represents an objective response lasting continuously for 6 months per independent assessment, the secondary end points included the OS and ORR. In regard to the T-VEC injection, the first dose was given at 10^6 pfu/mL (to seroconvert HSV-seronegative patients). Subsequent T-VEC doses of 10^8 pfu/mL were administered three weeks after the first dose and then once every 2 weeks. GM-CSF 125 $\mu g/m^2$ was administered subcutaneously once daily for 14 days in 28-day cycles [8]. In the final report of this study in 2019, a significantly higher DRR was reported with T-VEC (19.3%) than GM-CSF (1.4%). Similarly, the ORR was greater in the T-VEC (31.5%) than GM-CSF (6.4%) treatment. Fifty patients (16.9%) and one (0.7%) patient in the T-VEC and GM-CSF arms, respectively, achieved CR. The median

OS in the T-VEC arm reached 23.3 months (95% CI, 19.5–29.6) versus 18.9 months with GM-CSF (95% CI, 16.0–23.7). The toxicity profile was acceptable, with the most common adverse events (AEs) including fatigue, chills, pyrexia, nausea, and influenza-like illness. While the incidence of these AEs was highest during the first three cycles, most AEs lasted 2–4 days and subsequently subsided over time [9]. Based on the data from the OPTIM study, T-VEC was officially approved by the FDA on 27 October 2015.

Furthermore, other clinical trials of T-VEC monotherapy have been conducted and have shown promising results in terms of their efficacy and safety. For example, a phase 1 study (NCT03064763) assessed the safety and effectiveness of T-VEC in Japanese patients with advanced stage melanomas that could not be surgically removed. The study found that T-VEC had a favorable safety profile, with no dose-limiting toxicities being observed, and the most common side effects were fever and chills. Most AEs were grade 1 or 2, which were consistent with those observed in the OPTIM trial [25].

3.2. *T-VEC Combinational Therapy for Melanoma*

3.2.1. Rationale for T-VEC Combinational Therapy

The current frontline therapies for melanoma include chemotherapy, targeted therapy, immune checkpoint inhibitors (ICIs), and virotherapy (i.e., T-VEC). The activating mutation of BRAF, the key serine threonine protein kinase in the RAS/RAF/MEK/ERK pathway, has been found in nearly 70% of melanomas, with the consequential activation of the downstream MEK and ERK signaling contributing to the dysregulated proliferation of melanoma cell growth [26]. Vemurafenib was the first BRAFi approved by the FDA for the treatment of BRAF V600 mutant melanoma, followed by dabrafenib and encorafenib. While the BRAFis all exhibited improved survival outcomes in melanoma patients compared to the traditional chemotherapies, the rapid development of drug resistance to the BRAFi monotherapy was reported. The combination therapy of BRAFi and MEKi was developed subsequently to reduce this resistance, which was proven to be remarkably effective in an array of clinical trials. For instance, in the coBRIM trial, the combination of vemurafenib and cobimetinib resulted in a remarkably improved median OS (22.3 months) and progression-free survival (PFS) (12.3 months) compared to that of the vemurafenib monotherapy (OS, 17.4 months; PFS, 7.2 months) [27]. Similarly, in the COMBI-d trial, treatment with a combinational therapy of trametinib and dabrafenib led to a significantly prolonged median OS (25.1 months vs. 18.7 months) and increased median PFS (11.0 months vs. 8.8 months) in comparison to the dabrafenib monotherapy [28].

Interactions between immune checkpoints and their ligands negatively influence T cell function and the subsequent immune responses against tumor antigens. ICIs, which block these immunosuppressive pathways, have been shown to effectively elevate the antitumor immune reactions in preclinical studies. Among the ICIs, the blockade of CTLA-4 and interaction between PD-1 and PD-L1 are the two most prominent. The development of monoclonal antibodies against CTLA-4 (e.g., ipilimumab) and PD-1 (e.g., nivolumab and pembrolizumab), along with the successful survival outcomes in clinical trials with advanced melanoma patients, has significantly transformed the melanoma treatment landscape. For instance, in the CheckMate067 trial, untreated unresectable stage III or stage IV patients were randomly segregated into ipilimumab, nivolumab, and nivolumab + ipilimumab treatment groups. With a 6.5-year follow-up period, remarkable improvements were reported in the median OS values (19.9 months with ipilimumab, 36.9 months with nivolumab, and 72.1 months with nivolumab + ipilimumab) and median treatment-free intervals (1.9 months, 2.3 months, and 27.6 months with ipilimumab, nivolumab, and nivolumab + ipilimumab, respectively). In addition, 43%, 74%, and 81% of the patients after ipilimumab, nivolumab, and nivolumab + ipilimumab treatment, respectively, received no further subsequent systemic therapy [29,30].

While T-VEC, ICIs, and targeted therapies exhibit remarkable success, the combination of T-VEC with ICIs or targeted therapies would be expected to have synergistic efficacy. It has been shown that T-VEC infection and replication in tumor cells can elevate the inflammatory state of the tumor microenvironment, which can further promote T cell influx and activation [31]. While the GM-CSF gene product facilitates the recruitment and activation of antigen presentation cells (APCs), the oncolysis of the tumor cells spreads the tumor-associated antigens, which increases the availability to APCs and T cell priming. As the immune responses can be reduced via the expression of immune checkpoints on the T cells, such as CTLA-4 and PD-1, the coadministration of ICIs can prevent T cell exhaustion and prolong T cell activation and expansion [32].

3.2.2. Clinical Trials of T-VEC Combinational Therapy for Melanoma

The first randomized trial assessing the efficacy of the combinational therapy of T-VEC and ICIs was reported by Chesney et al. One hundred and ninety-eight patients with unresectable stage IIIB to IV melanomas were randomly segregated into the T-VEC + ipilimumab (n = 98) or ipilimumab monotherapy (n = 100) group. The toxicity profile was reported as mild, and the AEs mainly included fatigue, chills, and diarrhea. While three patients in the combination therapy group had fatal AEs, none were related to the treatment itself. The objective response was reported as thirty-eight patients (39%) in the combination therapy group and 18 patients (18%) in the ipilimumab monotherapy group. The median time to response was 5.8 months in the T-VEC + ipilimumab group (n = 38), which was not estimable in the ipilimumab group (n = 18). The median PFS was 8.2 months in the duplet group and 6.4 months in the monotherapy group. While this study indicates that the combination has greater antitumor activity without additional safety concerns compared to ipilimumab, several interesting findings are noted. First, it was notable that both the injected lesion and visceral lesions decreased in size in response to treatment. In total, 52% of the patients receiving combination therapy and 23% of the patients receiving ipilimumab monotherapy had visceral lesions that responded to treatment. Second, the efficacy of the treatments was shown to be affected by the tumor staging and existence of BRAF mutations. The ORR in the combination therapy group was significantly higher for patients with low tumor staging (IIIB/IIIC/IVM1a) in comparison to high tumor staging (IVM1b and IVM1c) (44% vs. 33%). The ORR in the combination arm was 42% among BRAF wild-type patients, which was greater than that among BRAF mutation patients (34%) [10].

In the other trial, the MASTERKEY-265 trial (phase Ib/III study), T-VEC + pembrolizumab was evaluated versus pembrolizumab monotherapy. In the phase Ib study, 21 patients with unresectable stage IIIB-IVM1c melanoma with injectable, measurable lesions and no prior systemic treatment were enrolled and followed for 18.6 (17.7–20.8) months before the time of reporting. There were no severe toxicities reported in any of the 21 patients, with the most common AEs including fatigue, chills, and fever. With the combinational therapy, the confirmed objective response rate was 61.9% (95% CI, 38.4–81.9%), while the confirmed CR rate was 33.3% (95% CI, 14.6–57.0%). Moreover, the combination treatment led to >50% reductions in 82% of injected, 43% of non-injected non-visceral, and 33% of non-injected visceral lesions [11]. All twenty-one patients enrolled were off treatment as of the data cutoff (Mar 2, 2020). Among them, 6 died and 15 are in long-term follow-up. With a median follow-up time of 58.6 months, the CR rate was reported as 43% (9/21 patients); 92.3% of the responders (12/13) remained in response, including all 9 patients with a CR. While the median PFS and OS were not reached at the data cutoff point, the 4-year PFS and OS rates were estimated as 55.9% and 71.4%, respectively. No additional safety signals were ever detected [33].

The remarkable results of the phase 1b part of MASTERKEY-265 led to the phase III randomized, double-blind KEYNOTE-034 study. In this study, a total of 692 patients with unresectable stage III-IVM1c melanoma who were naive to anti-PD1 therapy were randomized 1:1 to a T-VEC + pembrolizumab or placebo + pembrolizumab treatment. With a median follow-up of 31.0 months, it was reported that the median PFS was 14.3 months for the T-VEC + pembrolizumab arm and 8.5 months for the placebo + pembrolizumab arm. While the median OS was not reached for the T-VEC + pembrolizumab arm, the OS of the placebo + pembrolizumab arm was 49.2 months. However, statistical significance was not expected with OS in the primary OS analysis. The ORRs were 48.6% for the T-VEC + pembrolizumab group and 41.3% for the placebo + pembrolizumab group. The CR rate was greater in the T-VEC + pembrolizumab arm in comparison to the placebo + pembrolizumab arm (17.9% vs. 11.6%). The DRRs were 42.2% in the T-VEC + pembrolizumab arm and 34.1% for the placebo + pembrolizumab arm. Importantly, the safety profiles were acceptable, without any unknown safety issues from each agent [12].

In addition to the abovementioned trials, several other clinical trials involving the T-VEC combination therapy are ongoing to further evaluate the systemic efficacy of T-VEC. For instance, in a phase II clinical trial (NCT#02965716), patients with unresectable stage IIIB-IV melanoma who did not respond to PD-1/PD-L1 blockade were treated with T-VEC + pembrolizumab. This study had been designed to evaluate the T cell infiltration into tumors, the T-cell receptor (TCR) clonality in tumors and in peripheral blood, and the tumor immune microenvironment after T-VEC + pembrolizumab combination treatment, which will hopefully provide more in-depth information on the mechanisms of T-VEC in tumor eradication [34].

4. T-VEC Treatment in Other Cutaneous Cancer Types

Along with the success of T-VEC in melanoma treatment, T-VEC monotherapy and combination therapies are under exploration in other cutaneous cancer types, such as Merkel cell carcinoma (MCC) and cutaneous squamous cell carcinoma (CSCC).

As an aggressive malignancy from cutaneous neuroendocrine cells, MCC typically presents on the sun-exposed areas in the elderly. The current FDA-approved treatment for MCC includes chemotherapy and ICIs, such as PD-1 or PD-L1 blockade. Recent clinical trials reported superior ORR and PFS values with PD-1/PD-L1 treatment in comparison to chemotherapy; however, the CR rate was low, and most patients progressed in less than 12 months [35]. In regard to these observations, T-VEC has been assessed for MCC therapy. In Westbrook et al., four patients with regionally advanced MCC were treated with T-VEC. All four patients achieved durable CRs, with a median PFS of more than 16 months without severe AEs. Moreover, the treatment with T-VEC prevented distant metastasis in these high-risk individuals [14]. In another study, Knackstedt et al. reported on the combination therapy of T-VEC and a PD-1/PD-L1 inhibitor in two patients with anti-PD-1 refractory MCC. While the radiotherapy and chemotherapy had been utilized with failure, the T-VEC and PD-1/PD-L1 inhibitor combination therapy led to CR in one patient and near-CR in another patient [15].

CSCC is another common cutaneous malignancy, which has a wide range of presentations from low-risk in situ disease to high-risk advanced metastatic tumors. Compared to melanoma, CSCC has a less aggressive clinical course but a significantly higher incidence rate [36]. The current treatment options mainly include PD-L1 inhibitors, chemotherapy, and EGFR inhibitors. A single-arm phase II trial of T-VEC (NCT03714828) was conducted in treating low-risk invasive CSCC. With the Simon 2-stage design being used and a total sample size of 20 patients, 7 patients were recruited for stage 1 and an additional 13 patients would be recruited if five or more subjects met the primary endpoint in stage 1. In the interim analysis of 7 patients, all achieved overall CR. All AEs were of grades 1–2 based on the NCI Common Terminology Criteria for Adverse Events v. 4.0 (CTCAE v. 4.0), with the most common AEs including transient fatigue, flu-like symptoms, and headaches. At the time of analysis, the mean time to response was 43.4 days and the duration of the ORR was 190 days [37]. While-T-VEC has shown remarkable success with a 100% CR in stage 1, a high response rate will be expected and assessed at the completion of the study.

Currently, several other clinical trials are ongoing for assessing the efficacy of T-VEC in treating these cutaneous malignancies. For instance, the combination of T-VEC and radiotherapy is being evaluated in MCC and melanoma in a phase II trial (NCT02819843) [16]. In another phase II trial (NCT02978625), a combination therapy of T-VEC and nivolumab is being assessed in MCC, CSCC, and basal cell carcinoma [38–41].

5. T-VEC Treatment Practices in City of Hope

City of Hope is a National Cancer Institute (NCI)-designated Comprehensive Cancer Center and a member of the National Comprehensive Cancer Network (NCCN) (Figure 1). At City of Hope, T-VEC treatment has been applied to patients with recurrent or metastatic melanoma, metastatic CSCC, and metastatic MCC. While a few patients complained of chills, fever, and fatigue a few hours after T-VEC injection and some edema at the injection site, these symptoms usually lasted less than 24 h. Extensive fibrosis has been observed after T-VEC injection, which prevented further intratumoral injections. Overall, the toxicity profile of T-VEC has been reported as mild and tolerable.

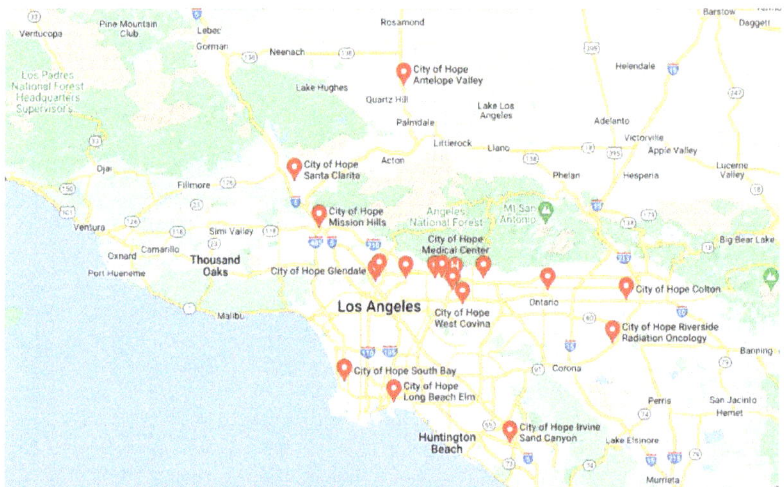

Figure 1. Map of City of Hope locations in Southern California. Teardrop with "H" represents COH main campus at Duarte, California, where the T-VEC treatment is performed. Other red teardrops represent 18 out of 27 campuses in City of Hope.

Among the melanoma patients under T-VEC treatment, nearly 32% of the patients were referred from other hospitals for either monotherapy or combination therapy. Overall, in comparison to T-VEC monotherapy, T-VEC + ICI combination therapies in which pembrolizumab was applied most frequently have resulted in higher CR rates, which indicates synergistically the more significant efficacy with the addition of ICIs. In light of this observation, we are currently undertaking preclinical studies that aim to explore melanoma treatment with the intratumoral injection of multi-drug combination therapies. Regarding the subsequent therapies following T-VEC, the PD-1 or CTLA-4 inhibitors as monotherapies or in combination and the BRAF or MEK inhibitors as monotherapies or in combination were most commonly administered. A small number of patients with metastatic CSCC and MCC were treated with either T-VEC monotherapy or combination therapies as off label treatments per the tumor board recommendations. While most patients with metastatic CSCC and MCC suffered from the progression of disease before eventually expiring, future trials on CSCC and MCC patients need to be conducted before the efficacy of T-VEC can be fully assessed in these two malignancies.

While most of the patients who were referred to City of Hope for T-VEC treatment lived within reasonable distance (less than 50 miles from City of Hope), several resided far away and even travelled four to five hours one way to receive treatment. Meanwhile, the regulations for the transportation, storage, and handling of T-VEC are cumbersome. For instance, T-VEC is usually stored frozen at −70 to −90 °C then thawed to a liquid state prior to preparation, which takes approximately 30 to 70 min in our experience. The pharmacy workflow must be adjusted so that trained technicians can prepare the syringes and the IV hood must be set aside for cleaning to reset the airflow. The main constraints include the lack of trained providers who can administer T-VEC, the freezer availability and capacity, and the biweekly scheduling. Additionally, insurance may not approve T-VEC for indications other than melanoma. All of these factors have limited the access of patients to T-VEC treatment.

6. Discussion and Future Directions

The landscape for cancer treatment has been rapidly evolving in the last few decades. With the advent of new drugs and combinations, the therapeutic options for patients have widely broadened and become more multidisciplinary. As the first OV approved by the FDA, T-VEC provides a new approach for cancer therapy regimens.

T-VEC was first studied in clinical trials of melanoma and demonstrated improved efficacy. For instance, in the phase III OPTIM study involving patients of unresected stages IIIB to IV melanoma, significantly higher ORR, DRR, and CR values with tolerable toxicity profiles were associated with the treatment of T-VEC in comparison to GM-CSF. With FDA approval, its application has been rapidly extended to the treatment of other cancer types. For instance, in a single-arm phase II trial where T-VEC was administered in patients with invasive CSCC, all 7 patients in stage 1 of the study achieved CR, with very mild AEs. Currently, T-VEC alone or as part of a combination therapy has been explored in clinical trials with a variety of cancers, such as MCC, CSCC, breast cancer, pancreatic cancer, colorectal cancer, and liver cancer. However, it is noteworthy that so far in most of the clinical trials, intratumoral injection remains the only option for virus administration. In fact, the intratumoral administration of T-VEC causes the direct lysis of tumor cells and increases the intratumoral infiltration of APCs and T cells, which leads to neoantigen recognition and strengthened tumor specificity. Moreover, intratumor injection protects the virus from the neutralizing antibodies and macrophage sequestration effect towards the virus. While the intratumoral route serves as a perfect means of eradication of locoregional cancers, it might not be effective with distant tumors that are inaccessible to direct injections or metastasized tumors that cannot be accurately located. Clinical trials are ongoing to evaluate the systemic route of OVs, which have demonstrated feasibility in systemic injections. More results are still needed to show the antitumor efficacy that can be achieved.

While ICIs have been commonly used in T-VEC combination therapies, other forms of treatment have also been evaluated. One example is a phase 1b clinical trial (NCT03088176) that will investigate the safety and tolerability of administering T-VEC locally, in conjunction with oral therapy with dabrafenib and trametinib. This study will be conducted with up to 20 patients with advanced melanoma who possess activating mutations in the BRAF gene. Another phase II trial (NCT02819843) intends to evaluate the effectiveness of T-VEC as a treatment for melanoma in conjunction with or without radiotherapy [42]. Interestingly, ongoing studies are investigating the potential benefits of using neoadjuvant T-VEC in patients with advanced, resectable melanoma. A phase 2 trial (NCT02211131) was conducted on 150 patients with resectable stage IIIB-IVM1a melanoma, who were randomized to receive T-VEC followed by surgery or surgery alone. The study found that the use of neoadjuvant T-VEC in combination with surgery resulted in a 25% reduction in the risk of disease recurrence compared to patients who received surgery alone [43]. Still, further research is needed to determine the best approach for utilizing T-VEC in combination with immunotherapy or other therapies for patients with advanced melanoma.

The success of T-VEC has amplified the interest of many researchers in cancer virotherapies. A number of other OVs have been designed and have undergone evaluation in preclinical and clinical studies as monotherapies or in combination with other systemic immunotherapies. For instance, TILT Biotherapeutics constructed TILT-123, an adenovirus engineered to express tumor necrosis factor (TNF)-α and IL-2. While its safety and biodistribution has been studied in mice and hamsters and it has been demonstrated to be safe in animals, the virus has been shown to induce rapid antitumor immune responses with viral replication restricted to the tumors and not normal tissues. With promising results, it is under evaluation in a phase 1 trial (NCT04217473) [44]. Another famous OV is Pexa-Vec (JX-594, Pexastimogene Devacirepvec) from SillaJen, a vaccinia virus genetically modified with thymidine kinase (TK) gene deletion and GM-CSF expression. While TK is essential for viral DNA production and has been overly expressed in the cancerous cells in comparison to in the normal cells, the deletion of the viral TK gene enables the OV to target the tumor cells more selectively while sparing the normal cells, which increases its tumor specificity. To date, JX-594 has been tested in a dozen clinical trials with many types of malignancies. All studies with JX-594 have shown excellent safety profile in more than 400 patients [45]. In general, most OVs follow the same principles regarding genetic modifications, which mainly include genetic alterations to limit pathogenicity, genomic deletions to enhance the tumor-specificity, and genomic additions to increase immune responses. While more virus species are being engineered and tested, more OVs are expected in the future to present even more options for cancer patients.

Currently, the genomic identification of cancer-promoting mutations can not only lead to drug development but can also provide information for individual patients to guide the treatment. While many drugs have shown remarkable efficacy in tumor suppression, it is still difficult to achieve a complete cure due to the refractoriness and high relapse rates of some tumors. Therefore, a combination of multiple therapies and new approaches is needed. Among the cancer therapies, the discovery of the immune checkpoints CTLA4, PD-1, and PDL1 and ICIs is a paramount achievement that has revolutionized the landscape of cancer treatment. The combination therapy of T-VEC and ICIs has, thus, appeared to be a very promising melanoma treatment approach. Indeed, many preclinical studies have provided evidence that supports the rationale of this combination. It has been shown that T-VEC can cause tumor regression and T cell infiltration, along with increased IFN-gamma and PD-L1 expression [46]. Liu et al. reported increased PD-L1 expression associated with OV monotherapy but better survival rates when combining OV with anti-PD-L1 in the mouse models [47]. The clinical trials of the combination therapy involving a variety of cancer types have also provided solid evidence for its success. At this time, there is still eager anticipation to see more trial results that may offer further insights on the future combination therapies for melanoma.

Author Contributions: T.Z. contributed to the literature review and writing. T.Z. and T.H.-T.J. contributed to the table design. Y.X. and T.Z. contributed to the conception and design. T.Z., J.H., K.H., Z.W., J.Y., H.Y. and Y.X. contributed to the proofreading and edition. All authors have read and agreed to the published version of the manuscript.

Funding: This research received no external funding.

Institutional Review Board Statement: Not applicable.

Informed Consent Statement: Not applicable.

Data Availability Statement: Not applicable.

Conflicts of Interest: All other authors declared no conflict of interest.

References

1. Kaushik, I.; Ramachandran, S.; Zabel, C.; Gaikwad, S.; Srivastava, S.K. The evolutionary legacy of immune checkpoint inhibitors. *Semin. Cancer Biol.* **2022**, *86 Pt 2*, 491–498. [CrossRef] [PubMed]
2. Naimi, A.; Mohammed, R.N.; Raji, A.; Chupradit, S.; Yumashev, A.V.; Suksatan, W.; Shalaby, M.N.; Thangavelu, L.; Kamrava, S.; Shomali, N.; et al. Tumor immunotherapies by immune checkpoint inhibitors (ICIs); the pros and cons. *Cell Commun. Signal.* **2022**, *20*, 44. [CrossRef]
3. Chaurasiya, S.; Fong, Y.; Warner, S.G. Oncolytic Virotherapy for Cancer: Clinical Experience. *Biomedicines* **2021**, *9*, 419. [CrossRef] [PubMed]
4. Toda, M.; Martuza, R.L.; Rabkin, S.D. Tumor growth inhibition by intratumoral inoculation of defective herpes simplex virus vectors expressing granulocyte-macrophage colony-stimulating factor. *Mol. Ther.* **2000**, *2*, 324–329. [CrossRef]
5. Kohlhapp, F.J.; Kaufman, H.L. Molecular Pathways: Mechanism of Action for Talimogene Laherparepvec, a New Oncolytic Virus Immunotherapy. *Clin. Cancer Res.* **2016**, *22*, 1048–1054. [CrossRef]

6. Hu, J.C.; Coffin, R.S.; Davis, C.J.; Graham, N.J.; Groves, N.; Guest, P.J.; Harrington, K.J.; James, N.D.; Love, C.A.; McNeish, I.; et al. A phase I study of OncoVEXGM-CSF, a second-generation oncolytic herpes simplex virus expressing granulocyte macrophage colony-stimulating factor. *Clin. Cancer Res.* **2006**, *12*, 6737–6747. [CrossRef] [PubMed]
7. Senzer, N.N.; Kaufman, H.L.; Amatruda, T.; Nemunaitis, M.; Reid, T.; Daniels, G.; Gonzalez, R.; Glaspy, J.; Whitman, E.; Harrington, K.; et al. Phase II clinical trial of a granulocyte-macrophage colony-stimulating factor-encoding, second-generation oncolytic herpesvirus in patients with unresectable metastatic melanoma. *J. Clin. Oncol.* **2009**, *27*, 5763–5771. [CrossRef] [PubMed]
8. Andtbacka, R.H.; Kaufman, H.L.; Collichio, F.; Amatruda, T.; Senzer, N.; Chesney, J.; Delman, K.A.; Spitler, L.E.; Puzanov, I.; Agarwala, S.S.; et al. Talimogene Laherparepvec Improves Durable Response Rate in Patients With Advanced Melanoma. *J. Clin. Oncol.* **2015**, *33*, 2780–2788. [CrossRef]
9. Andtbacka, R.H.I.; Collichio, F.; Harrington, K.J.; Middleton, M.R.; Downey, G.; Öhrling, K.; Kaufman, H.L. Final analyses of OPTiM: A randomized phase III trial of talimogene laherparepvec versus granulocyte-macrophage colony-stimulating factor in unresectable stage III-IV melanoma. *J. Immunother. Cancer* **2019**, *7*, 145. [CrossRef]
10. Chesney, J.; Puzanov, I.; Collichio, F.; Singh, P.; Milhem, M.M.; Glaspy, J.; Hamid, O.; Ross, M.; Friedlander, P.; Garbe, C.; et al. Randomized, Open-Label Phase II Study Evaluating the Efficacy and Safety of Talimogene Laherparepvec in Combination With Ipilimumab Versus Ipilimumab Alone in Patients With Advanced, Unresectable Melanoma. *J. Clin. Oncol.* **2018**, *36*, 1658–1667. [CrossRef]
11. Ribas, A.; Dummer, R.; Puzanov, I.; VanderWalde, A.; Andtbacka, R.H.I.; Michielin, O.; Olszanski, A.J.; Malvehy, J.; Cebon, J.; Fernandez, E.; et al. Oncolytic Virotherapy Promotes Intratumoral T Cell Infiltration and Improves Anti-PD-1 Immunotherapy. *Cell* **2017**, *170*, 1109–1119.e10. [CrossRef] [PubMed]
12. Ribas, A.; Chesney, J.; Long, G.V.; Kirkwood, J.M.; Dummer, R.; Puzanov, I.; Hoeller, C.; Gajewski, T.F.; Gutzmer, R.; Rutkowski, P.; et al. 1037O MASTERKEY-265: A phase III, randomized, placebo (Pbo)-controlled study of talimogene laherparepvec (T) plus pembrolizumab (P) for unresectable stage IIIB–IVM1c melanoma (MEL). *Ann. Oncol.* **2021**, *32*, S868–S869. [CrossRef]
13. Thompson, J.F.; Agarwala, S.S.; Smithers, B.M.; Ross, M.I.; Scoggins, C.R.; Coventry, B.J.; Neuhaus, S.J.; Minor, D.R.; Singer, J.M.; Wachter, E.A. Phase 2 Study of Intralesional PV-10 in Refractory Metastatic Melanoma. *Ann. Surg. Oncol.* **2015**, *22*, 2135–2142. [CrossRef]
14. Westbrook, B.C.; Norwood, T.G.; Terry, N.L.J.; McKee, S.B.; Conry, R.M. Talimogene laherparepvec induces durable response of regionally advanced Merkel cell carcinoma in 4 consecutive patients. *JAAD Case Rep.* **2019**, *5*, 782–786. [CrossRef]
15. Knackstedt, R.; Sussman, T.A.; McCahon, L.; Song, J.M.; Funchain, P.; Gastman, B. Pre-treated anti-PD-1 refractory Merkel cell carcinoma successfully treated with the combination of PD-1/PD-L1 axis inhibitors and TVEC: A report of two cases. *Ann. Oncol.* **2019**, *30*, 1399–1400. [CrossRef] [PubMed]
16. Shalhout, S.Z.; Kaufman, H.L.; Emerick, K.S.; Miller, D.M. Immunotherapy for Nonmelanoma Skin Cancer: Facts and Hopes. *Clin. Cancer Res.* **2022**, *28*, 2211–2220. [CrossRef] [PubMed]
17. Zhang, T.; Suryawanshi, Y.R.; Kordish, D.H.; Woyczesczyk, H.M.; Jeng, D.; Essani, K. Tanapoxvirus lacking a neuregulin-like gene regresses human melanoma tumors in nude mice. *Virus Genes* **2017**, *53*, 52–62. [CrossRef]
18. Viale, D.L.; Cafferata, E.G.; Gould, D.; Rotondaro, C.; Chernajovsky, Y.; Curiel, D.T.; Podhajcer, O.L.; Veronica Lopez, M. Therapeutic improvement of a stroma-targeted CRAd by incorporating motives responsive to the melanoma microenvironment. *J. Investig. Dermatol.* **2013**, *133*, 2576–2584. [CrossRef]
19. Vaha-Koskela, M.J.; Heikkila, J.E.; Hinkkanen, A.E. Oncolytic viruses in cancer therapy. *Cancer Lett.* **2007**, *254*, 178–216. [CrossRef]
20. Zhang, T.; Suryawanshi, Y.R.; Szymczyna, B.R.; Essani, K. Neutralization of matrix metalloproteinase-9 potentially enhances oncolytic efficacy of tanapox virus for melanoma therapy. *Med. Oncol.* **2017**, *34*, 129. [CrossRef]
21. Zhang, T.; Kordish, D.H.; Suryawanshi, Y.R.; Eversole, R.R.; Kohler, S.; Mackenzie, C.D.; Essani, K. Oncolytic Tanapoxvirus Expressing Interleukin-2 is Capable of Inducing the Regression of Human Melanoma Tumors in the Absence of T Cells. *Curr. Cancer Drug Targets* **2018**, *18*, 577–591. [CrossRef]
22. Liu, B.L.; Robinson, M.; Han, Z.Q.; Branston, R.H.; English, C.; Reay, P.; McGrath, Y.; Thomas, S.K.; Thornton, M.; Bullock, P.; et al. ICP34.5 deleted herpes simplex virus with enhanced oncolytic, immune stimulating, and anti-tumour properties. *Gene Ther.* **2003**, *10*, 292–303. [CrossRef] [PubMed]
23. Zhang, T.; Suryawanshi, Y.R.; Woyczesczyk, H.M.; Essani, K. Targeting Melanoma with Cancer-Killing Viruses. *Open Virol. J.* **2017**, *11*, 28–47. [CrossRef]
24. Hawkins, L.K.; Lemoine, N.R.; Kirn, D. Oncolytic biotherapy: A novel therapeutic plafform. *Lancet Oncol.* **2002**, *3*, 17–26. [CrossRef] [PubMed]
25. Yamazaki, N.; Koga, H.; Kojima, T.; Tsutsumida, A.; Namikawa, K.; Yi, M.; Mera, K.; Pickett-Gies, C. Early safety from a phase I, multicenter, open-label, dose de-escalation study of talimogene laherparepvec (T-VEC) in Japanese patients (pts) with unresectable stage IIIB-IV melanoma (MEL). *Ann. Oncol.* **2018**, *29*, ix107. [CrossRef]
26. Cancer Genome Atlas, N. Genomic Classification of Cutaneous Melanoma. *Cell* **2015**, *161*, 1681–1696. [CrossRef]
27. Ascierto, P.A.; McArthur, G.A.; Dreno, B.; Atkinson, V.; Liszkay, G.; Di Giacomo, A.M.; Mandala, M.; Demidov, L.; Stroyakovskiy, D.; Thomas, L.; et al. Cobimetinib combined with vemurafenib in advanced BRAF(V600)-mutant melanoma (coBRIM): Updated efficacy results from a randomised, double-blind, phase 3 trial. *Lancet Oncol.* **2016**, *17*, 1248–1260. [CrossRef]
28. Curti, B.D.; Faries, M.B. Recent Advances in the Treatment of Melanoma. *N. Engl. J. Med.* **2021**, *384*, 2229–2240. [CrossRef]

29. Tarhini, A.A.; Lee, S.J.; Hodi, F.S.; Rao, U.N.M.; Cohen, G.I.; Hamid, O.; Hutchins, L.F.; Sosman, J.A.; Kluger, H.M.; Eroglu, Z.; et al. Phase III Study of Adjuvant Ipilimumab (3 or 10 mg/kg) Versus High-Dose Interferon Alfa-2b for Resected High-Risk Melanoma: North American Intergroup E1609. *J. Clin. Oncol.* **2020**, *38*, 567–575. [CrossRef]
30. Eggermont, A.M.; Chiarion-Sileni, V.; Grob, J.J.; Dummer, R.; Wolchok, J.D.; Schmidt, H.; Hamid, O.; Robert, C.; Ascierto, P.A.; Richards, J.M.; et al. Prolonged Survival in Stage III Melanoma with Ipilimumab Adjuvant Therapy. *N. Engl. J. Med.* **2016**, *375*, 1845–1855. [CrossRef]
31. Sun, L.; Funchain, P.; Song, J.M.; Rayman, P.; Tannenbaum, C.; Ko, J.; McNamara, M.; Marcela Diaz-Montero, C.; Gastman, B. Talimogene Laherparepvec combined with anti-PD-1 based immunotherapy for unresectable stage III-IV melanoma: A case series. *J. Immunother. Cancer* **2018**, *6*, 36. [CrossRef] [PubMed]
32. Dummer, R.; Hoeller, C.; Gruter, I.P.; Michielin, O. Combining talimogene laherparepvec with immunotherapies in melanoma and other solid tumors. *Cancer Immunol. Immunother.* **2017**, *66*, 683–695. [CrossRef] [PubMed]
33. Long, G.; Dummer, R.; Johnson, D.; Michielin, O.; Martin-Algarra, S.; Treichel, S.; Chan, E.; Diede, S.; Ribas, A. 429 Long-term analysis of MASTERKEY-265 phase 1b trial of talimogene laherparepvec (T-VEC) plus pembrolizumab in patients with unresectable stage IIIB-IVM1c melanoma. *J. ImmunoTher. Cancer* **2020**, *8* (Suppl. S3), A261. [CrossRef]
34. Malvehy, J.; Samoylenko, I.; Schadendorf, D.; Gutzmer, R.; Grob, J.J.; Sacco, J.J.; Gorski, K.S.; Anderson, A.; Pickett, C.A.; Liu, K.; et al. Talimogene laherparepvec upregulates immune-cell populations in non-injected lesions: Findings from a phase II, multicenter, open-label study in patients with stage IIIB-IVM1c melanoma. *J. Immunother. Cancer* **2021**, *9*, e001621. [CrossRef] [PubMed]
35. Chan, I.S.; Bhatia, S.; Kaufman, H.L.; Lipson, E.J. Immunotherapy for Merkel cell carcinoma: A turning point in patient care. *J. Immunother. Cancer* **2018**, *6*, 23. [CrossRef]
36. Burns, C.; Kubicki, S.; Nguyen, Q.B.; Aboul-Fettouh, N.; Wilmas, K.M.; Chen, O.M.; Doan, H.Q.; Silapunt, S.; Migden, M.R. Advances in Cutaneous Squamous Cell Carcinoma Management. *Cancers* **2022**, *14*, 3653. [CrossRef]
37. Curiel, C.N.; Stratton, D.; Cui, H.; Roe, D.; Tiwari, H.A.; Sundararajan, S. A single arm phase 2 study of talimogene laherparepvec in patients with low-risk invasive cutaneous squamous cell cancer. Interim analysis. *J. Clin. Oncol.* **2022**, *40* (Suppl. 16), e21583. [CrossRef]
38. Kai, M.; Marx, A.N.; Liu, D.D.; Shen, Y.; Gao, H.; Reuben, J.M.; Whitman, G.; Krishnamurthy, S.; Ross, M.I.; Litton, J.K.; et al. A phase II study of talimogene laherparepvec for patients with inoperable locoregional recurrence of breast cancer. *Sci. Rep.* **2021**, *11*, 22242. [CrossRef]
39. Soliman, H.; Hogue, D.; Han, H.; Mooney, B.; Costa, R.; Lee, M.C.; Niell, B.; Williams, A.; Chau, A.; Falcon, S.; et al. A Phase I Trial of Talimogene Laherparepvec in Combination with Neoadjuvant Chemotherapy for the Treatment of Nonmetastatic Triple-Negative Breast Cancer. *Clin. Cancer Res.* **2021**, *27*, 1012–1018. [CrossRef]
40. Hecht, J.R.; Pless, M.; Cubillo, A.; Calvo, A.; Chon, H.J.; Liu, C.; Snyder, W.; Chan, E.; Chaney, M.F.; Chesney, J.A.; et al. Early safety from a phase I, multicenter, open-label clinical trial of talimogene laherparepvec (T-VEC) injected (inj) into liver tumors in combination with pembrolizumab (pem). *J. Clin. Oncol.* **2020**, *38* (Suppl. 15), 3015. [CrossRef]
41. Silk, A.W.; LeBoeuf, N.R.; Rabinowits, G.; Puzanov, I.; Burgess, M.A.; Devata, S.; Moore, D.; Goydos, J.S.; Chen, H.X.; Kaufman, H.; et al. A phase II study of talimogene laherparepvec followed by talimogene laherparepvec + nivolumab in refractory T cell and NK cell lymphomas, cutaneous squamous cell carcinoma, Merkel cell carcinoma, and other rare skin tumors (NCI #10057). *J. Clin. Oncol.* **2018**, *36* (Suppl. 5), TPS219. [CrossRef]
42. Cilento, M.A.; Klein, O.; Egan, E.; Roberts-Thomson, R. Talimogene laherparepvec resulting in near-complete response in a patient with treatment-refractory Merkel cell carcinoma. *Australas. J. Dermatol.* **2022**, *63*, e222–e225. [CrossRef] [PubMed]
43. Dummer, R.; Gyorki, D.E.; Hyngstrom, J.; Berger, A.C.; Conry, R.; Demidov, L.; Sharma, A.; Treichel, S.A.; Radcliffe, H.; Gorski, K.S.; et al. Neoadjuvant talimogene laherparepvec plus surgery versus surgery alone for resectable stage IIIB-IVM1a melanoma: A randomized, open-label, phase 2 trial. *Nat. Med.* **2021**, *27*, 1789–1796. [CrossRef] [PubMed]
44. Havunen, R.; Kalliokoski, R.; Siurala, M.; Sorsa, S.; Santos, J.M.; Cervera-Carrascon, V.; Anttila, M.; Hemminki, A. Cytokine-Coding Oncolytic Adenovirus TILT-123 Is Safe, Selective, and Effective as a Single Agent and in Combination with Immune Checkpoint Inhibitor Anti-PD-1. *Cells* **2021**, *10*, 246. [CrossRef]
45. Breitbach, C.J.; Bell, J.C.; Hwang, T.H.; Kirn, D.H.; Burke, J. The emerging therapeutic potential of the oncolytic immunotherapeutic Pexa-Vec (JX-594). *Oncolytic Virotherapy* **2015**, *4*, 25–31. [CrossRef]
46. Shibakita, M.; Tachibana, M.; Dhar, D.K.; Kotoh, T.; Kinugasa, S.; Kubota, H.; Masunaga, R.; Nagasue, N. Prognostic significance of Fas and Fas ligand expressions in human esophageal cancer. *Clin. Cancer Res.* **1999**, *5*, 2464–2469.
47. Liu, Z.; Ravindranathan, R.; Kalinski, P.; Guo, Z.S.; Bartlett, D.L. Rational combination of oncolytic vaccinia virus and PD-L1 blockade works synergistically to enhance therapeutic efficacy. *Nat. Commun.* **2017**, *8*, 14754. [CrossRef]

Journal of
Clinical Medicine

MDPI

Perspective

Integrating Early-Stage Drug Development with Clinical Networks; Challenges and Opportunities: The City of Hope Developing Experience

Miguel A. Villalona-Calero *, Jyoti Malhotra, Vincent Chung, Yan Xing, Stacy W. Gray, Heather Hampel, Stephen Gruber and Kevin McDonnell

City of Hope National Medical Center, Department of Medical Oncology, Duarte, CA 91010, USA
* Correspondence: mvillalona@coh.org

Abstract: Recent data suggest that patients with advanced cancer who participate in biomarker/genomically informed early-stage clinical trials experience clinical benefit. While most early-stage clinical trials are conducted in major academic centers, the majority of cancer patients in the United States are treated in community practices. Here, we describe ongoing efforts at the City of Hope Cancer Center to integrate our network community oncology clinical practices into our academic, centralized biomarker/genomic-driven, early-stage clinical trial program to build an understanding of the approaches that provide the benefits of early-stage clinical trial participation to community patients. Our efforts include three key initiatives: the development of a virtual "Refractory Disease" phase 1 trial matching televideo clinic, the construction of infrastructure to support the expansion of phase 1 clinical trials to a distant regional clinical satellite hub, and the implementation of an enterprise-wide precision medicine, germline, and somatic testing program. Our work at City of Hope may serve as an example to facilitate similar efforts at other institutions.

Keywords: integration; community clinical network; phase 1; genomic-driven; clinical trials

Citation: Villalona-Calero, M.A.; Malhotra, J.; Chung, V.; Xing, Y.; Gray, S.W.; Hampel, H.; Gruber, S.; McDonnell, K. Integrating Early-Stage Drug Development with Clinical Networks; Challenges and Opportunities: The City of Hope Developing Experience. *J. Clin. Med.* **2023**, *12*, 4061. https://doi.org/10.3390/jcm12124061

Academic Editor: Enrico Capobianco

Received: 21 April 2023
Revised: 29 May 2023
Accepted: 12 June 2023
Published: 15 June 2023

1. Introduction

Clinical development of a new molecular or biological entity is a long and costly process [1,2]. Advances in science and technology have enabled academic centers, biotechnology companies, and major pharmaceutical companies to access an ever-increasing number of agents with sufficient therapeutic potential to test in the clinic [3–5]. However, many drugs fail during drug development, either because of unacceptable toxicity, or lack of target effect [6,7]. Properly designed, executed, and analyzed early-stage clinical trials are fundamental for a new drug or combination of drugs to obtain eventual approval for marketing.

The complexity of performing early-stage trials and the burden posed to patients in terms of time, frequent travel, required procedures, and traditionally low response rates impacts referral patterns, and has historically limited the access of most patients to new anticancer agents until much later stages in the process of drug development [8,9]. Major academic centers and a few selected community practices currently share the majority of the responsibility for conducting these trials.

Clinicians as well as non-clinicians may question whether it is worthwhile for patients treated in community practices to pursue clinical trials with drugs being tested in early clinical drug development. Previous work in the field arguing against or encouraging referrals and participation is limited. Decoster et al. conducted a review of the antitumor activity and toxic deaths reported in single-agent phase I clinical trials in cancer patients using cytotoxic compounds between 1972 and 1987 [10]. A total of 6639 patients were accrued to 211 trials studying 87 compounds. There were 23 (0.3%) complete responders and 279 (4.2%) partial responders for an overall response rate (RR) of 4.5% among all entries.

Toxic deaths were rare and reported in only 31 patients (0.5% of the entire population). Similarly, Von Hoff et al., in 1991, reported their review of 228 phase 1 trials over a period of 14 years [11]. There were 75 complete and 432 partial responses recorded among 7960 patients for an overall objective RR of 6%.

In contrast, Chihara et al. reported on National Cancer Institute (NCI) sponsored phase 1 trials conducted between 2000 and 2019 [12]. The overall RR for all trials during the study period was 12.2% among 9325 patients and the complete RR was 2.7%. Overall response increased from 9.6% during the period 2000 to 2005 to 18% between 2013 and 2019, and complete RR from 2.5% to 4.3%. Overall RR for combination therapy was substantially higher than for monotherapy (15.8% vs. 3.5%). Furthermore, Chakiba et al. conducted a literature review of 224 phase 1 trials that were published from 1 January 2014 to 30 June 2015 [13]. The overall RR was 19.8%. Phase 1 trials employing an enrichment design (i.e., specific histologic characteristics, a specific biomarker, or both) were associated with a higher probability of clinical benefit, and a higher probability of an objective tumor response occurred among patients enrolled in phase 1 trials that included expansion cohorts.

Additional studies evaluated the impact of biomarker treatment strategies compared to an "all comers" approach in clinical trials. These analyses are limited in the phase 1 setting. Schwaederle et al. conducted a meta-analysis comparing patient outcomes in phase 1 studies that used a biomarker selection strategy with those that did not [14]. The analysis included trials performed between January 2011 and December 2013 and evaluated RR and progression-free survival (PFS). A total of 346 studies met the criteria for evaluation; 13,203 patients were treated within 351 study arms. Of these, 117 arms used a cytotoxic agent, whereas 234 arms used a targeted agent, with 57 (24.4%) being personalized. Non-personalized targeted agent arms had outcomes comparable with those that tested a cytotoxic agent. However, personalized arms using a genomic biomarker had a significantly higher median RR, 30% vs. 4.9% in the other arms, and a longer PFS, 5.7 months vs. 2.95 in the other arms. Furthermore, Mackley et al. [15] analyzed reports of 158 phase 1 trials published between January 2015 and July 2018; thus, not overlapping with the studies analyzed by Schwaederle and collaborators. The studies involved 6707 patients. The combined RR was 4%. Among the trials using tumor biomarkers as the eligibility criteria, the RR was higher: 12% vs. 4.9%. However, the same was true of trials focusing on single tumor type (13%) compared to multiple tumor types (3.8%). There were no treatment-related deaths, but the proportion of grade 3 to 4 toxicity was 13.2%.

Von Hoff et al. conducted a pilot study using molecular profiling (MP) of patients' tumors to find potential targets and select treatment based on these findings [16]. This group evaluated the premise that a substantial group of patients selected by this approach would experience improved clinical outcomes compared to their outcome with the immediate priorly administered treatment. The null hypothesis of $\leq$15% of this patient population having a PFS on MP-selected therapy/PFS on prior therapy of $\geq$1.3 was rejected. Eighteen of sixty-six patients (27%) had a PFS ratio of $\geq$1.3.

Important caveats in analyses of the clinical benefit of phase 1 trials in cancer patients, molecularly driven or not, include the multiplicity of tumor histological types usually accrued and the required multiple-dose evaluation steps for safety evaluation and regulatory agency mandates. Thus, even when a target is identified and there is an agent reasonably expected to result in efficacy, the range of doses tested include some below target inhibition and some unnecessarily toxic, beyond the requirement for target inhibition. Importantly, in order for a biological target to be clinically relevant and a molecularly targeted approach beneficial, potent drugs should be available that can interact successfully with the target without significant off-target toxicity.

Thus, overall, it is reasonable to conclude that it is worthwhile for an academic center to pursue systematic rational efforts to obtain promising targeted or immune-interacting agents, and to provide increased access to patients and care providers in community practices of these agents at the time when patients need them the most. That is, to provide these agents when their disease has become refractory to the available standard of care therapeutic approaches.

2. Current Challenges for Early-Stage Clinical Trials

Although the numbers of clinical trials with novel agents have increased, and expansion cohorts have become the routine, the number of clinical sites involved in a typical industry-sponsored clinical trial have exponentially increased, limiting the slots available to individual sites. This practice comports with corporate mandates to fill out available slots as quickly as possible in order to decrease the time to completion of the clinical trial.

Safety issues arising from the coordination of multiple sites have been partially offset by the institution of frequent investigators' calls and virtual meetings. However, these meetings often involve multiple time zones, substantial time demands and incomplete or stale data sets that may frustrate participants and result in attendee attrition. Unintentionally, these circumstances create competition among the sites for patients' slots, resulting in insufficient slots for patients who have time-sensitive needs for these investigational therapeutics.

From the participating sites' perspective, challenges include insufficient staff recruitment and retention, low capacity of treating units and hospitals (such as during the SARS-CoV-2 pandemic) [17,18], increasing clinical and regulatory demands on the clinicians–investigators' time, as well as prolonged time to activation of trials, frequent amendments, and reporting requirements [19]. Tracking down the genomic analyses of referred patients among the array of different analytic platforms not integrated into the medical records has also proven to be a difficult task.

From the community oncology perspective, and impacting community sites integration into clinical trials, referring patients to be placed on phase 1 waiting lists is, at a minimum, inconvenient and impractical. Furthermore, the necessity for patients to travel long distances for very frequent clinic visits, the requirement in many cases for tumor sample prescreening with uncertain outcome, together with multiple patients' procedures, and imaging while on trial, dampen the enthusiasm of previously motivated patients and families. An additional, sometimes insurmountable, challenge to community patients' participation is the approval process and financial limits of health management organizations (HMOs) [20].

3. City of Hope (COH) Community Oncology Practice Network

The COH Clinical Practice Network serves populations located in four Southern California counties: Los Angeles, Orange, San Bernardino/Inland Empire, and Riverside, encompassing 33,109 square miles and approximately 18 million residents (Figure 1). These sites were selected due to their demonstrated telehealth access, strong leadership commitment to quality care and research, and the diverse populations served. Over 30 community satellite practice sites with >150 physicians are distributed throughout our catchment area. The catchment area is one of the most diverse regions in the country (45% Hispanic/LatinX; 12% Asian/Pacific Islander; 6% Non-Hispanic Blacks) and includes a clinical practice site serving a low resource and socio-economically disadvantaged population in the California high-desert region. The majority of sites provide multidisciplinary cancer care. We share unique utilization of the EPIC electronic medical record, employment of COH tailored "Via" pathways, disease-focused tumor registries, as well as precision medicine genomics evaluation. Twenty percent of Duarte campus referrals for complex care or unique studies come from our community satellite practices. A Clinical Outpatient 190,000 Sq. Ft. hub capable of conducting all stages of clinical trials (Lennar Foundation Cancer Center) recently opened in Irvine, Orange County.

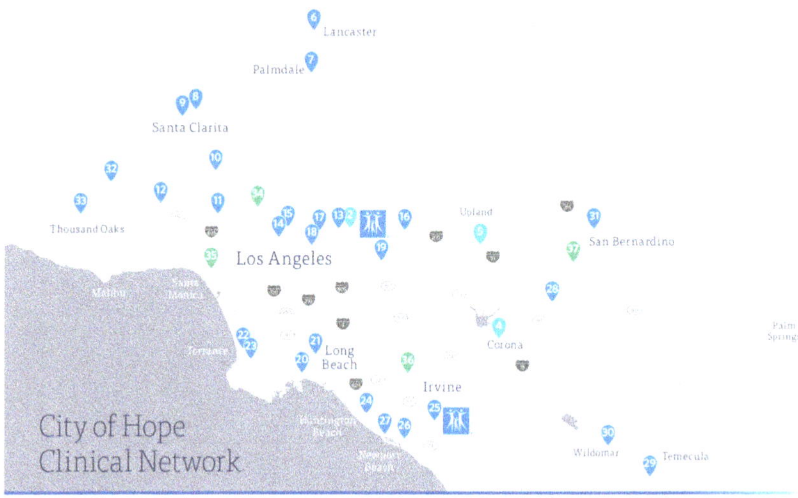

Figure 1. COH community practice sites.

The COH clinical practice network provides access to a very large cancer patient population (October 2020–September 2021: 104,378 unique patients; 359,679 completed appointments); (October 2021–September 2022: 107,870 unique patients; 367,812 completed appointments). The patients served are spread throughout a very large geographical territory and the network practices have different capabilities. Thus, concerted efforts are needed if we are to provide on-site access or directed channeling of patients to early-stage clinical investigations when appropriate.

4. Clinical Research Integration Opportunities

A number of initiatives have enhanced early-stage clinical investigation at COH. These include rapid clinical trial activation times (<90 days on average); uniform protocol templates; a single application form used by the protocol review committee, data safety monitoring board, and institutional review board; upstaffed regulatory start-up and contracting teams; specialized teams to build Epic Beacon and OnCore content; increased use of Master Clinical Trial Agreements with biotech and pharmaceutical companies that reduces the involvement of the general counsel; and participation as a leading academic organization in the NCI Early Therapeutics Clinical Trials Network.

Through an institution-sponsored Precision Medicine program, tumor whole exome and transcriptome sequencing together with germline gene panel sequencing are performed for City of Hope patients and their families. A digital infrastructure and informatics platform provides logistical and analytic support.

5. Ongoing COH Phase 1 Program Integrative Initiatives

5.1. Development of a Virtual "Refractory Disease" Phase 1 Trial—Matching Tele-Medicine Oncology Clinic

During this televideo-medicine-enabled clinic, interested patients referred by COH oncologists (community and academic) or regional HMO oncologists are evaluated. Referred patients have incurable tumors for which standard effective conventional therapies have been exhausted or are non-existent.

A full-time dedicated coordinator oversees the preparation of the clinic ahead of the scheduled appointment. The coordinator contacts referring physicians and patients and retrieves medical records which are made available to the phase 1 oncologist prior to the visit. The retrieval of all prior pathology data and reports and tumor genomic analyses is particularly challenging. In our experience, outside, often diverse, genomic analyses are frequently not incorporated into electronic medical records. If targetable alterations are identified during review, patients will be offered targeted treatment with therapies tailored to their tumor biology if not previously performed. In addition, in some cases, historical genomic analyses provide clues to potential targets amenable to not yet approved investigational agents when biological rationale exists supporting potential efficacy. If no such mutations are found (or while waiting for testing), available non-target specific phase 1 trials, such as those with novel mechanisms of action or immune-system-targeted therapies, are discussed. If the patient agrees to participate in a trial and they meet preliminary eligibility criteria during the virtual visit screening, the patients will visit the COH Duarte campus (or the Orange County clinical-research Hub, see below) to undergo examination and consent to the particular study; subsequently, they receive screening for trial initiation. However, if a tumor-specific later stage trial is available that better fits the patient, a recommendation and referral to the proper tumor-specific physician(s) is made, after obtaining approval from the referring physician.

These clinics (once a week with rotating phase 1 oncologists) were initiated during 2022 with great acceptance from patients and medical providers alike. Feedback communication with referring physician within 48 h is routinely performed. Patients are also provided access to our centralized germline and somatic genomic testing, if not previously performed (see below).

5.2. Expansion of Phase 1 Trials to City of Hope—Orange County

The large community oncology practice network at COH has increased our catchment area, enabled the enrollment of patients from diverse ethnicities and backgrounds as well as providing the patients the opportunity to seek specialized clinical care closer to home. However, early-phase trials can only be opened at sites that have the capabilities to conduct these trials. These requirements include the handling of research samples that may involve frequent and long hours of pharmacokinetic sampling and specialized on-site research pharmacy and radiology. Patients identified through the COH community network eligible for early-phase trials still must travel to the main campus in Duarte multiple times a month.

One recent development in the COH enterprise has been the opening of the City of Hope Orange County Lennar Foundation Cancer Center in Irvine, Orange County, in August 2022.

The Orange County Cancer Center offers all comprehensive cancer services with a dedicated clinical research unit and allows the conduct of on-site early-phase cancer trials. Within the County, this cancer center serves as a 'hub' offering a full array of clinical operations serving as the 'spokes' or regional community sites within Orange County. These 'spoke' sites typically provide only clinical services and late-phase trials. As the new cancer center is located within 25 miles from each of these regional sites (Figure 2), patients can be easily routed to the hub for early-phase trials as well as more specialized services.

Adding early-phase trials to a second campus beyond the academic Duarte campus created a number of challenges: increased regulatory processing, the establishment of an efficient clinical workflow for each trial, and the coordination of Orange County laboratories with laboratories at the Duarte campus for sample processing and shipping. Dedicated efforts have succeeded in ensuring that most services required for conducting early-phase trials are now operational at the Orange County site, such as contracting, institutional board review, data safety monitoring, budget review, and an electronic health system as well as phase I disease team management.

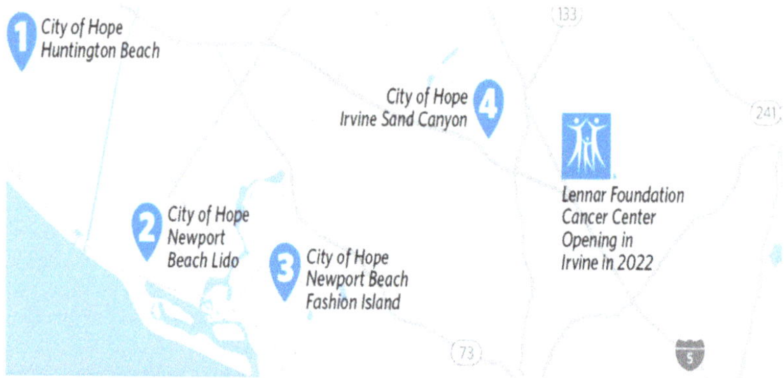

Figure 2. Map of COH Orange County clinical sites.

5.3. Precision Medicine Network Initiative

The efficient and successful conduct of early-stage drug trials depends critically upon the investigators' ability to identify patients whose tumor mutational profiles match the molecular target of an investigational drug [21–24]. Limiting the success of early-stage drug trials, patients frequently experience difficulty in obtaining tumor sequencing, trialists often lack ready access to completed studies, and clinicians may experience challenges in interpreting the often disparate, dense, and abstruse tumor sequencing reports [25–29]. These limitations may compound in the community oncology setting due to inconsistent tumor sequencing practices, inadequate administrative structure, and lack of advanced molecular expertise [30]. Helping to ameliorate these limitations and accelerate the progress of early-stage drug trials for its community oncology practice partners, City of Hope created the Center for Precision Medicine.

6. The City of Hope Center for Precision Medicine

The advent of targeted cancer therapies ushered in the era of precision medicine in clinical oncology [31]. City of Hope (COH), recognizing the unparalleled promise of precision medicine, established the COH Center for Precision Medicine (COH-CPM) in 2020 [32]. The COH-CPM aims to harness genomic-driven insights to pioneer personalized prevention and treatments towards improving the outcomes and quality of life for patients and their families. To accomplish its mission, the COH-CPM initiated the INSPIRE (Implementing Next-generation Sequencing for Precision Intervention and Risk Evaluation) study, a universal access investigation open to all patients at COH with a personal and/or family history of cancer.

COH-INSPIRE participants receive germline genetic assessment through testing with a custom 155 cancer gene panel [33]. Tumors of patients with an available cancer specimen undergo somatic tumor-normal whole exome and whole transcriptome sequencing [34,35]. To conduct the INSPIRE study, COH-CPM relies on an expert team of enrollment specialists, genetic counselors, and cancer genetic physicians who facilitate the participation and clinical management of patients. Since its inauguration, the INSPIRE study has experienced tremendous success with enrollment of nearly 15,000 patients. INSPIRE patients whose sequencing results pose complex genetic, genomic, and clinical questions receive in-depth review at 2 weekly clinical case conferences: a Genetics Case Conference and a Precision Oncology Tumor Board (POTB).

6.1. COH-CPM Clinical Case Conferences

Genetic counselors and cancer genetics physicians conduct the Genetics Case Conference with the aim of resolving challenging problems related to germline findings and genetic risk. The INSPIRE study has observed that nearly 1 in 5 (2654/14,346 [18.5%])

patients carry a germline pathogenic variant requiring clinical review and management; this significantly elevated pathogenicity rate ensures a full volume of complex case reviews but also, more clinically significant, validates a universal access model of precision medicine availability.

6.2. Precision Oncology Tumor Board

The second weekly conference, the POTB, complements the Genetics Case Conference. The POTB provides interpretation, targeted therapeutic insights, and clinical trial eligibility information related to tumor whole exome and transcriptome sequencing results. The POTB enlists the expertise of a multidisciplinary team comprising, among others, medical and surgical oncologists, genomic scientists, genetic counselors, and computational biologists. To date, the POTB has completed over 150 deep-dive analyses to help the tangible delivery of precision medicine to COH patients, positively impacting their treatment, health, and well-being.

7. Centralized Logistical Operation of COH-INSPIRE

Inaugural COH-CPM INSPIRE activities focused on optimizing precision medicine operations at the central, academic COH Duarte campus. Initial efforts sought to design, implement, and iteratively improve four core precision medicine operations: patient enrollment, specimen processing, germline and somatic tumor next-generation sequencing, and clinical management. Most patients enroll in the INSPIRE study through assigned study consenters stationed in the oncology subspecialty clinic where they receive treatment. Consenters may also enroll a patient remotely via a televideo protocol should in-person consenting prove infeasible. For germline DNA assessment, patients provide blood samples to a central specimen collection laboratory situated on campus. The COH Pathology department assumes responsibility for retrieving fresh-frozen, paraffin-embedded tumor blocks for somatic sequencing. Furthermore, the COH Pathology department oversees delivery of the germline and somatic tumor specimens to commercial NCS vendors who perform CLIA/CAP grade germline panel and somatic whole exome and transcriptome sequencing.

Vendors typically complete the sequencing of specimens within 10–14 days of specimen receipt and deliver test reports directly to the COH electronic medical record as primary BAM, FASTQ, and VCF sequencing files. A centralized electronic data warehouse, POSEIDON, receives copies of the primary BAM, FASTQ and VCF sequencing files [36]. Data analysts and computational biologists have access to these data files for downstream analysis. A clinical team of genetic counselors and cancer genetic physicians review all sequencing results; this team identifies patients requiring clinical management and/or further in-depth analyses. Optimization of these core operational activities has enabled high volume patient participation in the INSPIRE study at the COH Duarte campus.

8. Expansion of INSPIRE to the COH Community Oncology Network

Iterative improvements in the INSPIRE protocol established an efficiently functioning, fully interoperable, and incrementally more agile precision medicine workflow. Since initial optimization, COH-CPM has continued expansion of INSPIRE with serial introduction of the study across COH community oncology practices. To achieve streamlined integration of INSPIRE across the community practice enterprise, COH-CPM adopted a "hub-and-spoke" mode of operational logistics. Administration and pathology processing operations remain anchored at the Duarte campus hub, while specimen collection and in situ INSPIRE consenting takes place nodally at the community oncology clinic spokes. As with INSPIRE cases originating at the Duarte campus, all community oncology INSPIRE cases qualify for review through the Genetics Case Conference and POTB. Precision Medicine teams organize and conduct reviews at the central Duarte campus with remote televideo participation of community practices.

In 2021, COH-CPM successfully commenced network community INSPIRE participation. Among the first community oncology sites to participate, the COH Upland clinic, located in San Bernadino County, approximately 30 miles east of the central Duarte campus, serves a racially and ethnically diverse, historically underserved patient population. To date, over 1000 Upland community oncology patients have completed germline and somatic sequencing through the INSPIRE study. To expedite timely and convenient genetics care at the Upland clinic, a cancer genetics-trained surgical oncologist provides in-person services to INSPIRE patients requiring ancillary management.

COH-CPM has recently expanded INSPIRE to 20 COH community oncology practices; ongoing expansion efforts continue towards achieving full participation of all COH community oncology practices. Full expansion of INSPIRE promises the diverse community oncology patient population not only facilitated access to precision medicine resources but also widened avenues for potentially clinically impactful early drug development participation.

9. Leveraging INSPIRE to Accelerate Early Drug Development across the COH Community Oncology Network

Over the past 2 decades, the number of clinical trials requiring genomically informed biomarker qualification has increased exponentially from 15% to greater than 50% [37,38]. Often utilized qualifying biomarkers include, among others, germline and somatic pathogenic genetic variants and fusions, tumor mutational burden, homologous recombination deficiency, microsatellite instability, checkpoint inhibitor protein expression, and hormone receptor states [39]. INSPIRE directly or secondarily (through downstream analyses) has the ability to assess the gamut of these biomarkers; moreover, standard INSPIRE review protocols identify patients eligible for clinical trial enrollment based upon their genomic biomarker profile. These embedded review protocols permit proficient, routine identification of clinical trial-eligible patients and undergird a high-volume clinical trial selection process.

Expansion of INSPIRE across the COH community network affords the early drug development program four transformational opportunities to 1—increase early-stage clinical trial enrollment; 2—optimize investigational drug and clinical trial matching; 3—improve drug response rates; and 4—enhance healthcare equity for COH oncology community participants.

The INSPIRE study promotes not only improved quantity of enrollment, but also higher quality. INSPIRE's comprehensive assessment of germline, somatic whole exome and transcriptome alterations allows more specific drug matching and, consequently, more precise enrollment into early-stage drug development trials. More precise enrollment and optimized drug matching predicts improved drug response rates and, consequently, accelerated drug development [40,41]. Conversely, failure to "select the right drug for the right patient" may result in a lack of therapeutic efficacy and abandonment of further drug development efforts [42–44].

Historically underserved populations frequently demonstrate compromised awareness of genetics and genomics and the impact that these areas of medicine may have on their health and oncology treatment options [45,46]. The INSPIRE study proactively enrolls community oncology practice patients, many of whom present from marginalized and underserved regions of the COH catchment area. INSPIRE enrollment specialists and genetic counselors directly engage these patients, explaining the rationale and process of genetic and genomic testing; subsequently, patients meet with genetic counselors and cancer genetics physicians to review their test results and discuss health implications for them and their families. These direct interactions often represent the patients' first awareness of precision medicine and its promises. In tandem with increasing awareness, INSPIRE provides concrete access to the resources of precision medicine testing and offers a navigable clinical pathway to enter early-stage drug development studies [47].

Community participation in the INSPIRE study promotes patient health agency and autonomy. Knowledge of individual genetic alterations, both germline and somatic, empowers patients to make informed decisions regarding therapeutic options as well as clinical trial participation. Informed engagement with genetic specialists and early-stage drug development trialist moves the patient closer to the ideal of authentic shared decision-making in their cancer care [48,49].

10. Conclusions

Overcoming several significant practical challenges, our ongoing initiatives at COH have allowed our community clinical network to leverage the ever-transforming landscape of genetics, genomic, and novel therapies that is impacting cancer care. Our program has achieved excellent clinician and patient satisfaction. To date, the program has performed >12,000 germline and >6000 somatic tumor unique patient tests for a diverse patient population comprising 47% Hispanics/Latinx, 29% API, 12% African Americans, and 1% Native Americans.

Precision medicine-based innovation and discovery provide new opportunities for the early-stage drug development bench to clinic translational programs. In turn, these programs offer the COH community oncology patient population access to promising treatments.

We believe that the COH experience of developing and implementing a hub and spoke, early drug development program can serve as a model for other community oncology practices nationwide. Only with community integration can novel therapeutic discoveries reach their true clinical potential. However, such integration requires careful consideration and thoughtful deliberation regarding value versus cost, understanding not only therapeutic dividends but also the societal and ethical benefits of providing advanced genomic oncology care to underserved, disadvantaged populations.

Funding: This research received no external funding.

Institutional Review Board Statement: Not applicable.

Informed Consent Statement: Not applicable.

Data Availability Statement: Not applicable.

Conflicts of Interest: The authors declare no conflict of interest.

References

1. Dickson, M.; Gagnon, J.P. Key factors in the rising cost of new drug discovery and development. *Nat. Rev. Drug Discov.* **2004**, *3*, 417–429. [CrossRef]
2. DiMasi, J.A.; Grabowski, H.G.; Hansen, R.W. Innovation in the pharmaceutical industry: New estimates of R&D costs. *J. Health Econ.* **2016**, *47*, 20–33. [PubMed]
3. Mak, K.-K.; Pichika, M.R. Artificial intelligence in drug development: Present status and future prospects. *Drug Discov. Today* **2019**, *24*, 773–780. [CrossRef] [PubMed]
4. Ingber, D.E. Human organs-on-chips for disease modelling, drug development and personalized medicine. *Nat. Rev. Genet.* **2022**, *23*, 467–491. [CrossRef] [PubMed]
5. Mennen, S.M.; Alhambra, C.; Allen, C.L.; Barberis, M.; Berritt, S.; Brandt, T.A.; Campbell, A.D.; Castañón, J.; Cherney, A.H.; Christensen, M.; et al. The evolution of high-throughput experimentation in pharmaceutical development and perspectives on the future. *Org. Process Res. Dev.* **2019**, *23*, 1213–1242. [CrossRef]
6. Parasrampuria, D.A.; Benet, L.Z.; Sharma, A. Why drugs fail in late stages of development: Case study analyses from the last decade and recommendations. *AAPS J.* **2018**, *20*, 46. [CrossRef]
7. Lin, A.; Giuliano, C.J.; Palladino, A.; John, K.M.; Abramowicz, C.; Yuan, M.L.; Sausville, E.L.; Lukow, D.A.; Liu, L.; Chait, A.R.; et al. Off-target toxicity is a common mechanism of action of cancer drugs undergoing clinical trials. *Sci. Transl. Med.* **2019**, *11*, eaaw8412. [CrossRef]
8. Unger, J.M.; Vaidya, R.; Hershman, D.L.; Minasian, L.M.; Fleury, E.M. Systematic review and meta-analysis of the magnitude of structural, clinical, and physician and patient barriers to cancer clinical trial participation. *JNCI J. Natl. Cancer Inst.* **2019**, *111*, 245–255. [CrossRef]
9. Levit, L.A.; Byatt, L.; Lyss, A.P.; Paskett, E.D.; Levit, K.; Kirkwood, K.; Schenkel, C.; Schilsky, R.L. Closing the rural cancer care gap: Three institutional approaches. *JCO Oncol. Pract.* **2020**, *16*, 422–430. [CrossRef]
10. Decoster, G.; Stein, G.; Holdener, E.E. Responses and toxic deaths in phase I clinical trials. *Ann. Oncol.* **1990**, *1*, 175–181. [CrossRef]
11. Von Hoff, D.D.; Turner, J. Response rates, duration of response, and dose response effects in phase I studies of antineoplastics. *Investig. New Drugs* **1991**, *9*, 115–122. [CrossRef]

12. Chihara, D.; Lin, R.; Flowers, C.R.; Finnigan, S.R.; Cordes, L.M.; Fukuda, Y.; Huang, E.P.; Rubinstein, L.V.; Nastoupil, L.J.; Ivy, S.P.; et al. Early drug development in solid tumours: Analysis of National Cancer Institute-sponsored phase 1 trials. *Lancet* **2022**, *400*, 512–521. [CrossRef] [PubMed]
13. Chakiba, C.; Grellety, T.; Bellera, C.; Italiano, A. Encouraging Trends in Modern Phase 1 Oncology Trials. *N. Engl. J. Med.* **2018**, *378*, 2242–2243. [CrossRef] [PubMed]
14. Schwaederle, M.; Zhao, M.; Lee, J.; Lazar, V.; Leyland-Jones, B.; Schilsky, R.; Mendelsohn, J.; Kurzrock, R. Association of Biomarker-Based Treatment Strategies with response rates and progression free survival in refractory malignant neoplasms: A meta-analysis. *JAMA Oncol.* **2016**, *2*, 1452–1459. [CrossRef] [PubMed]
15. Mackley, M.; Fernandez, N.R.; Fletcher, B.; Woolcott, C.G.; Fernandez, C.V. Revisiting risk and benefit in early oncology trials in the era of precision medicien: A systematic review and meta-analysis of phase 1 trials of targeted single-agent anticancer therapies. *JCO Precis. Oncol.* **2021**, *5*, 17–26. [CrossRef]
16. Von Hoff, D.D.; Stephenson, J.J.S., Jr.; Rosen, P.; Loesch, D.M.; Borad, M.J.; Anthony, S.; Jameson, G.S.; Brown, S.; Cantafio, N.; Richards, D.A.; et al. Pilot study using molecular profiling of patients'tumors to find potential targets and select treatments for their refractory cancers. *J. Clin. Oncol.* **2010**, *28*, 4877–4883. [CrossRef]
17. Sathian, B.; Asim, M.; Banerjee, I.; Pizarro, A.B.; Roy, B.; Van Teijlingen, E.R.; Nascimento, I.J.B.D.; Alhamad, H.K. Impact of COVID-19 on clinical trials and clinical research: A systematic review. *Nepal. J. Epidemiol.* **2020**, *10*, 878–887. [CrossRef] [PubMed]
18. Lasch, F.; Psarelli, E.; Herold, R.; Mattsson, A.; Guizzaro, L.; Pétavy, F.; Schiel, A. The Impact of COVID-19 on the Initiation of Clinical Trials in Europe and the United States. *Clin Pharm. Ther.* **2022**, *111*, 1093–1102. [CrossRef]
19. Lee, C.; Werner, T.L.; Deal, A.M.; Krise-Confair, C.J.; Bentz, T.A.; Cummings, T.M.; Grant, S.C.; Lee, A.B.; Moehle, J.; Moffett, K.; et al. Clinical trial metrics: The complexity of conducting clinical trials in North American Cancer Centers. *JCO Oncol. Pract.* **2021**, *17*, e77–e93. [CrossRef]
20. Mendelsohn, J.; Moses, H.L.; Nass, S.J. *A National Cancer Clinical Trials System for the 21st Century: Reinvigorating the NCI Cooperative Group Program*; National Academies Press: Washington, DC, USA, 2010.
21. Goodsaid, F. Challenges of biomarkers in drug discovery and development. *Expert Opin. Drug Discov.* **2012**, *7*, 457–461. [CrossRef]
22. Rothwell, D.G.; Ayub, M.; Cook, N.; Thistlethwaite, F.; Carter, L.; Dean, E.; Smith, N.; Villa, S.; Dransfield, J.; Clipson, A.; et al. Utility of ctDNA to support patient selection for early phase clinical trials: The TARGET study. *Nat. Med.* **2019**, *25*, 738–743. [CrossRef] [PubMed]
23. Dienstmann, R.; Serpico, D.; Rodon, J.; Saura, C.; Macarulla, T.; Elez, E.; Alsina, M.; Capdevila, J.; Perez-Garcia, J.; Sánchez-Ollé, G.; et al. Molecular profiling of patients with colorectal cancer and matched targeted therapy in phase I clinical trials. *Mol. Cancer Ther.* **2012**, *11*, 2062–2071. [CrossRef]
24. Siu, L.L.; Conley, B.A.; Boerner, S.; LoRusso, P.M. Next-Generation Sequencing to Guide Clinical Trials. *Clin. Cancer Res.* **2015**, *21*, 4536–4544. [CrossRef]
25. Mateo, J.; Steuten, L.; Aftimos, P.; André, F.; Davies, M.; Garralda, E.; Geissler, J.; Husereau, D.; Martinez-Lopez, I.; Normanno, N.; et al. Delivering precision oncology to patients with cancer. *Nat. Med.* **2022**, *28*, 658–665. [CrossRef] [PubMed]
26. Kaminski, A.; Szamreta, E.A.; Shah, R.; Ning, N.; Aggarwal, J.; Hussain, A.; Adeboyeje, G. Barriers to next-generation sequencing despite increased utilization: U.S. physician survey results. *J. Clin. Oncol.* **2021**, *39* (Suppl. 15), e18754. [CrossRef]
27. Johansen Taber, K.A.; Dickinson, B.D.; Wilson, M. The promise and challenges of next-generation genome sequencing for clinical care. *JAMA Intern. Med.* **2014**, *174*, 275–280. [CrossRef]
28. Moor, J.; Gray, S.W.; Mitchell, S.A.; Klabunde, C.N.; Freedman, A.N. Oncologist Confidence in Genomic Testing and Implications for Using Multimarker Tumor Panel Tests in Practice. *JCO Precis. Oncol.* **2020**, *4*, 620–631. [CrossRef]
29. Roberts, M.C.; Spees, L.P.; Freedman, A.N.; Klein, W.M.P.; Das, I.P.; Butler, E.N.; de Moor, J.S. Oncologist-Reported Reasons for Not Ordering Multimarker Tumor Panels: Results from a Nationally Representative Survey. *JCO Precis. Oncol.* **2021**, *5*, 701–709. [CrossRef]
30. Ashbury, F.D.; Thompson, K.; Williams, C.; Williams, K. Challenges adopting next-generation sequencing in community oncology practice. *Curr. Opin. Oncol.* **2021**, *33*, 507–512. [CrossRef]
31. Bedard, P.L.; Hyman, D.M.; Davids, M.S.; Siu, L.L. Small molecules, big impact: 20 years of targeted therapy in oncology. *Lancet* **2020**, *395*, 1078–1088. [CrossRef]
32. Precision Medicine for All: Cancer Centre Makes Genomic Analysis a Top Priority. Available online: https://www.nature.com/articles/d42473-022-00259-y (accessed on 1 March 2023).
33. Solomon, I. INSPIRE Study (Implementing Next-generation Sequencing for Precision Intervention and Risk Evaluation): Scaling Return of Genomic Results. In Proceedings of the Conference: Precision Prevention, Early Detection, and Interception of Cancer, Austin, TX, USA, 1 January 2022.
34. McDonnell, K.; Hong, C.; Bonner, J.D.; Lindsey, S.S.; Solomon, I.; Hampel, H.; Park, W.; Idos, G.; Gray, S.W.; Gruber, S.B. Germline mutational landscape of non-highly penetrant Fanconi anemia genes unveiled from sequencing of 5,044 patients with solid tumor cancer. *J. Clin. Oncol.* **2022**, *40* (Suppl. 16), 10521. [CrossRef]
35. Kruper, L. Reappraising the Fanconi Anemia DNA repair pathway in breast cancer risk and precision intervention: Insights and opportunities from the City of Hope INSPIRE study. In Proceedings of the San Antonio Breast Cancer Symposium, San Antonio, TX, USA, 6–10 December 2022.

36. Courdy, S.; Hulse, M.; Nadaf, S.; Mao, A.; Pozhitkov, A.; Berger, S.; Chang, J.; Achuthan, S.; Kancharla, C.; Kunz, I.; et al. The City of Hope POSEIDON enterprise-wide platform for real-world data and evidence in cancer. *J. Clin. Oncol.* **2021**, *39* (Suppl. 15), e18813. [CrossRef]
37. Vadas, A.; Bilodeau, T.J.; Oza, C. Special Report: The Evolution of Biomarker Use in Clinical Trials for Cancer Treatments. Available online: https://www.thejournalofprecisionmedicine.com/the-journal-of-precision-medicine/special-report-the-evolution-of-biomarker-use-in-clinical-trials-for-cancer-treatments/ (accessed on 1 March 2023).
38. Goulart, B.H.; Clark, J.W.; Pien, H.H.; Roberts, T.G.; Finkelstein, S.N.; Chabner, B.A. Trends in the use and role of biomarkers in phase I oncology trials. *Clin. Cancer Res.* **2007**, *13 Pt 1*, 6719–6726. [CrossRef] [PubMed]
39. Malone, E.R.; Oliva, M.; Sabatini, P.J.B.; Stockley, T.; Siu, L.L. Molecular profiling for precision cancer therapies. *Genome Med.* **2020**, *12*, 8. [CrossRef]
40. Jardim, D.L.; Schwaederle, M.; Hong, D.S.; Kurzrock, R. An appraisal of drug development timelines in the Era of precision oncology. *Oncotarget* **2016**, *7*, 53037–53046. [CrossRef]
41. Collins, F.S.; Varmus, H. A new initiative on precision medicine. *N. Engl. J. Med.* **2015**, *372*, 793–795. [CrossRef]
42. Hamburg, M.A.; Collins, F.S. The path to personalized medicine. *N. Engl. J. Med.* **2010**, *363*, 301–304. [CrossRef]
43. Kelloff, G.J.; Sigman, C.C. Cancer biomarkers: Selecting the right drug for the right patient. *Nat. Rev. Drug Discov.* **2012**, *11*, 201–214. [CrossRef]
44. Kiriiri, G.K.; Njogu, P.M.; Mwangi, A.N. Exploring different approaches to improve the success of drug discovery and development projects: A review. *Future J. Pharm. Sci.* **2020**, *6*, 27. [CrossRef]
45. Giri, V.N.; Shimada, A.; Leader, A.E. Predictors of Population Awareness of Cancer Genetic Tests: Implications for Enhancing Equity in Engaging in Cancer Prevention and Precision Medicine. *JCO Precis Oncol.* **2021**, *5*, 1699–1708. [CrossRef]
46. Catz, D.S.; Green, N.S.; Tobin, J.N.; Lloyd-Puryear, M.A.; Kyler, P.; Umemoto, A.; Cernoch, J.; Brown, R.; Wolman, F. Attitudes about genetics in underserved, culturally diverse populations. *Community Genet.* **2005**, *8*, 161–172. [CrossRef] [PubMed]
47. Khoury, M.J.; Bowen, S.; Dotson, W.D.; Drzymalla, E.; Green, R.F.; Goldstein, R.; Kolor, K.; Liburd, L.C.; Sperling, L.S.; Bunnell, R. Health equity in the implementation of genomics and precision medicine: A public health imperative. *Genet. Med.* **2022**, *24*, 1630–1639. [CrossRef] [PubMed]
48. Elwyn, G.; Cochran, N.; Pignone, M. Shared Decision Making-The Importance of Diagnosing Preferences. *JAMA Intern. Med.* **2017**, *177*, 1239–1240. [CrossRef] [PubMed]
49. Blasimme, A.; Vayena, E. Becoming partners, retaining autonomy: Ethical considerations on the development of precision medicine. *BMC Med. Ethics* **2016**, *17*, 67. [CrossRef]

Journal of
Clinical Medicine

MDPI

Perspective

Integrative Oncology and the Clinical Care Network: Challenges and Opportunities

George Semeniuk [1,†], Bahareh Bahadini [1,†], Eugene Ahn [1,2], Jasmine Zain [1], Jessica Cheng [1], Ameish Govindarajan [1], Judy Rose [1] and Richard T. Lee [1,*]

1 City of Hope Comprehensive Cancer Center, Duarte, CA 91010, USA; gsemeniuk@coh.org (G.S.); bbahadini@coh.org (B.B.); eugene.ahn@ctca-hope.com (E.A.); jazain@coh.org (J.Z.); jescheng@coh.org (J.C.); agovindarajan@coh.org (A.G.); jrose@coh.org (J.R.)
2 City of Hope Chicago, Zion, IL 60099, USA
* Correspondence: richlee@coh.org
† These authors contributed equally to this work.

Abstract: Integrative oncology is a new and growing field of cancer care. Integrative oncology is a patient-centered, evidence-based field of comprehensive cancer care that utilizes integrative therapies such as mind-body practices, acupuncture, massage, music therapy, nutrition, and exercise in collaboration with conventional cancer treatments. Patient interest and utilization has been growing over the past two decades. Clinical research has shown the benefits of these approaches to improving symptom management and quality of life, and is now being incorporated into national guidelines from the National Comprehensive Cancer Network (NCCN) and American Society for Clinical Oncology (ASCO). The availability of these services at cancer centers is growing, although the structure and implementation of integrative oncology remains highly variable. This article discusses the benefits of integrative oncology and provides an overview of the current state of integrative oncology programs nationwide. Current challenges and opportunities for cancer centers to provide integrative services is reviewed in the areas of programmatic structure, clinical service, education, and research.

Keywords: integrative oncology; clinical care network; implementation

Citation: Semeniuk, G.; Bahadini, B.; Ahn, E.; Zain, J.; Cheng, J.; Govindarajan, A.; Rose, J.; Lee, R.T. Integrative Oncology and the Clinical Care Network: Challenges and Opportunities. *J. Clin. Med.* **2023**, *12*, 3946. https://doi.org/10.3390/jcm12123946

Academic Editors: Prakash Kulkarni and Ravi Salgia

Received: 14 April 2023
Revised: 29 May 2023
Accepted: 31 May 2023
Published: 9 June 2023

1. Introduction

Over 19 million people around the world were diagnosed with cancer and almost 10 million died from cancer in 2020 [1]. By 2040, new cases and death totals are expected to reach approximately 28 million and 16 million, respectively [2]. Cancer treatment alone costs the world approximately USD 1.2 trillion dollars annually. According to the National Center for Complementary and Integrative Health (NCCIH), "if a non-mainstream approach is used together with conventional medicine, it's considered complimentary", while "if a non-mainstream approach is used in place of conventional medicine, it's considered alternative" [3]. Therefore, integrative medicine is practiced in combination with conventional cancer care, not as an "alternative". Thus, these types of therapies are commonly termed integrative, complementary, and alternative medicine (ICAM) and have become increasingly popular in Western medicine. The Academic Consortium for Integrative Medicine and Health describes this approach as one that "reaffirms the importance of the relationship between practitioner and patient, focuses on the whole person, is informed by evidence, and makes use of all appropriate therapeutic and lifestyle approaches, healthcare professionals and disciplines to achieve optimal health and healing". The NCCIH divides ICAM modalities into three main categories: nutrition, psychological, and physical. Nutrition approaches focus on food as medicine and encompass special diets, dietary patterns, and natural products (i.e., dietary supplements such as vitamins, minerals, herbs, and botanicals). Psychological aspects include mindfulness and spirituality, while physical modalities include massage and spinal manipulation, and increasingly popular is the combination of psychological and physical approaches known as mind and body practices, including acupuncture, massage therapy, mindfulness, meditation, music therapy, and yoga (see Figure 1 from NCCIH) [3].

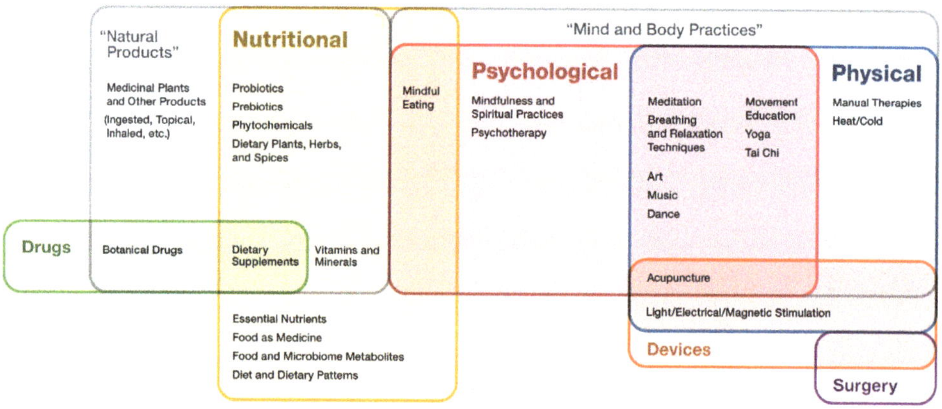

Figure 1. NCCIH ICAM approaches.

As this applies to cancer care, The Society of Integrative Oncology (SIO) defines integrative medicine as a "patient centered, evidence-informed field of cancer care that utilizes mind and body practices, natural products, and/or lifestyle modifications from different traditions alongside conventional cancer treatments" [4]. In practice, integrative medicine is a multidisciplinary approach to treating patients with efforts to combine the benefits of integrative and conventional modalities that have been shown to be safe and effective. These types of therapies are provided by a different group of professional caregivers, with different educational backgrounds, credentials, and regulatory oversight that varies by therapy, geography, and culture.

The use of ICAM has become increasingly popular. According to the 2012 National Health Interview Survey, 59 million Americans over the age of four used at least one complementary health approach, equating to USD 30.2 billion dollars in out-of-pocket expenses [5]. Dietary supplements and yoga appeared to be the most popular modalities in 2017, with 57.6% and 14.2% of adults reporting their usage, respectively [6,7]. People use integrative medicine for a wide array of diseases and conditions. According to surveys, a majority of patients use integrative medicine most commonly for back pain or back problems, head or chest colds, neck pain or neck problems, joint pain or stiffness, and anxiety or depression [8]. As more patients are using integrative medicine, so are practitioners. According to a national survey of medical oncologists, one third reported using ICAM and nearly two thirds had recommended some form of ICAM to patients [9].

Cancer patients and survivors in particular, increasingly look to ICAM in attempt to decrease recurrence, manage the side effects of treatment, and manage and treat other comorbidities [10]. More specifically, approximately one third of cancer patients reported using ICAM in the last 12 months in 2019, with herbal supplements being the most commonly used modality. Of the cancer patients using ICAM, 29.3% did not disclose their ICAM use to their physicians [11]. This high use of herbal supplements, in combination with lack of disclosure to their healthcare providers, is worrisome not only because dietary supplements are not thoroughly regulated, and thus may have quality and safety issues, but also because many have drug interactions and the potential for negative outcomes. Therefore, it is important for physicians to discuss ICAM with their patients to ensure they are using ICAM in safe and effective ways. This also requires educating physicians on the various ICAM modalities, such as their safety, efficacy, and effects on the body and on other conventional treatments.

Despite the growing efforts to provide conventional cancer care globally, and similar attempts to increase the use of ICAM into Western medicine practices, a systematic integration of these two approaches is still missing. This is where the field of integrative oncology can serve as a bridge between the two models and help both patients and practitioners find solutions to navigate between the two systems. This article focuses on integrative oncology and the current state of practice, with an emphasis on clinical care network as most patients are treated at community centers near home rather than large academic cancer centers.

2. Benefits of Integrative Oncology

Growing evidence indicates the important role of integrative oncology on outcomes in cancer care, with most studies indicating improvement in symptoms and quality of life including pain, nausea/vomiting, anxiety, hot flashes, insomnia, neuropathy, and dry mouth [12]. As a result, integrative therapies are now being incorporated in national guidelines from the National Comprehensive Cancer Network and the American Society for Clinical Oncology [13–15].

The mind-body connection is an important aspect of integrative oncology, as emphasized in the recent Institute of Medicine (IOM) report "Cancer Care for the Whole Patient" (IOM) [16]. In this comprehensive report, it is mentioned that "cancer care today often provides state-of-the-science biomedical treatment but fails to address the psychological and social (psychosocial) problems associated with the illness. These problems—including... anxiety, depression or other emotional problems—cause additional suffering, weaken adherence to prescribed treatments, and threaten patients' return to health". Extensive research has documented that mind–body interventions appear to address many of the issues mentioned in the IOM report.

Mindfulness meditation is the practice of bringing one's attention to the present via thoughts, feelings, emotions, and physical presence with openness. One of the most well studied techniques is Mindfulness-Based Stress Reduction (MSBR), which can be taught via an 8-week course [17,18], and a variety of other techniques have also been studied such as yoga meditation and Tibetan meditation. The MBSR training is generally a 2.5 h session each week for 8 weeks, along with an 8 h retreat. A clinical trial of 229 women found that MSBR significantly improved total mood, quality of life, and well-being compared to the control group, in women who had undergone breast cancer treatment [19]. Another randomized controlled trial examining breast cancer survivors found mindfulness awareness practices to significantly reduce stress, pro-inflammatory gene expression, inflammatory signaling, fatigue, sleep disturbance, and vasomotor symptoms, though these effects, aside from cancer-related distress, did not persist at a 3 month follow-up [20]. Six other studies found that meditation can also significantly decrease anxiety, fatigue, and depression in cancer patients [21–26], as well as lead to stress reduction and improve overall quality of life [21,27]. Anderson et al. demonstrated a survival advantage among breast cancer patients randomized to a psychological intervention with more than 10 years of follow-up [28]. Although the exact mechanism behind this effect remains unclear, the authors have proposed several theories. Psychological therapies such as Cognitive Emotional Behavior Therapy (CEBT), Cognitive Behavioral Therapy (CBT), and supportive-expressive therapy aim to alleviate anxiety and depression commonly associated with cancer treatments. These negative emotional states are believed to suppress the immune and neuroendocrine systems, which may impact survival [29]. Patients are taught problem-solving skills, cognitive flexibility, and relaxation techniques to better cope with stressful situations. However, a recent meta-analysis of psychosocial interventions showed only minimal short-term improvements in survival, with individually delivered interventions failing to show any survival benefits [30]. ASCO/SIO guidelines recommend meditation to reduce anxiety (Grade A), treat mood disturbance and depressive symptoms (Grade A), and improve quality of life (Grade A) [31].

level support within the cancer center to ensure long term viability. Many examples of programs being started with philanthropy investment, only later to slowly disappear, illustrate a common story among integrative medicine/oncology programs, which speaks to the importance of institutional support, but also programmatic leadership. Having oncologists either leading or establishing strong partnerships is critical to understanding the nuances with modern cancer care, as well as developing trust among the cancer center staff by being an advocate from within. A series of guidelines have been developed by SIO and endorsed by ASCO for highlighting evidence-based integrative therapies in breast cancer and more recently, integrative approaches to cancer pain [14,15]. Additional practice guidelines are currently under development by SIO to facilitate the incorporation of integrative therapies in an evidence-based approach.

Integrative medicine is patient-specific and based on a patient's goals, values, cultures, and philosophy of health. Practice guidelines are often generic and difficult to adapt to individual patients. A universal, one-size-fits-all approach to implementation of these evidence-based guidelines is impossible and therefore personalization and cultural tailoring are necessary for each cancer center. Additionally, the limited published data is available discussing how to establish integrative oncology programs. Based on our knowledge and experience, challenges to implementing a community network integrative oncology program include a lack of financial resources, clinical delivery due to limited numbers of integrative oncology trained practitioners and leaders, as well as difficulties advocating and collaborating with other departments (e.g., palliative care physicians and social workers), and developing an educational forum for disease-specific cancers.

Financial: Funding and budgetary constraints pose the number one challenge of a community integrative network program. Increased cancer survival rates, diminishing payments and reimbursements, and expanding and aging populations make obtaining enough funding to support a network expansion of an existing integrative program difficult. Without more rigorous scientific evidence in large, reproducible, randomized controlled studies, it will be difficult to justify the costs of payments from Medicare and insurers for many integrative supplements and practices. Most patients will have high out-of-pocket costs already limiting their access to standard care in addition to receiving integrative therapies, which has been identified in surveys of patients [60,61]. Developing a financially feasible integrative oncology program that expands patient access is challenging, but necessary. Insurance will typically cover integrative oncology visits with an MD or NP. Still, visits with other integrative care providers, such as naturopaths and acupuncturists, often incur high out-of-pocket costs. Additionally, racial disparities and lower socioeconomic patients are often under-represented in integrative oncology programs due to financial constraints, time away from work, transportation, and lack of childcare. Ideally, programs should offer a sliding scale fee structure that allows patients of limited financial resources to still be able to receive integrative therapies as medically indicated.

The financial model also needs to be constantly evaluated for sustainability, as philanthropy is unpredictable. Achieving defined budgetary support from the healthcare system is paramount as the integrative oncology program matures. The clinical services themselves also provide significant revenue in which to sustain the program, although this alone is rarely enough. Thus, successful programs often draw from all three sources of funding: institutional, philanthropy, and clinical revenue. Programs should also focus on the return on investment (ROI) for healthcare systems, which may come in the form of increased patient satisfaction, improved quality of life, decreased cost or healthcare utilization, and market differentiation. Thus, the benefit of integrative programs goes beyond the revenue it is able to generate, which is not always understood by senior leaders.

Clinical Services: Clinical delivery of programs also varies between centers. Three types of services are generally found: outpatient, inpatient, and group. Mostly commonly, these services include integrative oncology consultation, acupuncture, massage, mind–body medicine, music therapy, nutrition, and exercise counseling. Careful consideration should be made in the referral process and, if possible, adherence to routine procedures for referral to other services such as supportive/palliative care or physical therapy. Where to deliver care is also a major challenge, given many cancer centers have outgrown the available clinical space. Ideally, these services should be provided within the cancer center so they can be both accessible and visible to patients. Some programs have begun to deliver services simultaneously while patients receive chemotherapy in the infusion suites, which then obviates the need for separate space and appointment.

Many comprehensive cancer centers have associated integrative oncology programs with their cancer services. These programs are intricately linked in proximity to large academic centers. However, incorporating an integrative oncology program into multiple community network sites poses many challenges. The demand for integrative oncology is increasing, but having all the available integrative services for patients at each network site will be complex and problematic. Barriers to developing and expanding such a program to the community include financial limitations, lack of skilled and certified integrative practitioners, spatial constraints at each site, and adaptation of innovative technology.

Additionally, the development and expansion of integrative oncology programs to academic network sites depends on the availability of a skilled workforce. Currently, most oncologists do not have an integrative oncology background. Only a minority have received formal integrative medicine training and board certification and dedicated therapists who focus only on oncology patients are scarce. Recruiting enough experienced integrative providers that could provide care at these individual sites makes it challenging to meet the high demands of interested patients. Developing an integrative oncology program in the community also requires increasing square footage space, including dedicated exam rooms, quiet areas for services such as meditation, and procedure rooms for massage and acupuncture treatments. Most importantly, merging integrative therapies into a cancer patient's care plan will require the endorsement and adoption of the whole care team, including medical, radiation, and surgical oncologists, which requires a culture change for most places as these clinicians may not be aware of the value of integrative therapies. This also entails making sure services are available for referral by the treatment team.

Currently in community practices, the most readily available integrative oncology care model entails the patient seeing a conventional oncologist while also being managed in parallel by an integrative medicine provider who may or may not have had formal oncology training. Obviously, there are risks with this arrangement, especially if there is no collaboration or discussion. It is recommended in these settings to ensure there is a dialogue between the oncologist and the integrative medicine provider, including the sharing of medical records to reduce the chances of serious drug interactions and use of contraindicated interventions. Through this dialogue, there is an opportunity for the oncologist to gauge their level of comfort with the integrative medicine provider, and ensure the integrative medicine provider is using similar evidence-based standards as the oncologist. Not only will this promote safety and increase the knowledge base of the integrative medicine provider and oncologist, but a constructive relationship will eventually lead to referrals from the integrative medicine provider as well due to the mutual respect. Likewise, such interactions will help the oncologist identify practices that are marketed as integrative but are actually alternative. Sometimes patients may choose to seek out alternative therapies while also receiving conventional cancer treatment, and patients are encouraged to at least inform their oncology team so they are aware of the potential risks, even if they do not agree with the combined approach. The optimal situation is one in which an established working relationship exists between the oncology team and the integrative practitioners, and includes a shared medical record system for enhanced communication. Additionally, oncologists should consider developing a network of preferred providers in preparation for when patients inquire about receiving integrative therapies.

Telehealth has been rapidly adopted across several health systems in the United States and around the world during the COVID-19 pandemic, and as patients are more isolated from family and community they attempt to explore strategies and interventions to successfully manage their symptoms, incorporate a healthy lifestyle, and improve their overall health. Telehealth has had its own unique opportunities and challenges in oncology. Considering that cancer patients more frequently use ICAM than the general population, to meet patient's needs, integrative oncology programs have become more widely available in several cancer centers [58]. Despite more integrative oncology programs have become available in cancer centers around the nation, access to such programs is still limited to major medical centers. Even when a cancer program has an established integrative oncology program, access may be further limited due to time constraints and challenges with the coordination of care with multiple appointment requirements and geographic barriers. Adopting telemedicine in integrative oncology may therefore help reduce some of these challenges. According to one study at University of Texas MD Anderson, looking at this paradigm shift during the pandemic, telehealth integrative oncology consultations were provided to 509 new patients from 21 April 2020 to 21 October 2020 compared to 842 new patients in person during the same period in 2019 [62]. In this study, it was concluded that delivering integrative oncology consultations using telehealth is feasible and meets patient's needs. These patients reported lower symptom burden and more concerns about lifestyle, herbs, and supplements. Additional studies have also shown the ability of telehealth to effectively deliver diet and exercise interventions for patients with cancer [63–66]. We believe that use of the telehealth platform to provide integrative oncology consultations is beneficial if in-person consultation is not feasible to counsel patients on healthy lifestyle and address any questions regarding natural products or other integrative modalities.

Another option to consider is a group medical visit (GMV) model to increase access to specialized integrative oncology care. This model has been tested successfully in other community settings such as UCSF Integrative Oncology [67]. GMV aims to present integrative oncology information to groups of individuals, not to serve as a support group, except for informal patient interactions. Patients are divided into one of three appropriate series: patients on active treatment or maintenance treatment of their specific cancer (tumor-specific), patients finished with treatment and in remission (tumor agnostic), and any cancer patient with metastatic disease on treatment (tumor agnostic). Each series consists of three meetings: session 1 covers nutrition and cancer, session 2 covers cannabis and other supplements, and session 3 covers acupuncture and stress reduction therapies. All three meetings last 2 h, including didactic content presented by the physician, time for patient questions, and individual consultations in a nearby room. Groups include up to eight patients, and each is allowed to bring one guest.

The program for GMV integrative oncology visits has been demonstrated previously to be financially viable. It is more efficient for a provider to bill for more patients receiving shorter individual consultations as part of the GMV than to spend that time in private, 1 h consultations. The revenue from group visits significantly exceeds the revenue potential when compared to the time spent on individual visits. When similar models were implemented in other communities, patient and provider satisfaction surveys were high and rated with scores indicating the enjoyment and high value of sessions.

Education and Research: The lack of awareness of integrative therapies has been noted by both patients and clinicians and is a barrier to utilization [60]. Therefore, the success of any integrative oncology program relies on the ability to inform patients and clinicians about the potential benefits of these approaches. A strategy for clinicians is to provide experiential opportunities for them to receive clinical services such as acupuncture or meditation. The services could be provided as part of a routine meeting such as a department/division meeting or even an in-service meeting for nurses. Another important aspect of training is the development of specialists in integrative oncology, including medical oncology, radiation oncology, and surgical oncology as well as integrative therapists specializing in cancer–acupuncture, massage, and mediation. Given the growing demand by patients, and the lack of clinicians to staff this demand, it is essential that comprehensive cancer centers consider this part of the long-term mission. Along with the SIO, only four academic institutions provide formal integrative oncology training for healthcare providers: MD Anderson, Memorial Sloan-Kettering, University of Arizona, and the University of Michigan [12].

To successfully provide integrative oncology in the community, patients need to be aware of the services offered and have an easy way to access the program promptly. Education is vital to helping patients consider integrative therapies as part of their treatment. One strategy is to utilize current infrastructure to help educate patients. Many cancer centers are exploring the use of patient navigators familiar with all the services offered to assist the patient with reinforcing and detailing their goals and treatment plan. Having the patient navigator be familiar with available integrative services is a significant opportunity to engage patients, as the navigator facilitates timely appointments, obtains medical records, and completes pre-authorizations. Many patients already find and use outside resources to help them with this process. Third-party companies and concierge physician services are often hired to advocate for the patient by reviewing medical research on their behalf, helping them locate practitioners and clinical trials, and offering integrative solutions for their problems with discussions on supplements, medical cannabis, nutrition, and lifestyle changes.

Although there are a growing number of large, randomized clinical trials of integrative therapies, there continues to be a relatively weak foundation of positive trials in oncology to compel insurance companies of both the clinical and financial benefit to cover these services. Clinical programs should consider incorporating routine collection of patient outcomes to demonstrate the real-world value of integrative interventions. These should also be accompanied by some financial analysis as well, given many of these benefits may come in the form of decreased length of stay or decreased use of healthcare services such as emergency room visits [68]. If infrastructure and funding allows, integrative oncology programs should be providing patients the opportunity to participate in clinical trials to, in turn, help produce more critical data to understand how to best use these approaches to improve patient care.

5. Conclusions

Integrative oncology is growing in cancer care, both in patient interest as well as in the positive clinical impact it provides. Benefits of early integrative oncology support may include optimization of symptom management and improved quality of life, as well as a potentially enhanced ability to deliver chemotherapy due to the mitigation of side effects by integrative oncology approaches. The complexity of cancer treatment regimens and the potential of complementary and integrative care, to either enhance or interfere with treatment, underscores the need for more integrative oncology providers. It is important that all practitioners involved in the care of cancer patients have the knowledge and skills to recognize the benefits of integrative oncology. Cancer centers, both at comprehensive cancer centers as well as community network sites, should provide opportunities for patients to utilize these therapies as part of the treatment plan. Programs require support from senior leadership and integrative oncology programs, and need to be culturally sensitive to meet the unique needs of their patient population. A global commitment to research is critical to advancing evidence-based integrative oncology programs.

Author Contributions: Conceptualization G.S., B.B. and R.T.L.; investigation, G.S., B.B. and R.T.L.; writing—original draft preparation, G.S., B.B. and R.T.L.; writing—review and editing, G.S., B.B., E.A., J.Z., J.C., A.G., J.R. and R.T.L.; supervision, R.T.L. All authors have read and agreed to the published version of the manuscript.

Funding: This research received no external funding.

Institutional Review Board Statement: Not applicable.

Informed Consent Statement: Not applicable.

Data Availability Statement: Not applicable.

Conflicts of Interest: The authors declare no conflict of interest.

References

1. International Agency for Research on Cancer. Latest Global Cancer Data: Cancer Burden Rises to 19.3 Million New Cases and 10.0 Million Cancer Deaths in 2020. Available online: https://www.iarc.who.int/wp-content/uploads/2020/12/pr292_E.pdf (accessed on 1 March 2023).
2. Sung, H.; Ferlay, J.; Siegel, R.L.; Laversanne, M.; Soerjomataram, I.; Jemal, A.; Bray, F. Global Cancer Statistics 2020: GLOBOCAN Estimates of Incidence and Mortality Worldwide for 36 Cancers in 185 Countries. *CA Cancer J. Clin.* **2021**, *71*, 209–249. [CrossRef]
3. *Complementary, Alternative, or Integrative Health: What's in a Name*; National Center for Complementary and Integrative Health: Bethesda, MD, USA, 2021.
4. Witt, C.M.; Balneaves, L.G.; Cardoso, M.J.; Cohen, L.; Greenlee, H.; Johnstone, P.; Kucuk, O.; Mailman, J.; Mao, J.J. A Comprehensive Definition for Integrative Oncology. *J. Natl. Cancer Inst. Monogr.* **2017**, *2017*, lgx012. [CrossRef]
5. Nahin, R.L.; Barnes, P.M.; Stussman, B.J. *Expenditures on Complementary Health Approaches: United States, 2012*; National Health Status Reports no. 95; CDC: New York, NY, USA, 2016; pp. 1–11.
6. Clarke, T.C.; Barnes, P.M.; Black, L.I.; Stussman, B.J.; Nahin, R.L. *Use of Yoga, Meditation, and Chiropractors Among U.S. Adults Aged 18 and Over, National Health Interview Survey, 2012 and 2017*; National Center for Health Statistics: Hyattsville, MD, USA, 2018.
7. Mishra, S.; Stierman, B.; Gahche, J.J.; Potischman, N. *Dietary Supplement Use Among Adults: United States, 2017–2018*; National Center for Health Statistics: Hyattsville, MD, USA, 2021.
8. Barnes, P.M.; Powell-Griner, E.; McFann, K.; Nahin, R.L. Complementary and alternative medicine use among adults: United States, 2002. *Adv. Data* **2004**, *2*, 1–19. [CrossRef]
9. Lee, R.T.; Barbo, A.; Lopez, G.; Melhem-Bertrandt, A.; Lin, H.; Olopade, O.I.; Curlin, F.A. National survey of US oncologists' knowledge, attitudes, and practice patterns regarding herb and supplement use by patients with cancer. *J. Clin. Oncol.* **2014**, *32*, 4095–4101. [CrossRef] [PubMed]
10. Mao, J.J.; Palmer, C.S.; Healy, K.E.; Desai, K.; Amsterdam, J. Complementary and alternative medicine use among cancer survivors: A population-based study. *J. Cancer Surviv.* **2011**, *5*, 8–17. [CrossRef]
11. Sanford, N.N.; Sher, D.J.; Ahn, C.; Aizer, A.A.; Mahal, B.A. Prevalence and Nondisclosure of Complementary and Alternative Medicine Use in Patients With Cancer and Cancer Survivors in the United States. *JAMA Oncol.* **2019**, *5*, 735–737. [CrossRef]
12. Latte-Naor, S.; Mao, J.J. Putting Integrative Oncology Into Practice: Concepts and Approaches. *J. Oncol. Pract.* **2019**, *15*, 7–14. [CrossRef]
13. Dans, M.; Kutner, J.S.; Agarwal, R.; Baker, J.N.; Bauman, J.R.; Beck, A.C.; Campbell, T.C.; Carey, E.C.; Case, A.A.; Dalal, S.; et al. NCCN Guidelines(R) Insights: Palliative Care, Version 2.2021. *J. Natl. Compr. Canc Netw.* **2021**, *19*, 780–788. [CrossRef] [PubMed]
14. Lyman, G.H.; Greenlee, H.; Bohlke, K.; Bao, T.; DeMichele, A.M.; Deng, G.E.; Fouladbakhsh, J.M.; Gil, B.; Hershman, D.L.; Mansfield, S.; et al. Integrative Therapies During and After Breast Cancer Treatment: ASCO Endorsement of the SIO Clinical Practice Guideline. *J. Clin. Oncol.* **2018**, *36*, 2647–2655. [CrossRef]
15. Mao, J.J.; Ismaila, N.; Bao, T.; Barton, D.; Ben-Arye, E.; Garland, E.L.; Greenlee, H.; Leblanc, T.; Lee, R.T.; Lopez, A.M.; et al. Integrative Medicine for Pain Management in Oncology: Society for Integrative Oncology-ASCO Guideline. *J. Clin. Oncol.* **2022**, *40*, 3998–4024. [CrossRef] [PubMed]
16. Adler, N.E.; Page, A.E.K. (Eds.) *Cancer Care for the Whole Patient: Meeting Psychosocial Health Needs*; The National Academies Collection: Reports funded by National Institutes of Health; National Academies Press: Washington, DC, USA, 2008.
17. Niazi, A.K.; Niazi, S.K. Mindfulness-based stress reduction: A non-pharmacological approach for chronic illnesses. *N. Am. J. Med. Sci.* **2011**, *3*, 20–23. [CrossRef] [PubMed]
18. Gu, J.; Strauss, C.; Bond, R.; Cavanagh, K. How do mindfulness-based cognitive therapy and mindfulness-based stress reduction improve mental health and wellbeing? A systematic review and meta-analysis of mediation studies. *Clin. Psychol. Rev.* **2015**, *37*, 1–12. [CrossRef]
19. Hoffman, C.J.; Ersser, S.J.; Hopkinson, J.B.; Nicholls, P.G.; Harrington, J.E.; Thomas, P.W. Effectiveness of mindfulness-based stress reduction in mood, breast- and endocrine-related quality of life, and well-being in stage 0 to III breast cancer: A randomized, controlled trial. *J. Clin. Oncol.* **2012**, *30*, 1335–1342. [CrossRef] [PubMed]
20. Bower, J.E.; Crosswell, A.D.; Stanton, A.L.; Crespi, C.M.; Winston, D.; Arevalo, J.; Ma, J.; Cole, S.W.; Ganz, P.A. Mindfulness meditation for younger breast cancer survivors: A randomized controlled trial. *Cancer* **2015**, *121*, 1231–1240. [CrossRef] [PubMed]
21. Greenlee, H.; DuPont-Reyes, M.J.; Balneaves, L.G.; Carlson, L.E.; Cohen, M.R.; Deng, G.; Johnson, J.A.; Mumber, M.; Seely, D.; Zick, S.M.; et al. Clinical practice guidelines on the evidence-based use of integrative therapies during and after breast cancer treatment. *CA Cancer J. Clin.* **2017**, *67*, 194–232. [CrossRef] [PubMed]
22. Piet, J.; Wurtzen, H.; Zachariae, R. The effect of mindfulness-based therapy on symptoms of anxiety and depression in adult cancer patients and survivors: A systematic review and meta-analysis. *J. Consult. Clin. Psychol.* **2012**, *80*, 1007–1020. [CrossRef] [PubMed]
23. Lengacher, C.A.; Johnson-Mallard, V.; Post-White, J.; Moscoso, M.S.; Jacobsen, P.B.; Klein, T.W.; Widen, R.H.; Fitzgerald, S.G.; Shelton, M.M.; Barta, M.; et al. Randomized controlled trial of mindfulness-based stress reduction (MBSR) for survivors of breast cancer. *Psychooncology* **2009**, *18*, 1261–1272. [CrossRef]

24. Zainal, N.Z.; Booth, S.; Huppert, F.A. The efficacy of mindfulness-based stress reduction on mental health of breast cancer patients: A meta-analysis. *Psychooncology* **2013**, *22*, 1457–1465. [CrossRef]
25. Cramer, H.; Lauche, R.; Paul, A.; Dobos, G. Mindfulness-based stress reduction for breast cancer—A systematic review and meta-analysis. *Curr. Oncol.* **2012**, *19*, e343–e352. [CrossRef]
26. Gok Metin, Z.; Karadas, C.; Izgu, N.; Ozdemir, L.; Demirci, U. Effects of progressive muscle relaxation and mindfulness meditation on fatigue, coping styles, and quality of life in early breast cancer patients: An assessor blinded, three-arm, randomized controlled trial. *Eur. J. Oncol. Nurs.* **2019**, *42*, 116–125. [CrossRef]
27. Musial, F.; Bussing, A.; Heusser, P.; Choi, K.E.; Ostermann, T. Mindfulness-based stress reduction for integrative cancer care: A summary of evidence. *Forsch. Komplement.* **2011**, *18*, 192–202. [CrossRef]
28. Andersen, B.L.; Yang, H.C.; Farrar, W.B.; Golden-Kreutz, D.M.; Emery, C.F.; Thornton, L.M.; Young, D.C.; Carson, W.E., III. Psychologic intervention improves survival for breast cancer patients: A randomized clinical trial. *Cancer* **2008**, *113*, 3450–3458. [CrossRef] [PubMed]
29. Antoni, M.H.; Lutgendorf, S.K.; Cole, S.W.; Dhabhar, F.S.; Sephton, S.E.; McDonald, P.G.; Stefanek, M.; Sood, A.K. The influence of bio-behavioural factors on tumour biology: Pathways and mechanisms. *Nat. Rev. Cancer* **2006**, *6*, 240–248. [CrossRef]
30. Fu, W.W.; Popovic, M.; Agarwal, A.; Milakovic, M.; Fu, T.S.; McDonald, R.; Fu, G.; Lam, M.; Chow, R.; Cheon, S.; et al. The impact of psychosocial intervention on survival in cancer: A meta-analysis. *Ann. Palliat. Med.* **2016**, *5*, 93–106. [CrossRef] [PubMed]
31. Lyman, G.H.; Bohlke, K.; Cohen, L. Integrative Therapies During and After Breast Cancer Treatment: ASCO Endorsement of the SIO Clinical Practice Guideline Summary. *J. Oncol. Pract.* **2018**, *14*, 495–499. [CrossRef] [PubMed]
32. Lin, K.Y.; Hu, Y.T.; Chang, K.J.; Lin, H.F.; Tsauo, J.Y. Effects of yoga on psychological health, quality of life, and physical health of patients with cancer: A meta-analysis. *Evid. Based Complement. Alternat Med.* **2011**, *2011*, 659876. [CrossRef] [PubMed]
33. Danhauer, S.C.; Addington, E.L.; Sohl, S.J.; Chaoul, A.; Cohen, L. Review of yoga therapy during cancer treatment. *Support. Care Cancer* **2017**, *25*, 1357–1372. [CrossRef] [PubMed]
34. Cramer, H.; Lauche, R.; Klose, P.; Lange, S.; Langhorst, J.; Dobos, G.J. Yoga for improving health-related quality of life, mental health and cancer-related symptoms in women diagnosed with breast cancer. *Cochrane Database Syst. Rev.* **2017**, *1*, CD010802. [CrossRef]
35. Hershman, D.L.; Unger, J.M.; Greenlee, H.; Capodice, J.L.; Lew, D.L.; Darke, A.K.; Kengla, A.T.; Melnik, M.K.; Jorgensen, C.W.; Kreisle, W.H.; et al. Effect of Acupuncture vs Sham Acupuncture or Waitlist Control on Joint Pain Related to Aromatase Inhibitors among Women with Early-Stage Breast Cancer: A Randomized Clinical Trial. *JAMA* **2018**, *320*, 167–176. [CrossRef]
36. Shen, J.; Wenger, N.; Glaspy, J.; Hays, R.D.; Albert, P.S.; Choi, C.; Shekelle, P.G. Electroacupuncture for control of myeloablative chemotherapy-induced emesis: A randomized controlled trial. *JAMA* **2000**, *284*, 2755–2761. [CrossRef]
37. He, Y.; Guo, X.; May, B.H.; Zhang, A.L.; Liu, Y.; Lu, C.; Mao, J.J.; Xue, C.C.; Zhang, H. Clinical Evidence for Association of Acupuncture and Acupressure With Improved Cancer Pain: A Systematic Review and Meta-Analysis. *JAMA Oncol.* **2020**, *6*, 271–278. [CrossRef] [PubMed]
38. Dong, B.; Lin, L.; Chen, Q.; Qi, Y.; Wang, F.; Qian, K.; Tian, L. Wrist-ankle acupuncture has a positive effect on cancer pain: A meta-analysis. *BMC Complement. Med. Ther.* **2021**, *21*, 24. [CrossRef] [PubMed]
39. Tao, W.W.; Jiang, H.; Tao, X.M.; Jiang, P.; Sha, L.Y.; Sun, X.C. Effects of Acupuncture, Tuina, Tai Chi, Qigong, and Traditional Chinese Medicine Five-Element Music Therapy on Symptom Management and Quality of Life for Cancer Patients: A Meta-Analysis. *J. Pain. Symptom Manag.* **2016**, *51*, 728–747. [CrossRef]
40. Chiu, H.Y.; Hsieh, Y.J.; Tsai, P.S. Systematic review and meta-analysis of acupuncture to reduce cancer-related pain. *Eur. J. Cancer Care* **2017**, *26*, e12457. [CrossRef] [PubMed]
41. Choi, T.Y.; Lee, M.S.; Kim, T.H.; Zaslawski, C.; Ernst, E. Acupuncture for the treatment of cancer pain: A systematic review of randomised clinical trials. *Support. Care Cancer* **2012**, *20*, 1147–1158. [CrossRef] [PubMed]
42. Wu, X.; Chung, V.C.; Hui, E.P.; Ziea, E.T.; Ng, B.F.; Ho, R.S.; Tsoi, K.K.; Wong, S.Y.; Wu, J.C. Effectiveness of acupuncture and related therapies for palliative care of cancer: Overview of systematic reviews. *Sci. Rep.* **2015**, *5*, 16776. [CrossRef]
43. Paley, C.A.; Johnson, M.I.; Tashani, O.A.; Bagnall, A.M. Acupuncture for cancer pain in adults. *Cochrane Database Syst. Rev.* **2015**, *10*, CD007753. [CrossRef]
44. Kutner, J.S.; Smith, M.C.; Corbin, L.; Hemphill, L.; Benton, K.; Mellis, B.K.; Beaty, B.; Felton, S.; Yamashita, T.E.; Bryant, L.L.; et al. Massage therapy versus simple touch to improve pain and mood in patients with advanced cancer: A randomized trial. *Ann. Intern. Med.* **2008**, *149*, 369–379. [CrossRef]
45. Cassileth, B.R.; Vickers, A.J. Massage therapy for symptom control: Outcome study at a major cancer center. *J. Pain. Symptom Manag.* **2004**, *28*, 244–249. [CrossRef]
46. Lee, S.H.; Kim, J.Y.; Yeo, S.; Kim, S.H.; Lim, S. Meta-Analysis of Massage Therapy on Cancer Pain. *Integr. Cancer Ther.* **2015**, *14*, 297–304. [CrossRef]
47. Kinkead, B.; Schettler, P.J.; Larson, E.R.; Carroll, D.; Sharenko, M.; Nettles, J.; Edwards, S.A.; Miller, A.H.; Torres, M.A.; Dunlop, B.W.; et al. Massage therapy decreases cancer-related fatigue: Results from a randomized early phase trial. *Cancer* **2018**, *124*, 546–554. [CrossRef]
48. Pan, Y.Q.; Yang, K.H.; Wang, Y.L.; Zhang, L.P.; Liang, H.Q. Massage interventions and treatment-related side effects of breast cancer: A systematic review and meta-analysis. *Int. J. Clin. Oncol.* **2014**, *19*, 829–841. [CrossRef]

49. Pierce, J.P.; Natarajan, L.; Caan, B.J.; Parker, B.A.; Greenberg, E.R.; Flatt, S.W.; Rock, C.L.; Kealey, S.; Al-Delaimy, W.K.; Bardwell, W.A.; et al. Influence of a diet very high in vegetables, fruit, and fiber and low in fat on prognosis following treatment for breast cancer: The Women's Healthy Eating and Living (WHEL) randomized trial. *JAMA* **2007**, *298*, 289–298. [CrossRef]
50. Chlebowski, R.T.; Aragaki, A.K.; Anderson, G.L.; Pan, K.; Neuhouser, M.L.; Manson, J.E.; Thomson, C.A.; Mossavar-Rahmani, Y.; Lane, D.S.; Johnson, K.C.; et al. Dietary Modification and Breast Cancer Mortality: Long-Term Follow-Up of the Women's Health Initiative Randomized Trial. *J. Clin. Oncol.* **2020**, *38*, 1419–1428. [CrossRef]
51. Rock, C.L.; Thomson, C.A.; Sullivan, K.R.; Howe, C.L.; Kushi, L.H.; Caan, B.J.; Neuhouser, M.L.; Bandera, E.V.; Wang, Y.; Robien, K.; et al. American Cancer Society nutrition and physical activity guideline for cancer survivors. *CA Cancer J. Clin.* **2022**, *72*, 230–262. [CrossRef] [PubMed]
52. Campbell, K.L.; Winters-Stone, K.M.; Wiskemann, J.; May, A.M.; Schwartz, A.L.; Courneya, K.S.; Zucker, D.S.; Matthews, C.E.; Ligibel, J.A.; Gerber, L.H.; et al. Exercise Guidelines for Cancer Survivors: Consensus Statement from International Multidisciplinary Roundtable. *Med. Sci. Sports Exerc.* **2019**, *51*, 2375–2390. [CrossRef] [PubMed]
53. Ligibel, J.A.; Bohlke, K.; May, A.M.; Clinton, S.K.; Demark-Wahnefried, W.; Gilchrist, S.C.; Irwin, M.L.; Late, M.; Mansfield, S.; Marshall, T.F.; et al. Exercise, Diet, and Weight Management During Cancer Treatment: ASCO Guideline. *J. Clin. Oncol.* **2022**, *40*, 2491–2507. [CrossRef] [PubMed]
54. Van Blarigan, E.L.; Fuchs, C.S.; Niedzwiecki, D.; Zhang, S.; Saltz, L.B.; Mayer, R.J.; Mowat, R.B.; Whittom, R.; Hantel, A.; Benson, A.; et al. Association of Survival With Adherence to the American Cancer Society Nutrition and Physical Activity Guidelines for Cancer Survivors After Colon Cancer Diagnosis: The CALGB 89803/Alliance Trial. *JAMA Oncol.* **2018**, *4*, 783–790. [CrossRef]
55. Lee, R.T.; Kwon, N.; Wu, J.; To, C.; To, S.; Szmulewitz, R.; Tchekmedyian, R.; Holmes, H.M.; Olopade, O.I.; Stadler, W.M.; et al. Prevalence of potential interactions of medications, including herbs and supplements, before, during, and after chemotherapy in patients with breast and prostate cancer. *Cancer* **2021**, *127*, 1827–1835. [CrossRef]
56. Yeung, K.S.; Gubili, J.; Mao, J.J. Herb-Drug Interactions in Cancer Care. *Oncology* **2018**, *32*, 516–520. [PubMed]
57. Newmaster, S.G.; Grguric, M.; Shanmughanandhan, D.; Ramalingam, S.; Ragupathy, S. DNA barcoding detects contamination and substitution in North American herbal products. *BMC Med.* **2013**, *11*, 222. [CrossRef]
58. Brauer, J.A.; El Sehamy, A.; Metz, J.M.; Mao, J.J. Complementary and alternative medicine and supportive care at leading cancer centers: A systematic analysis of websites. *J. Altern. Complement. Med.* **2010**, *16*, 183–186. [CrossRef] [PubMed]
59. Yun, H.; Sun, L.; Mao, J.J. Growth of Integrative Medicine at Leading Cancer Centers between 2009 and 2016: A Systematic Analysis of NCI-Designated Comprehensive Cancer Center Websites. *J. Natl. Cancer Inst. Monogr.* **2017**, *2017*, lgx004. [CrossRef] [PubMed]
60. Jiang, C.; Larbi, O.; Feyes, D.; Wang, G.M.; Momotaz, H.; Li, M.; Daunov, K.; Daly, B.; Mazanec, S.; Rodgers-Melnick, S.; et al. A survey of cancer patients, caregivers, and providers regarding familiarity, importance, and utilization of supportive and integrative oncology services. *Support. Care Cancer* **2021**, *29*, 5777–5785. [CrossRef] [PubMed]
61. Larbi, O.M.; Jiang, C.; McLane, B.; Wang, G.M.; Daunov, K.; Hobson, S.M.; Daly, B.; Mazanec, S.R.; Feyes, D.; Rodgers-Melnick, S.; et al. Interest and Willingness to Pay for Integrative Therapies of Patients With Cancer and Caregivers. *JCO Oncol. Pract.* **2021**, *17*, e1622–e1630. [CrossRef] [PubMed]
62. Narayanan, S.; Lopez, G.; Powers-James, C.; Fellman, B.M.; Chunduru, A.; Li, Y.; Bruera, E.; Cohen, L. Integrative Oncology Consultations Delivered via Telehealth in 2020 and In-Person in 2019: Paradigm Shift During the COVID-19 World Pandemic. *Integr. Cancer Ther.* **2021**, *20*, 1534735421999101. [CrossRef]
63. Valle, C.G.; Diamond, M.A.; Heiling, H.M.; Deal, A.M.; Hales, D.P.; Nezami, B.T.; Pinto, B.M.; LaRose, J.G.; Rini, C.M.; Tate, D.F. Effect of an mHealth intervention on physical activity outcomes among young adult cancer survivors: The IMPACT randomized controlled trial. *Cancer* **2023**, *129*, 461–472. [CrossRef] [PubMed]
64. Reeves, M.M.; Terranova, C.O.; Winkler, E.A.H.; McCarthy, N.; Hickman, I.J.; Ware, R.S.; Lawler, S.P.; Eakin, E.G.; Demark-Wahnefried, W. Effect of a Remotely Delivered Weight Loss Intervention in Early-Stage Breast Cancer: Randomized Controlled Trial. *Nutrients* **2021**, *13*, 4091. [CrossRef]
65. Lee, K.; Shamunee, J.; Lindenfeld, L.; Ross, E.; Hageman, L.; Sedrak, M.S.; Wong, F.L.; Nakamura, R.; Forman, S.J.; Bhatia, S.; et al. Feasibility of implementing a supervised telehealth exercise intervention in frail survivors of hematopoietic cell transplantation: A pilot randomized trial. *BMC Cancer* **2023**, *23*, 390. [CrossRef]
66. Finley, D.J.; Stevens, C.J.; Emond, J.A.; Batsis, J.A.; Fay, K.A.; Darabos, C.; Sacks, O.A.; Cook, S.B.; Lyons, K.D. Potential effectiveness of a surgeon-delivered exercise prescription and an activity tracker on pre-operative exercise adherence and aerobic capacity of lung cancer patients. *Surg. Oncol.* **2021**, *37*, 101525. [CrossRef]
67. Thompson-Lastad, A.; Atreya, C.E.; Chao, M.T.; Pollak, C.; Dhruva, A.; Santana, T.; Abrams, D.I. Improving Access to Integrative Oncology Through Group Medical Visits: A Pilot Implementation Project. *J. Altern. Complement. Med.* **2019**, *25*, 733–739. [CrossRef] [PubMed]
68. Dusek, J.A.; Griffin, K.H.; Finch, M.D.; Rivard, R.L.; Watson, D. Cost Savings from Reducing Pain Through the Delivery of Integrative Medicine Program to Hospitalized Patients. *J. Altern. Complement. Med.* **2018**, *24*, 557–563. [CrossRef] [PubMed]

Journal of
Clinical Medicine

MDPI

Case Report

Complete Pathologic Response to PARP Inhibitor Olaparib in a Patient with Stage IVB Recurrent Endometrioid Endometrial Adenocarcinoma

Rosemary Noel Senguttuvan [1], Christina Wei [2], Mustafa Raoof [1], Thanh H. Dellinger [1] and Edward Wenge Wang [3,*]

1 Department of Surgery, City of Hope Comprehensive Cancer Center (COH), Duarte, CA 91010, USA; rsenguttuvan@coh.org (R.N.S.); mraoof@coh.org (M.R.); tdellinger@coh.org (T.H.D.)
2 Department of Pathology, City of Hope Comprehensive Cancer Center (COH), Duarte, CA 91010, USA; cwei@coh.org
3 Department of Medical Oncology, City of Hope Comprehensive Cancer Center (COH), Duarte, CA 91010, USA
* Correspondence: edwang@coh.org

Abstract: Treatment for endometrial cancer is rapidly evolving with the increased use and integration of somatic tumor RNA sequencing in clinical practice. There is a paucity of data regarding PARP inhibition in endometrial cancer given that mutations in homologous recombination genes are rare, and currently no FDA approval exists. A 50-year-old gravida 1 para 1 woman with a diagnosis of stage IVB poorly differentiated endometrioid endometrial adenocarcinoma presented to our comprehensive cancer center. Following surgical staging, she was placed on adjuvant chemotherapy with carboplatin/paclitaxel which was held multiple times due to poor performance status and complications. CT scan of the abdomen and pelvis following cycles 3 of adjuvant chemotherapy showed recurrent progressive disease. She received one cycle of liposomal doxorubicin but discontinued it due to severe cutaneous toxicity. Based on the BRIP1 mutation identified, the patient was placed on compassionate use of Olaparib in January 2020. Imaging during this surveillance period showed a significant decrease in hepatic, peritoneal, and extraperitoneal metastases, and eventually the patient had a clinical complete response in a year. The most recent CT A/P in December 2022 showed no sites of active recurrent or metastatic disease in the abdomen or pelvis. We present a unique case of a patient with recurrent stage IVB poorly differentiated endometrioid endometrial adenocarcinoma with multiple somatic gene mutations including BRIP1, who had a pathologic complete response following compassionate use of Olaparib for 3 years. To our knowledge, this is the first reported case of high grade endometrioid endometrial cancer that has shown a pathologic complete response to a PARP inhibitor.

Keywords: endometrial cancer; PARP inhibition; pathologic complete response; Olaparib; BRIP1 mutation; ATM mutation; RAD51C mutation; POLE mutation; TMB high

Citation: Senguttuvan, R.N.; Wei, C.; Raoof, M.; Dellinger, T.H.; Wang, E.W. Complete Pathologic Response to PARP Inhibitor Olaparib in a Patient with Stage IVB Recurrent Endometrioid Endometrial Adenocarcinoma. *J. Clin. Med.* **2023**, *12*, 3839. https://doi.org/10.3390/jcm12113839

Academic Editor: Jacek Szamatowicz

Received: 10 May 2023
Revised: 28 May 2023
Accepted: 2 June 2023
Published: 4 June 2023

1. Introduction

In 2022, an estimated 65,950 cases of uterine cancer were diagnosed in the United States with 12,550 women succumbing to the disease [1]. The incidence of uterine cancer has continued to rise in the United States over the last 10 years [1]. Certain factors known to negatively affect prognosis at the time of diagnosis include higher FIGO grade, extensive lymphovascular space invasion, invasion of the outer third of the myometrium, non-endometrioid histologic subtypes, and loss of p53 expression [2,3]. Treatment of early-stage disease involves upfront surgical staging with adjuvant chemotherapy and radiotherapy depending on final pathologic stage along with consideration of the presence of high-risk histological features. Chemotherapy and radiation following surgery for patients at advanced stages is recommended but carries significant short and long-term toxicities [3]. Thus, the investigation of alternative, less-toxic therapy is of interest.

Treatment for endometrial cancer is rapidly evolving with the increased uptake and integration of somatic tumor RNA sequencing and germline testing into clinical practice. The newly released 2023 National Comprehensive Cancer Network (NCCN) guidelines for uterine cancer include molecular tumor profiling into the four distinct molecular subtypes of endometrial cancer established by the Cancer Genome Atlas Research Network: (1) hypermutation in the exonuclease domain of DNA polymerase-ε (POLEmut); (2) mismatch repair deficiency, which confers microsatellite instability (MMRd); (3) mutations in TP53; and (4) tumors with none of the aforementioned classifications ('no specific molecular profile' or 'NSMP') [4,5]. Clinical trials are ongoing to determine the optimal treatment algorithm for each molecular category [6].

In ovarian cancer, the inhibition of poly (ADP-ribose) polymerase (PARP1) has resulted in superior clinical outcomes in patients that have germline or somatic mutations in BRCA1, BRCA2, and DNA homologous repair deficiency-related genes in frontline [7–9] and recurrent [10–12] settings. However, there is a paucity of data regarding use of PARP inhibition in endometrial cancer given that mutations in BRCA1, BRCA2 and other homologous recombination genes are rare [13]. There is currently no FDA approved use of PARP inhibitors in endometrial cancer. Various clinical trials are ongoing to elucidate the utility of PARP inhibitors in advanced or recurrent endometrial cancer, though data have yet to emerge [14,15].

We describe a case of a sustained complete pathologic response to Olaparib in a patient with stage IVB poorly differentiated endometrioid endometrial adenocarcinoma with multiple somatic mutations in the pathways of DNA repair, signal transduction, and metabolism.

2. Case Presentation

2.1. History of Presenting Illness

A 50-year-old gravida 1 para 1 woman with a diagnosis of stage IVB poorly differentiated endometrioid endometrial adenocarcinoma presented to the Gynecologic Oncology Department at City of Hope Comprehensive Cancer Center (COH) in mid-2019 for evaluation and treatment. Prior to presenting to COH, the patient had uterine leiomyomas diagnosed twenty years prior with subsequent menorrhagia, anemia, and multiple hospital visits. She had a negative endometrial biopsy on record from 2018. Following worsening bleeding in 2019, the patient underwent another endometrial biopsy, which identified a high-grade endometrial carcinoma. Pelvic ultrasound at this time showed the uterus with multilobulated appearance secondary to multiple leiomyomas (14 × 9.5 × 11 cm) with poor discernment of the endometrium. CT A/P without contrast prior to her presentation at COH showed a large multi-fibroid uterus (17.7 × 10.2 × 14.7 cm) and prominent retroperitoneal lymph nodes, progressively increased in size (18 mm × 15 mm), which is concerning for metastatic disease.

Her past medical history is significant for systemic lupus erythematosus, diagnosed at 16 years of age, and complicated with renal failure, requiring hemodialysis and renal transplant into the left lower pelvis at age 25, followed by long-term immunosuppression with prednisone. Her family medical history included a cousin with endometrial cancer of unknown age and was otherwise noncontributory. The patient had no history of smoking, alcohol, or drug use.

Physical exam at initial presentation was remarkable for a 20-week sized uterus. She was counseled regarding her diagnosis and elected to undergo primary surgery. She thus underwent an exploratory laparotomy, modified radical hysterectomy, bilateral salpingo-oophorectomy, right pelvic lymphadenectomy, and infra-gastric omentectomy in June 2019.

2.2. Surgical Findings

The patient had a globularly enlarged uterus measuring approximately 20 cm, with a tumor extending to the serosa, parametria, posterior lower uterine segment, and cervix, requiring a modified radical hysterectomy. The adnexa had no gross pathology. The posterior pelvic cul-de-sac was obliterated by the tumor. There were significant adhesions of the left pelvic kidney to the left lateral aspect of the uterus and parametrium, requiring extensive lysis of adhesions. An enlarged right common iliac lymph node was present. The omentum appeared grossly normal. The upper abdomen was within normal limits. At the end of the procedure, the patient was optimally cytoreduced.

2.3. Pathology

The hysterectomy specimen consisted of a 1251-g uterine corpus containing a 12.7 cm endometrial mass diffusely involving the myometrium, extending past the lower uterine segment into the anterior and posterior cervix. The tumor involved the left parametrium and cul de sac peritoneum, bilateral fallopian tubes and ovaries, and the omentum. Macrometastatic carcinoma (3.4 cm tumor deposit) was found involving the right common iliac lymph node. Microscopic examination of the endometrial tumor demonstrated a highly infiltrative, poorly differentiated carcinoma with extensive lymphovascular space invasion (Figure 1). Interestingly, tumor infiltrating lymphocytes were minimal in quantity and were not a prominent feature of the tumor. On balance, the immunomorphologic finding was consistent with a FIGO stage IV high-grade endometrial carcinoma, endometrioid type, FIGO grade 3. The patient was staged as stage IVB. An incidental serous tubal intraepithelial carcinoma was noted in one fallopian tube.

2.4. Immunohistochemistry (IHC)

The tumor immunoprofile demonstrated positivity for pan-cytokeratin, PAX8, and ER (60%, moderate intensity). TP53 IHC showed a wild-type staining pattern. Mismatch repair protein analysis demonstrated loss of nuclear expression for MSH6 and intact nuclear expression for MLH1, MSH2, and PMS2.

2.5. Somatic Tumor Testing

GEM ExTRA® somatic tumor testing [16] was performed on the surgical specimen identifying alterations shown in Table 1, notably mutations in BRIP1 (c.2098-2A>G), ATM (R23*), RAD51C (R370*), POLE (V411L), MTOR (A469T), PTEN (R233*), TP53 (N131I), MSH2 (E580*), and MSH6 (E641*). The tumor had a high tumor mutation burden (344 Mut/Mb), but microsatellite instability (MSI) was stable. Germline testing was performed using Invitae Germline Precision Medicine American College of Medical Genetics (ACMG)® Panels 22081I-VT0013 and 22081I-VT0014 and was uninformative [17].

When interpreted in conjunction with the molecular profile (POLE mutation p.V411L, high TMB, microsatellite stable), the molecular classification best fits the POLE-ultramutated molecular subtype in the context of a multiple-classifier (MSH6 deficiency by IHC, POLE mutation by molecular sequencing, and microsatellite stable status). The presence of TP53 mutation may be interpreted as a passenger mutation in this context.

2.6. Adjuvant Therapy and First Recurrence

Adjuvant chemotherapy with weekly carboplatin and paclitaxel was initiated but paused several times due to poor performance status and complications. CT scan of the abdomen and pelvis following three cycles of adjuvant chemotherapy showed recurrent progressive disease with new hepatic metastases, abdominopelvic peritoneal and mesenteric carcinomatosis, left supraclavicular, retroperitoneal, and pelvic nodal metastasis, and metastatic deposits within the abdominal wall. The patient was then switched to liposomal doxorubicin and received one cycle, but this was discontinued due to severe cutaneous toxicity necessitating hospitalization.

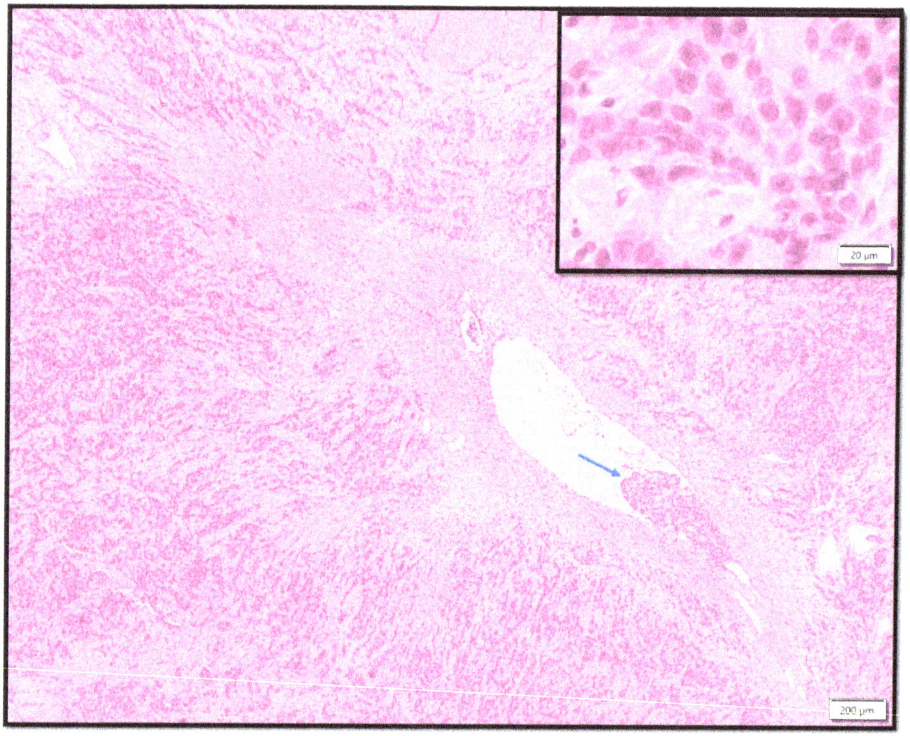

Figure 1. The histologic section of the uterine tumor showed poorly differentiated neoplastic cells with destructive growth pattern, diffusely infiltrating throughout the myometrium. The lower magnification image (4×) showed the presence of lymphovascular space invasion (arrow). The inset shows a high magnification view (40×) of the tumor cells, with high grade cytologic features including nucleomegaly, irregular nuclear contour, stippled to clumped chromatin, and variably prominent nucleoli. Scattered mitotic activity and numerous apoptotic cells can be seen, signifying high proliferative activity.

2.7. *Compassionate Treatment with Olaparib*

Based on the BRIP1 mutation identified, the patient was placed on compassionate use of Olaparib in January 2020, with the dose adjusted to 200 mg, twice daily. Imaging during this surveillance period showed a significant decrease in hepatic, peritoneal, and extraperitoneal metastases, and eventually the patient had a clinically complete response in a year.

The patient continued Olaparib until November 2022, when she experienced abdominal pain secondary to a ventral midline abdominal wall hernia with a resulting partial small bowel obstruction. After failed conservative treatment, the patient underwent a diagnostic laparoscopy with conversion to laparotomy, incisional/ventral hernia repair plus mesh implantation for recurrent obstruction and bowel incarceration within the hernia. Ileocecectomy/colectomy was also performed at this time due to plaque-like deposits noted on the serosal surface of the small bowel and mesentery, which are concerning for recurrent disease. Nodules in the omentum and transverse colon mesentery were resected. Pathology was negative for malignancy, and the transition point for the small bowel obstruction was thus determined to be due to scarring from the tumor treatment on the mesentery of the terminal ileum and right colon.

Table 1. Tumor genomic alterations identified by GEM ExTRA® somatic tumor testing.

Gene Name	Gene Symbol	Mutation
AT-rich interactive domain-containing protein 1A	ARID1A	E1767* H1380Y
ATM serine/threonine kinase	ATM	R23* c.2921 + 1G>A
BRCA1 interacting protein C-terminal helicase 1	BRIP1	c.2098-2A>G
Cyclin-dependent kinase inhibitor 2A	CDKN2A	D74N
F-Box and WD repeat domain containing 7	FBXW7	R479Q
MutL homolog 3	MLH3	K703fs
MutS homolog 2	MSH2	E580*
MutS homolog 6	MSH6	E641*)
Mammalian target of rapamycin	MTOR	A469T
Neurofibromin 1	NF1	W221*
Phosphatidylinositol-4,5-bisphosphate 3-kinase catalytic subunit alpha	PIK3CA	E545D R348*
DNA polymerase epsilon catalytic subunit	POLE	V411L
Phosphatase and tensin homolog	PTEN	R233* T277A
RAD51 paralog C	RAD51C	R370*
Tumor protein P53	TP53	N131I
Tuberous sclerosis complex 2	TSC2	G1157*
Tumor mutational burden	TMB	High 344 mut/mB
Microsatellite instability	MSI	Stable

The most recent CT A/P in December 2022 showed no sites of active recurrent or metastatic disease in the abdomen or pelvis. Figure 2A shows regression of peritoneal carcinomatosis by CT A/P at the time of progression in October 2019 in comparison to the most recent imaging in December 2022. Figure 2B is representative of the regression of one of the patient's hepatic lesions. Though not shown, all other abdominal and hepatic metastases demonstrated complete regression. Olaparib was discontinued in January 2023, after 3 years. The patient is currently considered to have a complete response without radiographic and pathologic evidence of disease.

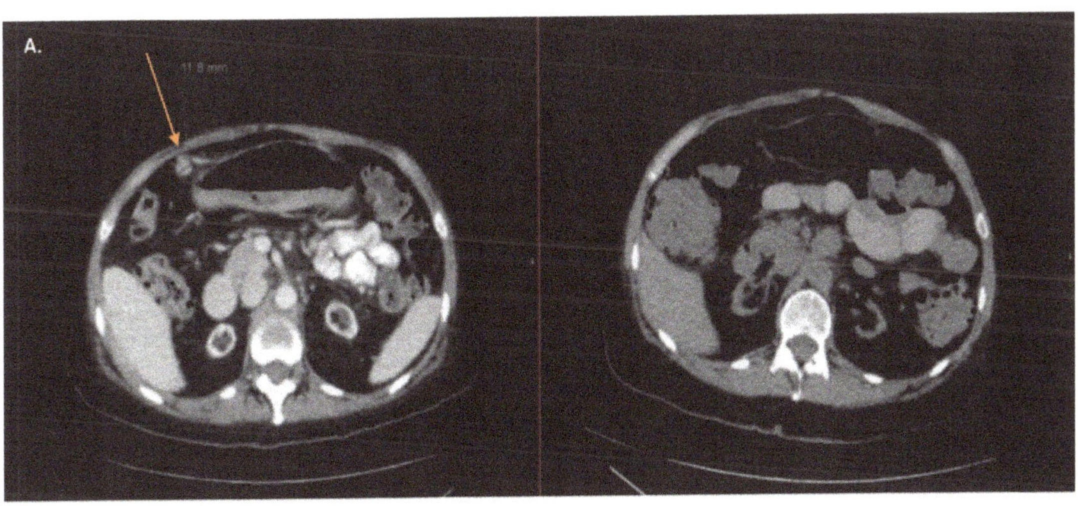

Figure 2. *Cont.*

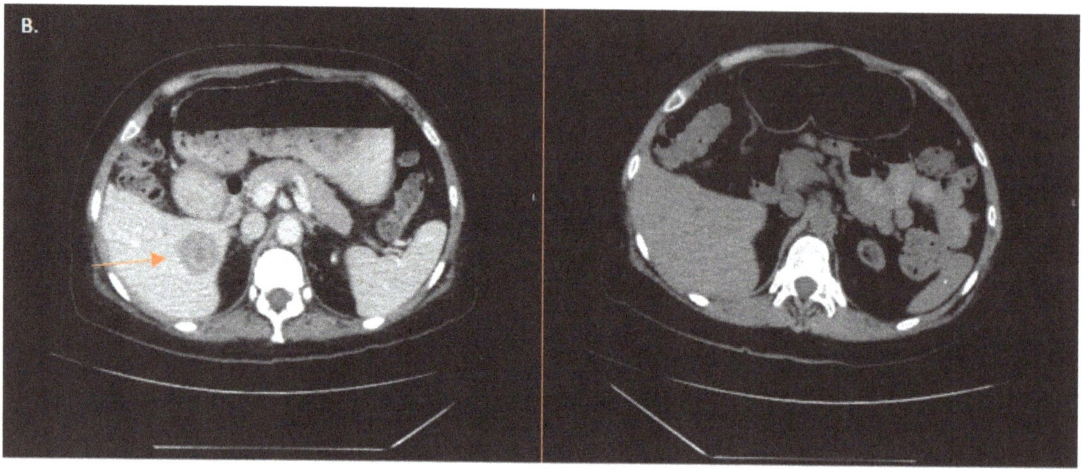

Figure 2. (**A**,**B**): CT comparison demonstrating regression of abdoinopelvic metastases represented by arrow. Axial views of metastases in October 2019 compared with corresponding axial views demonstrating complete radiographic regression of abdominopelvic metastases in December 2022.

3. Discussion

We present a unique case of a patient with recurrent stage IVB poorly differentiated endometrioid endometrial adenocarcinoma with multiple somatic gene mutations including BRIP1, who had a pathologic complete response following compassionate use of Olaparib for 3 years. To our knowledge, this is the first reported case of high grade endometrioid endometrial cancer that has shown a pathologic complete response to a PARP inhibitor. One case of high grade serous endometrial carcinoma with clinical radiographic response to Olaparib is noted in the literature [18], however, no pathologic confirmation was reported.

PARP inhibition increases both progression-free survival and overall survival in patients with BRCA-deficient and homologous recombination deficient ovarian cancer [7–12]. Heeke and colleagues reported that 34.4% of endometrial cancers possess molecular aberrations of genes involved in the homologous recombination pathway [13]. BRIP1 (BRCA1 interacting helicase 1) is actively involved in the homologous recombination pathway by interacting with the BRCT repeats of BRCA1 [19]. A mutation of BRIP1 is associated with Fanconi anemia and breast cancer [20,21]; however, only 0.14% of endometrial cancers have BRIP1 mutations. This makes the study of BRCA-mutated and homologous recombination deficient mutated endometrial cancer patients difficult given the low incidence; however, it is worth investigation if clinical outcomes such as those reported in the ovarian cancer literature can be achieved with PARP inhibitor use in patients with endometrial cancer who harbor these germline and somatic mutations. There are ongoing clinical trials seeking to answer this question [14]. Other gene mutations in the DNA repair pathway including POLE, ATM, and RAD51C may also contribute to the significant response to PARP inhibition [22].

The classification of endometrial cancer is moving away from traditional type I and type II classification system and towards molecular categorization [4]. As the prognostic utility of molecular subtyping has been elucidated, the investigation of the optimal treatment for each of the four molecular subtypes originally determined by the Cancer Genome Atlas Research Network is ongoing in the overarching Refining Adjuvant Treatment In Endometrial Cancer (RAINBO) umbrella program [6]. This molecular classification works well for patients who neatly fit into each category; however, in cases such as ours where multiple genomic alterations exist that fit into more than one of the four molecular subcategories, the approach to treatment becomes challenging.

Our patient's genomic analysis indicated her tumor was MMRd, POLE ultramutated, and possessed a passenger TP53 mutation with a resulting high tumor mutation burden. Given these multiple mutations, it is unclear which prognostic molecular subcategory she would fall into. This presents challenges to the clinician given the immense genomic heterogeneity, and an individualized approach to the specific molecular profile of the tumor should be undertaken to optimize clinical outcomes. Patients with endometrial cancer who have genomic alternations in homologous recombination related downstream genes should have a shared decision-making discussion with their provider regarding the potential benefits of PARP inhibitor therapy. Such decisions should be undertaken in collaboration with a multidisciplinary team consisting of the patient's medical oncologist, surgeon, nursing team, and pharmacist to provide optimal team-based care.

Author Contributions: Conceptualization, E.W.W. and R.N.S.; Methodology, E.W.W. and R.N.S.; Validiation, C.W., M.R., T.H.D. and E.W.W.; Writing—original draft preparation, R.N.S.; Writing—review and editing, C.W., M.R., T.H.D. and E.W.W.; Visualization, E.W.W.; Supervision, E.W.W.; Project administration, E.W.W. All authors have read and agreed to the published version of the manuscript.

Funding: This research received no external funding.

Institutional Review Board Statement: Not applicable.

Informed Consent Statement: Informed consent was obtained from the patient described in the study.

Data Availability Statement: Not applicable.

Acknowledgments: The authors thank Nicola Welch, CMPP (Whipbird Communications) for providing medical writing and editorial support.

Conflicts of Interest: The authors declare no conflict of interest.

References

1. Cancer Stat Facts: Uterine Cancer. 2022. Available online: https://seer.cancer.gov/statfacts/html/corp.html (accessed on 17 April 2023).
2. Vargas, R.; Rauh-Hain, J.A.; Clemmer, J.; Clark, R.M.; Goodman, A.; Growdon, W.B.; Schorge, J.O.; del Carmen, M.G.; Horowitz, N.S.; Boruta, D.M. Tumor size, depth of invasion, and histologic grade as prognostic factors of lymph node involvement in endometrial cancer: A SEER analysis. *Gynecol. Oncol.* **2014**, *133*, 216–220. [CrossRef] [PubMed]
3. de Boer, S.M.; Powell, M.E.; Mileshkin, L.; Katsaros, D.; Bessette, P.; Haie-Meder, C.; Tubiana-Mathieu, N. Adjuvant chemoradiotherapy versus radiotherapy alone in women with high-risk endometrial cancer (PORTEC-3): Patterns of recurrence and post-hoc survival analysis of a randomised phase 3 trial. *Lancet Oncol.* **2019**, *20*, 1273–1285. [CrossRef] [PubMed]
4. Levine, D.A.; The Cancer Genome Atlas Research Network. Integrated genomic characterization of endometrial carcinoma. *Nature* **2013**, *497*, 67–73. [CrossRef] [PubMed]
5. Uterine Neoplasms. Available online: https://www.nccn.org/professionals/physician_gls/pdf/uterine.pdf (accessed on 17 April 2023).
6. Consortium, R.R. Refining adjuvant treatment in endometrial cancer based on molecular features: The RAINBO clinical trial program. *Int. J. Gynecol. Cancer* **2023**, *33*, 109–117. [CrossRef] [PubMed]
7. Moore, K.; Colombo, N.; Scambia, G.; Kim, B.G.; Oaknin, A.; Friedlander, M.; DiSilvestro, P. Maintenance Olaparib in Patients with Newly Diagnosed Advanced Ovarian Cancer. *N. Engl. J. Med.* **2018**, *379*, 2495–2505. [CrossRef] [PubMed]
8. González-Martín, A.; Pothuri, B.; Vergote, I.; DePont Christensen, R.; Graybill, W.; Mirza, M.R.; Monk, B.J. Niraparib in Patients with Newly Diagnosed Advanced Ovarian Cancer. *N. Engl. J. Med.* **2019**, *381*, 2391–2402. [CrossRef] [PubMed]
9. Monk, B.J.; Parkinson, C.; Lim, M.C.; O'Malley, D.M.; Oaknin, A.; Wilson, M.K.; Kristeleit, R.S. A Randomized, Phase III Trial to Evaluate Rucaparib Monotherapy as Maintenance Treatment in Patients with Newly Diagnosed Ovarian Cancer (ATHENA–MONO/GOG-3020/ENGOT-ov45). *J. Clin. Oncol.* **2022**, *40*, 3952–3964. [CrossRef] [PubMed]
10. Poveda, A.; Floquet, A.; Ledermann, J.A.; Asher, R.; Penson, R.T.; Oza, A.M.; Hegg, R. Olaparib tablets as maintenance therapy in patients with platinum-sensitive relapsed ovarian cancer and a BRCA1/2 mutation (SOLO2/ENGOT-Ov21): A final analysis of a double-blind, randomised, placebo-controlled, phase 3 trial. *Lancet Oncol.* **2021**, *22*, 620–631. [CrossRef] [PubMed]
11. Mirza, M.R.; Monk, B.J.; Herrstedt, J.; Oza, A.M.; Mahner, S.; Redondo, A.; Matulonis, U.A. Niraparib Maintenance Therapy in Platinum-Sensitive, Recurrent Ovarian Cancer. *N. Engl. J. Med.* **2016**, *375*, 2154–2164. [CrossRef] [PubMed]
12. Kristeleit, R.; Lisyanskaya, A.; Fedenko, A.; Dvorkin, M.; de Melo, A.C.; Shparyk, Y.; Oza, A.M. Rucaparib versus standard-of-care chemotherapy in patients with relapsed ovarian cancer and a deleterious BRCA1 or BRCA2 mutation (ARIEL4): An international, open-label, randomised, phase 3 trial. *Lancet Oncol.* **2022**, *23*, 465–478. [CrossRef] [PubMed]

13. Heeke, A.L.; Pishvaian, M.J.; Lynce, F.; Xiu, J.; Brody, J.R.; Chen, W.-J.; Baker, T.M.; Marshall, J.L.; Isaacs, C. Prevalence of Homologous Recombination-Related Gene Mutations Across Multiple Cancer Types. *JCO Precis Oncol.* **2018**, *2018*, 1–13. [CrossRef] [PubMed]
14. Joly, F.; Brachet, P.E.; Lacourtoisie, S.A.; Asselain, B.; Floquet, A.; Collard, O.; Alexandre, J. Multicentre randomized phase II trial of olaparib as maintenance therapy in platinum-sensitive advanced endometrial carcinoma: The GINECO-UTOLA study. *J. Clin. Oncol.* **2020**, *38* (Suppl. 15), TPS6109. [CrossRef]
15. Musacchio, L.; Caruso, G.; Pisano, C.; Cecere, S.C.; Di Napoli, M.; Attademo, L.; Tambaro, R.; Russo, D.; Califano, D.; Palaia, I.; et al. PARP Inhibitors in Endometrial Cancer: Current Status and Perspectives. *Cancer Manag. Res.* **2020**, *12*, 6123–6135. [CrossRef] [PubMed]
16. White, T.; Szelinger, S.; LoBello, J.; King, A.; Aldrich, J.; Garinger, N.; Halbert, M.; Richholt, R.F.; Mastrian, S.D.; Babb, C.; et al. Analytic validation and clinical utilization of the comprehensive genomic profiling test, GEM ExTra(®). *Oncotarget* **2021**, *12*, 726–739. [CrossRef] [PubMed]
17. Rehm, H.L.; Bale, S.J.; Bayrak-Toydemir, P.; Berg, J.S.; Brown, K.K.; Deignan, J.L.; Friez, M.J.; Funke, B.H.; Hegde, M.R.; Lyon, E. ACMG clinical laboratory standards for next-generation sequencing. *Genet. Med.* **2013**, *15*, 733–747. [CrossRef] [PubMed]
18. Nakamura, K.; Aimono, E.; Tanishima, S.; Imai, M.; Nagatsuma, A.K.; Hayashi, H.; Yoshimura, Y.; Nakayama, K.; Kyo, S.; Nishihara, H. Olaparib Monotherapy for BRIP1-Mutated High-Grade Serous Endometrial Cancer. *JCO Precis. Oncol.* **2020**, *4*, 283–290. [CrossRef] [PubMed]
19. Yu, X.; Chini, C.C.S.; He, M.; Mer, G.; Chen, J. The BRCT domain is a phospho-protein binding domain. *Science* **2003**, *302*, 639–642. [CrossRef] [PubMed]
20. Ouhtit, A.; Gupta, I.; Shaikh, Z. BRIP1, a potential candidate gene in development of non-BRCA1/2 breast cancer. *Front. Biosci. Elite Ed.* **2016**, *8*, 289–298. [CrossRef] [PubMed]
21. Magrin, L.; Fanale, D.; Brando, C.; Fiorino, A.; Corsini, L.R.; Sciacchitano, R.; Filorizzo, C.; Dimino, A.; Russo, A.; Bazan, V. POLE, POLD1, and NTHL1: The last but not the least hereditary cancer-predisposing genes. *Oncogene* **2021**, *40*, 5893–5901. [CrossRef] [PubMed]
22. Abe, A.; Imoto, I.; Ueki, A.; Nomura, H.; Kanao, H. Moderate-Risk Genes for Hereditary Ovarian Cancers Involved in the Homologous Recombination Repair Pathway. *Int. J. Mol. Sci.* **2022**, *23*, 11790. [CrossRef] [PubMed]

Journal of
Clinical Medicine

MDPI

Brief Report

The Midwest Sarcoma Trials Partnership: Bridging Academic and Community Networks in a Collaborative Approach to Sarcoma

Natalie K. Heater [1], Scott Okuno [2], Steven Robinson [2], Steven Attia [3], Mahesh Seetharam [4], Brittany L. Siontis [2], Janet Yoon [5], Sant Chawla [6], Mohammed M. Milhem [7], Varun Monga [7], Keith Skubitz [8], John Charlson [9], Angela C. Hirbe [10], Mia C. Weiss [10], Brian Van Tine [10] and Mark Agulnik [5,*]

1 Department of Medicine, Northwestern University, Chicago, IL 60611, USA
2 Department of Medical Oncology, Mayo Clinic, Rochester, MN 55905, USA
3 Department of Hematology and Oncology, Mayo Clinic, Jacksonville, FL 32224, USA
4 Department of Hematology and Oncology, Mayo Clinic, Phoenix, AZ 85054, USA
5 City of Hope Medical Center, Duarte, CA 91010, USA
6 Sarcoma Oncology Center, Santa Monica, CA 90403, USA
7 Department of Internal Medicine, Division of Hematology, Oncology and Blood and Marrow Transplantation, University of Iowa, Iowa City, IA 52242, USA
8 Masonic Cancer Center, University of Minnesota, Minneapolis, MN 55455, USA
9 Department of Hematology/Oncology, Medical College of Wisconsin, Milwaukee, WI 53226, USA
10 Division of Oncology, Department of Medicine, Washington University School of Medicine, St. Louis, MO 63130, USA
* Correspondence: magulnik@coh.org

Abstract: The treatment of sarcoma necessitates a collaborative approach, given its rarity and complex management. At a single institution, multidisciplinary teams of specialists determine and execute treatment plans involving surgical, radiation, and medical management. Treatment guidelines for systemic therapies in advanced or nonresectable soft tissue sarcoma have advanced in recent years as new immunotherapies and targeted therapies become available. Collaboration between institutions is necessary to facilitate accrual to clinical trials. Here, we describe the success of the Midwest Sarcoma Trials Partnership (MWSTP) in creating a network encompassing large academic centers and local community sites. We propose a new model utilizing online platforms to expand the reach of clinical expertise for the treatment of advanced soft tissue sarcoma.

Keywords: advanced soft tissue sarcoma; sarcoma treatment; Midwest Sarcoma Trials Partnership; targeted therapy; multidisciplinary tumor board; collaboration

Citation: Heater, N.K.; Okuno, S.; Robinson, S.; Attia, S.; Seetharam, M.; Siontis, B.L.; Yoon, J.; Chawla, S.; Milhem, M.M.; Monga, V.; et al. The Midwest Sarcoma Trials Partnership: Bridging Academic and Community Networks in a Collaborative Approach to Sarcoma. *J. Clin. Med.* **2023**, *12*, 2561. https://doi.org/10.3390/jcm12072561

Academic Editor: Hong Shen

Received: 9 December 2022
Revised: 27 February 2023
Accepted: 23 March 2023
Published: 29 March 2023

1. Sarcoma and the Complexities of Treatment

Sarcomas are a rare, heterogenous group of malignant tumors that arise from tissues of mesenchymal origin, comprising approximately 1% of all diagnosed malignancies worldwide, with an incidence of approximately 13,000 cases per year in the United States [1]. Sarcoma affects patients across the lifespan and demographic spectrum. Over 100 histologic subtypes have been identified, with the majority originating from soft tissue (80%) and the remainder originating from bone [2]. Sarcoma primarily spreads hematogenously [3]. Over 20 different genetic syndromes have been shown to harbor a predisposition toward developing sarcoma [4,5]. Prognosis depends on histologic subtype, depth of invasion, grade, and tumor size. Mortality in sarcoma is regrettably high, with 5-year overall survivorship ranging from 43% to 73%, although reports from national cancer databases may not be fully accurate in their survival rates for rare tumors such as sarcoma [6,7].

The standard of care in managing sarcomas requires a collaborative approach by a multidisciplinary tumor board (MTB), comprising of radiologists, pathologists, geneticists, surgical oncologists, radiation oncologists, orthopedic oncologists, and medical oncologists, to determine the optimal management. Treatment is often multimodal. For patients with localized or oligometastatic disease, the standard treatment is complete surgical resection, with some patients benefiting from radiation therapy and/or systemic treatment (chemotherapy, immunotherapy, and targeted therapy). Systemic treatment is typically palliative for nonresectable or widely metastatic sarcoma, where the overall survival is a dismal 12–14 months [8].

Given the poor prognosis of sarcoma without definitive surgical management, clinical trials are necessary to further delineate therapies for nonresectable or widely metastatic sarcoma. Current guidelines from the National Comprehensive Cancer Network (NCCN) recommend anthracycline or gemcitabine-based chemotherapies as first-line therapy for nonresectable disease if a clinical trial is not available [9]. The average progression-free survival (PFS) after these first-line agents is 4–6 months [8]. Patients receive a median of three different systemic treatments, with variable benefits of treatment after third-line therapies [10]. Over the past decade, immunotherapies and targeted therapies have enhanced treatment options for advanced sarcoma [11–13].

2. Collaboration between Academic and Community Programs in Sarcoma

The National Cancer Institute (NCI) is working to bridge the gap in access to research and clinical trials between academic and community cancer centers. Across the United States, 64 hospitals have been named Designated Cancer Centers (DCCs) and receive funding from the NCI to conduct studies to enhance patient care. The vast majority of these DCCs are affiliated with university medical centers. Many DCCs collaborate with local community sites or establish satellite clinics to form larger networks, which have succeeded in increasing community access to clinical trials. For example, the City of Hope encompasses 27 sites across 5 counties in Southern California. Community sites have been shown to contribute up to one-third of total clinical trial accruals across DCC-associated networks [14]. The availability of clinical trials for patients who initially present to a community site associated with the City of Hope has led to increased access to clinical trials for patients of diverse backgrounds [15].

For patients with sarcoma, larger DCC-associated networks can improve access to MTBs for patients who present to local partners. This carries many benefits for patient care, especially the ability to expedite the referral process for treatment planning while continuing patient care at local sites.

However, even a large DCC-associated network generally will not be large enough to support a clinical trial in rare cancers, such as sarcoma, without patient accrual from outside the network. Even then, given the heterogeneity across subtypes of sarcoma, studies investigating individual subtypes are limited by patient accrual, whether at local community sites or at large academic centers. Thus, sarcoma experts now recognize the need for a high level of communication and collaboration across multiple DCC-associated networks in order to increase patient accrual and, thereby, improve the quality of evidence available for individual subtypes.

The concept of clinical trial alliances in oncology is not novel. Various groups, such as Alliance, ECOG-ACRIN and SWOG, have all formed collaborations for cancer research, although without a specific focus on sarcoma. The Sarcoma Alliance for Research through Collaboration (SARC) is the largest sarcoma-specific clinical trial collective. The SARC was founded in 2003 by five sarcoma experts; today, it encompasses 85 cancer centers in the United States, along with 6 international institutions. The SARC has completed 15 clinical trials and has 8 open trials as of 2022, and it curates a sarcoma-specific database hosting the prospective data from these trials [7]. In addition, the SARC has partnered with the NCI to provide funding opportunities for researchers and to bring sarcoma experts together through semiannual meetings.

3. The City of Hope and the Midwest Sarcoma Trials Partnership

The City of Hope is moving sarcoma research ahead through collaborative initiatives, such as the Midwest Sarcoma Trials Partnership (MWSTP). The MWSTP was established in 2012 with the goal of improving the care of patients with sarcoma by increasing patient accrual to clinical trials. The majority of MWSTP anchor sites also belong to the SARC, highlighting the intertwined nature of collaboration in sarcoma. The seven original member institutions are the Mayo Clinic (including locations in Arizona and Florida), the University of Minnesota, the University of Wisconsin, the Medical College of Wisconsin, the University of Iowa, the Washington University in St. Louis, and the Northwestern University. As a result of leadership changes, the City of Hope Comprehensive Cancer Center joined the partnership in 2020. MWSTP members meet monthly to discuss open clinical trials and encourage collaboration across member sites. Combining expertise across eight states and thousands of patients, the MWSTP enables the development of investigator-initiated trials with access to patients across multiple health care systems. Additionally, it provides a forum for physicians from the MWSTP network to collaborate on retrospective review of data for the purpose of publishing treatment experiences across these centers (Figure 1).

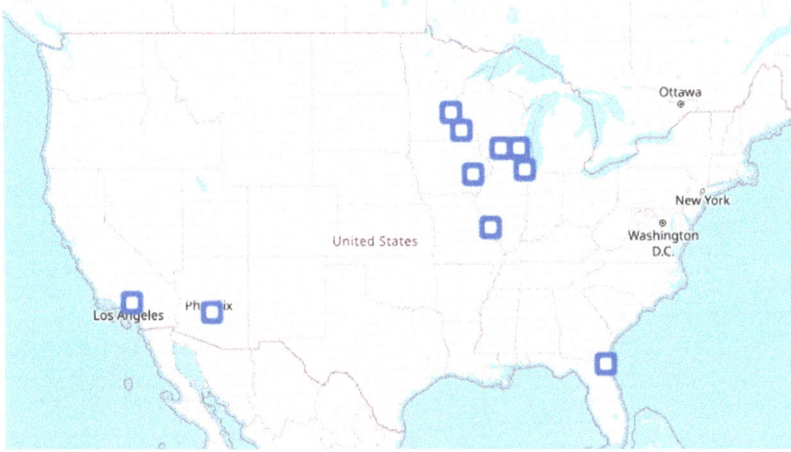

Figure 1. MWSTP Anchor Site Locations.

4. MWSTP's Impact on Soft Tissue Sarcoma Research

Since 2012, the Midwest Sarcoma Trials Partnership has completed four clinical trials and is currently accruing patients to three active clinical trials investigating chemotherapy and targeted therapies for the treatment of sarcoma, specifically soft tissue sarcoma. These phase II studies have recruited across the eight anchor sites and have allowed the expansion of clinical trial availability. The results of the MWSTP's research have significantly influenced current NCCN guidelines for sarcoma treatment.

The MWSTP's trial with regorafenib has led to new treatment options for patients with metastatic angiosarcoma, a rare and aggressive variant of sarcoma arising from blood and lymphatic vessels [16,17]. There is a paucity of angiosarcoma-specific data, with the first dedicated phase II trial occurring in 2008 [18]. In order to increase access to this trial, the MWSTP collaborated with two non-MWSTP sites, Sarcoma Oncology Group (Santa Monica, CA, USA) and Mercy Health (Janesville, WI, USA).

Regorafenib is a small-molecule inhibitor with activity against VEGFR 1-3, PDGFb, RET, and KIT [19]. In the MWSTP's 2021 phase II trial involving 31 patients from across the expanded MWSTP network, regorafenib showed activity against previously treated metastatic angiosarcoma, with an overall response rate of 17.4% and a median PFS of 5.5 months. Two patients had a complete response, two patients had a partial response, and ten patients had stable disease, for an overall clinical benefit rate of 60.8% [20]. Based on this study, current NCCN guidelines now include regorafenib as a recommended agent in the treatment of angiosarcoma [9] (Figure 2).

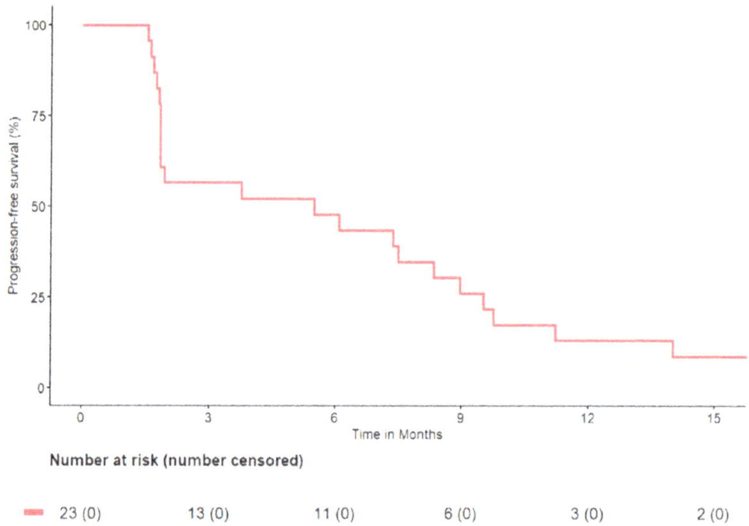

Figure 2. Kaplan–Meier curve of progression-free survival in patients receiving at least 2 cycles of regorafenib [20].

Another targeted therapy, pazopanib, is a small-molecule tyrosine kinase inhibitor of VEGFR and PDGFa/b that has single-agent activity in non-adipocytic soft tissue sarcoma [21]. The MWSTP's first trial of treatment-naïve patients with advanced sarcoma studied pazopanib as a first-line treatment for patients who were determined to be unsuitable for doxorubicin chemotherapy. The primary endpoint at 16 weeks was met, with 39% of patients achieving clinical benefit (complete response, partial response, or stable disease). Secondary endpoints included PFS, overall survival (OS), and quality of life. PFS was 3.67 months, and OS was 14.16 months. Side effects were similar to prior studies with pazopanib, with no appreciable decrease in quality of life [22]. Pazopanib has since been shown to be non-inferior to doxorubicin in the front-line setting, highlighting the future of this targeted therapy as a mainstay of treatment for metastatic sarcoma [12].

The MWSTP also studied tivozanib, a small-molecule tyrosine kinase inhibitor with activity against VEGFR1-3, PDGFa/b, and cKIT. In a phase II trial involving 58 patients with previously treated soft tissue sarcoma, tivozanib was well tolerated, with 36% of patients exhibiting PFS at the primary endpoint of 4 months, and a median PFS of 3.5 months [23]. Response to tivozanib did not correlate with the genetic expression of VEGFR1-3, PDGFa, or PDGFb as measured with immunohistochemical staining of tumor tissue. Although there are currently no ongoing trials of tivozanib in sarcoma, as of March 2021, tivozanib met FDA approval for treatment of relapsed/refractory renal cell carcinoma, another malignancy that is known to spread hematogenously [24].

Given the success of pazopanib as a single agent, the MWSTP conducted the first trial of pazopanib in combination with topotecan, a cytotoxic chemotherapy, in patients with previously treated advanced non-adipocytic sarcoma. Unfortunately, this phase II trial did not meet its primary endpoint of 66% of patients exhibiting PFS at 12 weeks. Higher rates of grade 3 or 4 toxicities, including hematologic toxicity and hypertension, were observed in this study in comparison to prior studies with pazopanib or topotecan as a single agent [25]. Thus, the combination of pazopanib and topotecan did not move forward to phase III trials. Of note, the trial enrolled an osteosarcoma cohort. Utilizing efficacy benchmarks, a threshold of 11 out of 36 potentially enrolled patients with PFS greater than 20 weeks was needed in order to demonstrate efficacy. In our study, this level was exceeded with a PFS rate of 45.5% at six months, indicating a high likelihood of efficacy in the treatment of this disease or an effect from pazopanib alone.

The open trials of the MWSTP include several promising studies. A phase II trial of abemaciclib for the treatment of sarcoma with cyclin-dependent kinase (CDK) pathway alteration has been opened since 2019 [26]. A phase II trial of temolozomide with cabozantinib in advanced sarcoma has recently completed accrual [27]. In 2022, a phase I clinical trial of NOX66 plus doxorubicin in anthracycline-naïve patients with sarcoma opened, with results being expected in 2024 [28]. In 2023, the MWSTP will open a study of lubinectedin with radiation for the treatment of retroperitoneal soft tissue sarcoma of the extremity.

The impact of the MWSTP extends beyond the conduct of clinical trials; it utilizes retrospective reviews to study community issues that impact future patient care. The MWSTP studied the administration of anthracyclines and/or ifosfamide in pregnancy-associated sarcomas [29]. In this multi-institutional study of treatment regimens for sarcomas during pregnancy, a high rate of fetal demise was seen only in patients receiving both doxorubicin and ifosfamide, especially when the treatment was initiated earlier in the second trimester. While limited by the small sample size, this review encompassed the largest study to date of sarcoma patients who received anthracyclines and/or ifosfamide during pregnancy. Future endeavors toward building an international registry of sarcoma patients would allow further investigations into this topic.

The MWSTP also conducted a retrospective review to report the safety, efficacy, and prognostic factors related to checkpoint inhibitors in soft tissue sarcoma. The results confirm the activity and safety of anti-PD-1 therapy in advanced sarcoma [30]. A notable response rate was observed in undifferentiated pleomorphic sarcoma and leiomyosarcoma subtypes. This study expands the knowledge base beyond what is currently available from clinical trials involving checkpoint inhibitors in metastatic sarcoma.

5. Conclusions

The inherent nature of sarcoma requires a multidisciplinary and collaborative approach to treatment. Due to its rarity, access to clinical expertise is necessary. At a single institution level, whether academic or community-based, MTBs are the cornerstone of management for patients in a local geographic area. Sarcoma networks, such as the MWSTP, allow cross-communications across MTBs and coordination of clinical trials across multiple anchor sites. The success of the MWSTP shows that all participating sites, regardless of whether they are academic or community-based, and whether they are an anchor site or an affiliated center, can contribute to the enhancement of care for each individual patient they bring to the network.

As a positive outcome of the COVID-19 pandemic, telehealth has made virtual collaboration more attainable than ever, allowing the expansions of existing medical networks via online communication. In previous studies of sarcoma MTBs that moved to an online platform as a result of the pandemic, there was no perceived difference in quality of discussion compared to in-person meetings [31], while also having no measurable effect on overall survival [32]. The potential impact of utilizing technology to create virtual MTBs, which connect single-institution or single-network MTBs into one large MTB, could change our entire existing framework of sarcoma care. Patients with rare subtypes of sarcoma would have increased access to an MTB specific to their condition, thus utilizing cumulative expertise from clinicians across the country and the world. The future of sarcoma care lies in increasing cooperation between existing sarcoma networks to improve access to clinical trials for all patients with sarcoma.

Author Contributions: Conceptualization, N.K.H. and M.A.; Writing – original draft, N.K.H. and M.A.; Writing – review & editing, S.O., S.R., S.A., M.S., B.L.S., J.Y., S.C., M.M.M., V.M., K.S., J.C., A.C.H., M.C.W., B.V.T. and M.A.; Supervision, M.A. All authors have read and agreed to the published version of the manuscript.

Funding: This research received no external funding.

Institutional Review Board Statement: Not applicable.

Informed Consent Statement: Not applicable.

Data Availability Statement: Data sharing not applicable. No new data were created or analyzed in this study. Data sharing is not applicable to this article.

Conflicts of Interest: The authors declare no conflict of interest.

References

1. Siegel, R.L.; Miller, K.D.; Fuchs, H.E.; Jemal, A. Cancer statistics, 2022. *CA Cancer J. Clin.* **2022**, *72*, 7–33. [CrossRef] [PubMed]
2. Grünewald, T.G.; Alonso, M.; Avnet, S.; Banito, A.; Burdach, S.; Cidre-Aranaz, F.; Di Pompo, G.; Distel, M.; Dorado-Garcia, H.; Garcia-Castro, J.; et al. Sarcoma treatment in the era of molecular medicine. *EMBO Mol. Med.* **2020**, *12*, e11131. [CrossRef]
3. Yoon, S.S.; Segal, N.H.; Olshen, A.B.; Brennan, M.F.; Singer, S. Circulating angiogenic factor levels correlate with extent of disease and risk of recurrence in patients with soft tissue sarcoma. *Ann. Oncol.* **2004**, *15*, 1261–1266. [CrossRef]
4. Farid, M.; Ngeow, J. Sarcomas Associated with Genetic Cancer Predisposition Syndromes: A Review. *Oncol.* **2016**, *21*, 1002–1013. [CrossRef]
5. Liu, L.; Dehner, C.; Grandhi, N.; Lyu, Y.; Borcherding, D.C.; Chrisinger, J.S.A.; Zhang, X.; Luo, J.; Tao, Y.; Parkes, A.; et al. The Impact of TSC-1 and -2 Mutations on Response to Therapy in Malignant PEComa: A Multicenter Retrospective Analysis. *Genes* **2022**, *13*, 1932. [CrossRef]
6. Hoven-Gondrie, M.L.; Bastiaannet, E.; Ho, V.K.; van Leeuwen, B.L.; Liefers, G.-J.; Hoekstra, H.J.; Suurmeijer, A.J.H. Worse Survival in Elderly Patients with Extremity Soft-Tissue Sarcoma. *Ann. Surg. Oncol.* **2016**, *23*, 2577–2585. [CrossRef] [PubMed]
7. Lyu, H.G.; Haider, A.H.; Landman, A.B.; Raut, C.P. The opportunities and shortcomings of using big data and national databases for sarcoma research. *Cancer* **2019**, *125*, 2926–2934. [CrossRef]
8. Judson, I.; Verweij, J.; Gelderblom, H.; Hartmann, J.T.; Schöffski, P.; Blay, J.-Y.; Kerst, J.M.; Sufliarsky, J.; Whelan, J.; Hohenberger, P.; et al. Doxorubicin alone versus intensified doxorubicin plus ifosfamide for first-line treatment of advanced or metastatic soft-tissue sarcoma: A randomised controlled phase 3 trial. *Lancet Oncol.* **2014**, *15*, 415–423. [CrossRef] [PubMed]
9. von Mehren, M.; Kane, J.M.; Agulnik, M.; Bui, M.M.; Carr-Ascher, J.; Choy, E.; Connelly, M.; Dry, S.; Ganjoo, K.N.; Gonzalez, R.J.; et al. Soft Tissue Sarcoma, Version 2.2022, NCCN Clinical Practice Guidelines in Oncology. *J. Natl. Compr. Cancer Netw.* **2022**, *20*, 815–833. [CrossRef]
10. Savina, M.; Le Cesne, A.; Blay, J.-Y.; Ray-Coquard, I.; Mir, O.; Toulmonde, M.; Cousin, S.; Terrier, P.; Ranchere-Vince, D.; Meeus, P.; et al. Patterns of care and outcomes of patients with METAstatic soft tissue SARComa in a real-life setting: The METASARC observational study. *BMC Med.* **2017**, *15*, 78. [CrossRef] [PubMed]
11. Saerens, M.; Brusselaers, N.; Rottey, S.; Decruyenaere, A.; Creytens, D.; Lapeire, L. Immune checkpoint inhibitors in treatment of soft-tissue sarcoma: A systematic review and meta-analysis. *Eur. J. Cancer* **2021**, *152*, 165–182. [CrossRef] [PubMed]
12. Grünwald, V.; Karch, A.; Schuler, M.; Schöffski, P.; Kopp, H.-G.; Bauer, S.; Kasper, B.; Lindner, L.H.; Chemnitz, J.-M.; Crysandt, M.; et al. Randomized Comparison of Pazopanib and Doxorubicin as First-Line Treatment in Patients with Metastatic Soft Tissue Sarcoma Age 60 Years or Older: Results of a German Intergroup Study. *J. Clin. Oncol.* **2020**, *38*, 3555–3564. [CrossRef] [PubMed]
13. Rytlewski, J.; Milhem, M.M.; Monga, V. Turning 'Cold' tumors 'Hot': Immunotherapies in sarcoma. *Ann. Transl. Med.* **2021**, *9*, 1039. [CrossRef]
14. Carpenter, W.R.; Fortune-Greeley, A.K.; Zullig, L.L.; Lee, S.-Y.; Weiner, B.J. Sustainability and performance of the National Cancer Institute's Community Clinical Oncology Program. *Contemp. Clin. Trials* **2012**, *33*, 46–54. [CrossRef]
15. Kim, D.J.; Otap, D.; Ruel, N.; Gupta, N.; Khan, N.; Dorff, T. NCI–Clinical Trial Accrual in a Community Network Affiliated with a Designated Cancer Center. *J. Clin. Med.* **2020**, *9*, 1970. [CrossRef]
16. Mehta, C.R.; Liu, L.; Theuer, C. An adaptive population enrichment phase III trial of TRC105 and pazopanib versus pazopanib alone in patients with advanced angiosarcoma (TAPPAS trial). *Ann. Oncol.* **2019**, *30*, 103–108. [CrossRef]
17. Sturm, E.C.; Marasco, I.S.; Katz, S.C. Multidisciplinary Management of Angiosarcoma—A Review. *J. Surg. Res.* **2021**, *257*, 213–220. [CrossRef] [PubMed]
18. Penel, N.; Bui, B.N.; Bay, J.-O.; Cupissol, D.; Ray-Coquard, I.; Piperno-Neumann, S.; Kerbrat, P.; Fournier, C.; Taieb, S.; Jimenez, M.; et al. Phase II Trial of Weekly Paclitaxel for Unresectable Angiosarcoma: The ANGIOTAX Study. *J. Clin. Oncol.* **2008**, *26*, 5269–5274. [CrossRef]
19. Wilhelm, S.M.; Dumas, J.; Adnane, L.; Lynch, M.; Carter, C.A.; Schütz, G.; Thierauch, K.-H.; Zopf, D. Regorafenib (BAY 73-4506): A new oral multikinase inhibitor of angiogenic, stromal and oncogenic receptor tyrosine kinases with potent preclinical antitumor activity. *Int. J. Cancer* **2011**, *129*, 245–255. [CrossRef]
20. Agulnik, M.; Schulte, B.; Robinson, S.; Hirbe, A.C.; Kozak, K.; Chawla, S.P.; Attia, S.; Rademaker, A.; Zhang, H.; Abbinanti, S.; et al. An open-label single-arm phase II study of regorafenib for the treatment of angiosarcoma. *Eur. J. Cancer* **2021**, *154*, 201–208. [CrossRef]

21. van der Graaf, W.T.; Blay, J.-Y.; Chawla, S.P.; Kim, D.-W.; Bui-Nguyen, B.; Casali, P.G.; Schöffski, P.; Aglietta, M.; Staddon, A.P.; Beppu, Y.; et al. Pazopanib for metastatic soft-tissue sarcoma (PALETTE): A randomised, double-blind, placebo-controlled phase 3 trial. *Lancet* **2012**, *379*, 1879–1886. [CrossRef] [PubMed]
22. Hirbe, A.C.; Eulo, V.; Moon, C.I.; Luo, J.; Myles, S.; Seetharam, M.; Toeniskoetter, J.; Kershner, T.; Haarberg, S.; Agulnik, M.; et al. A phase II study of pazopanib as front-line therapy in patients with non-resectable or metastatic soft-tissue sarcomas who are not candidates for chemotherapy. *Eur. J. Cancer* **2020**, *137*, 1–9. [CrossRef] [PubMed]
23. Agulnik, M.; Costa, R.; Milhem, M.; Rademaker, A.; Prunder, B.; Daniels, D.; Rhodes, B.; Humphreys, C.; Abbinanti, S.; Nye, L.; et al. A phase II study of tivozanib in patients with metastatic and nonresectable soft-tissue sarcomas. *Ann. Oncol.* **2017**, *28*, 121–127. [CrossRef]
24. Chang, E.; Weinstock, C.; Zhang, L.; Fiero, M.H.; Zhao, M.; Zahalka, E.; Ricks, T.K.; Zirkelbach, J.F.; Qiu, J.; Yu, J.; et al. FDA Approval Summary: Tivozanib for Relapsed or Refractory Renal Cell Carcinoma. *Clin. Cancer Res.* **2022**, *28*, 441–445. [CrossRef] [PubMed]
25. Schulte, B.; Mohindra, N.; Milhem, M.; Attia, S.; Robinson, S.; Monga, V.; Hirbe, A.C.; Oppelt, P.; Charlson, J.; Helenowski, I.; et al. Phase II study of pazopanib with oral topotecan in patients with metastatic and non-resectable soft tissue and bone sarcomas. *Br. J. Cancer* **2021**, *125*, 528–533. [CrossRef] [PubMed]
26. (U.S.). NLoM. Abemaciclib for Bone and Soft Tissue Sarcoma with Cyclin-Dependent Kinase (CDK) Pathway Alteration. Identifier NCT04040205 2019. Available online: https://clinicaltrials.gov/ct2/show/NCT04040205 (accessed on 20 November 2022).
27. Medicine NLo. Cabozantinib and Temozolomide for the Treatment of Unresectable or Metastatic Leiomyosarcoma or Other Soft Tissue Sarcoma. Identifier NCT04200443. Available online: https://clinicaltrials.gov/ct2/show/NCT04040205 (accessed on 20 November 2022).
28. (U.S.). NLoM. A Dose Escalation and Dose Expansion Study of NOX66 Plus Doxorubicin in Anthracycline-naïve, Adult Pa-tients with Soft Tissue Sarcoma. Identifier NCT05100628. Available online: https://clinicaltrials.gov/ct2/show/NCT05100628 (accessed on 20 November 2022).
29. Miller, D.; Livingston, J.A.; Park, Y.; Posey, K.; Godbole, S.; Skubitz, K.; Robinson, S.I.; Agulnik, M.; Davis, L.E.; Van Tine, B.A.; et al. Pregnancy outcomes related to the treatment of sarcomas with anthracyclines and/or ifosfamide during pregnancy. *Cancer Med.* **2022**, *11*, 3471–3478. [CrossRef]
30. Monga, V.; Skubitz, K.M.; Maliske, S.; Mott, S.L.; Dietz, H.; Hirbe, A.C.; Van Tine, B.A.; Oppelt, P.; Okuno, S.; Robinson, S.; et al. A Retrospective Analysis of the Efficacy of Immunotherapy in Metastatic Soft-Tissue Sarcomas. *Cancers* **2020**, *12*, 1873. [CrossRef]
31. Rajasekaran, R.B.; Whitwell, D.; Cosker, T.D.A.; Gibbons, C.L.M.H.; Carr, A. Will virtual multidisciplinary team meetings become the norm for musculoskeletal oncology care following the COVID-19 pandemic?—Experience from a tertiary sarcoma centre. *BMC Musculoskelet. Disord.* **2021**, *22*, 18. [CrossRef]
32. Pan, M.; Yu, J.; Sidhu, M.; Seto, T.; Fang, A. Impact of a Virtual Multidisciplinary Sarcoma Case Conference on Treatment Plan and Survival in a Large Integrated Healthcare System. *JCO Oncol. Pract.* **2021**, *17*, e1711–e1718. [CrossRef]

Journal of
Clinical Medicine

MDPI

Brief Report

Difficulties in Defining Oligometastatic Prostate Cancer: Implications for Clinical Trial Accrual and Community Practice Adoption of Metastasis-Directed Therapy Approaches

Tanya Barauskas Dorff [1,*], Saro Kasparian [1], Natasha Garg [2], Sandy Liu [3], Sumanta Kumar Pal [1], Jeffrey Wong [4] and Savita Dandapani [4]

[1] Department of Medical Oncology and Therapeutics Research, City of Hope Comprehensive Cancer Center, 1500 E. Duarte Rd. Pavillion #2250, Duarte, CA 91010, USA
[2] Department of Medical Oncology and Therapeutics Research, City of Hope, Upland, CA 91786, USA
[3] Department of Medical Oncology and Therapeutics Research, City of Hope Lennar Cancer Center, Irvine, CA 92618, USA
[4] Department of Radiation Oncology, City of Hope Comprehensive Cancer Center, Duarte, CA 91010, USA
* Correspondence: tdorff@coh.org; Tel.: +1-626-218-8231; Fax: +1-626-218-8233

Abstract: Background: Metastasis-directed therapy is widely utilized for oligometastatic prostate cancer patients, but standard imaging does not always identify metastases definitively and, even with PSMA PET, there may be equivocal findings. Not all clinicians have access to detailed imaging review, particularly outside of academic cancer centers, and PET scan access is also limited. We sought to understand how imaging interpretation impacted recruitment to a clinical trial for oligometastatic prostate cancer. Methods: IRB approval was obtained to review medical records from all patients screened for the institutional IRB-approved clinical trial for men with oligometastatic prostate cancer involving androgen deprivation plus stereotactic radiation to all metastatic sites, as well as radium223 (NCT03361735). Clinical trial inclusion required at least one bone metastatic lesion and no more than five total sites of metastasis, including soft tissue sites. Tumor board discussion records were reviewed, along with results from additional radiology studies ordered or confirmatory biopsies performed. Clinical characteristics such as PSA level and Gleason score were studied for association with likelihood of oligometastatic disease confirmation. Results: At the time of data analysis, 18 subjects were deemed eligible and 20 were not eligible. The most common reasons for ineligibility were no confirmed bone metastasis in 16 patients (59%) and too many metastatic sites in 3 (11%). The median PSA of eligible subjects was 3.28 (range 0.4–45.5), whereas the median PSA of those found to be ineligible was 10.45 (range 3.7–26.3) when there were too many metastases identified, and 2.7 (range 0.2–34.5) when metastases were unconfirmed. PET imaging (PSMA or fluciclovine PET) increased the number of metastases, while MRI resulted in downstaging to non-metastatic disease. Conclusions: This research suggests that additional imaging (i.e., at least two independent imaging modalities of a possible metastatic lesion) or tumor board adjudication of imaging findings may be critical to correctly identify patients appropriate for enrollment in oligometastatic protocols. This should be considered as trials of metastasis-directed therapy for oligometastatic prostate cancer accrue and results are translated to broader oncology practice.

Keywords: oligometastatic; prostate cancer; metastasis-directed therapy; imaging; PSMA PET

Citation: Dorff, T.B.; Kasparian, S.; Garg, N.; Liu, S.; Pal, S.K.; Wong, J.; Dandapani, S. Difficulties in Defining Oligometastatic Prostate Cancer: Implications for Clinical Trial Accrual and Community Practice Adoption of Metastasis-Directed Therapy Approaches. *J. Clin. Med.* **2023**, *12*, 2011. https://doi.org/10.3390/jcm12052011

Academic Editors: Cristian Fiori and Enrico Checcucci

Received: 12 January 2023
Revised: 14 February 2023
Accepted: 27 February 2023
Published: 3 March 2023

1. Introduction

Metastatic prostate cancer remains incurable despite recent improvements in outcomes with systemic therapy. Whereas previously any metastatic disease, even if only to pelvic lymph nodes, was felt to represent a disseminated disease state which should only be treated with systemic therapy, sophisticated analysis of metastases from an autopsy series identified that metastatic deposits can create additional metastases [1]. This raised the possibility that enhanced eradication of cancer at visible metastatic foci using focal radiation could prevent or reduce further cancer spread. Preliminary success with this approach has been observed. For instance, the ORIOLE trial [2] found improved progression-free survival when stereotactic ablative radiation (SABR) was used to treat oligometastatic sites (compared to observation). However, this study utilized PET scans to define oligometastases, and 19% of those treated with SABR had progression within 6 months, suggesting that additional occult metastatic sites existed.

Advances in imaging with prostate-cancer-specific PET tracers such as [^{11}C]Choline, [^{18}F]DCFPyL, and [^{68}Ga]Ga-PSMA-11 have resulted in improved sensitivity and specificity for detecting metastatic foci [3–5]. However, these are not widely available, and the majority of prostate cancer patents enrolling on clinical trials continue to undergo conventional imaging with CT and technetium bone scans to define their eligibility.

Our institution is recruiting subjects to an IRB-approved investigator-initiated clinical trial (NCT03361735) designed to enroll men with oligometastatic, castration-sensitive prostate cancer. Participants receive 9 months of androgen deprivation therapy, stereotactic ablative radiation to all metastatic sites, and 6 doses of radium223 (55 kBq/kg intravenously, once every 4 weeks × 6 doses). The protocol requires at least one bone metastasis and no more than five sites of metastasis for inclusion. During screening for the clinical trial, it was noted that review of imaging by the tumor board, or additional imaging studies ordered to verify the oligometastatic status, often led to a determination that the patient was ineligible. Based on this observation, we obtained IRB approval to evaluate the subjects who did not successfully enroll, with the goal of evaluating which imaging modalities were most helpful in confirming or refuting an oligometastatic state and determining whether any clinical characteristics should raise questions about the certainty of imaging findings. This information was felt to have the potential to benefit future clinical trials for oligometastatic prostate cancer patients, and also help practicing oncologists note imaging pitfalls when recommending metastasis-directed therapy for patients presenting to them with what appears to be oligometastatic disease.

2. Methods

After IRB approval via an amendment to the main clinical trial protocol, a retrospective chart review was performed on patients who signed consent for the clinical trial protocol (NCT03361735) or were being considered for enrollment based on tumor board records.

Treatment on the trial included androgen deprivation therapy for 9 months, SBRT to all metastatic sites, and radium223 infusions (55 kBq/kg IV, once every 4 weeks for 6 doses). The primary endpoint of this trial was time to treatment failure.

Baseline disease characteristics and the results of all imaging studies and biopsies that were performed as part of the eligibility determination were tabulated. For the purposes of this analysis, eligible men included those enrolled on the protocol, as well as those who were deemed eligible but declined participation, while ineligible men were those who were excluded from participation due to having more than the allowed number of metastases or a lack of confirmed metastatic disease after tumor board review or additional imaging or biopsy. Comparison between eligible and ineligible patient groups was performed using a *t*-test for continuous variables and chi-square for categorical variables.

3. Results

This study began in 2018. At the time of analysis, 18 subjects had been deemed eligible, while 20 others were deemed ineligible. Baseline characteristics are summarized in Table 1. The median PSA of eligible subjects was 3.28 (range 0.4–45.5), whereas the median PSA of those found to be ineligible was 10.45 (range 3.7–26.3) when there were too many metastases identified, and 2.7 (range 0.2–34.5) when no metastases could be confirmed. There was no difference in Gleason grade group between patients confirmed to be oligometastatic and those recategorized as either non-metastatic or having more than five metastatic sites.

Table 1. Characteristics of eligible and ineligible patients being evaluated for enrollment into the oligometastatic protocol.

	Eligible (n = 19)	Ineligible—Too Many Metastases (n = 6)	Ineligible—Not Metastatic (n = 13)	p Value ^
Median PSA * (range)	3.28 (0.4–45.5)	10.45 (3.7–26.3)	2.7 (0.2–34.5)	
Primary untreated	18.1 (9.1–45.5)	8.9 (4.6–13.1)	13.7 (4.8–34.5)	Many p = 0.057 Few p = 1.0
Primary treated	1.7 (0.4–27.5)	12 (3.7–26.3)	1.174 (0.2–2.3)	
Gleason grade group N (%)				
1	3 (16%)	0	1 (8%)	p = 0.17
2–3	8 (42%)	4 (67%)	4 (31%)	p = 0.37
4–5	8 (42%)	2 (33%)	8 (61%)	p = 0.14
Imaging modalities				
MRI	10	1	7	
PET (fluciclovine)	7	4	5	
PET (PSMA)	3	0	1	

* PSA at the time of eligibility assessment for enrollment on the clinical trial. ^ p values were calculated comparing eligible patients to those with too many ("Many") metastases and eligible patients to those with too few ("Few") metastases using Fisher's exact test for continuous variables and the chi-square test for the Gleason grade group comparing too many versus too few.

MRI was performed in 19 of the 38 patients (50%) and PET scans were performed in 21 of 38 patients (55%). Biopsy of a bone lesion was performed in four cases (one each: femur, iliac, ischium, and rib), and three of these patients had PSA <1 at the time. In all four cases, biopsy was negative for malignancy, which was used to determine that the patient was ineligible. One of these patients later developed metastasis in a different bone but did not appear to develop metastasis in the original biopsied area. Three subjects with PSA >10 who were initially suspected of having metastatic disease were deemed non-metastatic after tumor board imaging review and/or MRI. Two subjects with PSA <5 were found to have too many metastases to qualify, in both cases based on fluciclovine PET imaging.

Skull (n = 2) and femur (n = 3) findings were most commonly recategorized as non-metastatic on further imaging or further review by the tumor board, while acetabular lesions were more commonly confirmed (n = 2). Spine and pelvic findings were evenly divided between confirmed and unconfirmed patients.

4. Discussion

Metastasis-directed therapy holds significant promise for oligometastatic prostate cancer, but clinical trials in this space have utilized different imaging modalities to define their oligometastatic populations (Table 2). In this experience, there was a high ineligibility rate during screening for a therapeutic clinical trial for patients with oligometastatic prostate cancer, with subjects having either too many metastases or a lack of confirmed metastases. This was the result of a high degree of scrutiny and utilization of additional imaging and/or biopsy in order to confirm eligibility. It raises questions about how community oncologists can best adopt metastasis-directed therapy for oligometastatic prostate cancer patients without access to the resources available at a tertiary academic center.

Table 2. Imaging used to qualify patients for enrollment in select published oligometastatic prostate cancer clinical trials.

Study	# Metastases	Other Restrictions	Imaging Used to Define # of Metastases
POP-STAR [6]	1–3	Bone or LN only	[^{18}F]-NaF PET/CT
ORIOLE [2]	1–3	Asymptomatic, arose in the prior 6 months, ≤5 cm in long axis or ≤250 cm^2	Conventional imaging
STOMP [7]	1–3	Extracranial, negative MRI or biopsy of prostate bed even if choline PET negative in prostate bed	[^{11}C]Choline PET

signifies the number of metastases.

Variability in radiologic interpretation in cancer patients has been well documented, for instance when evaluating the RECIST response [8] and even when using conventional imaging, which most radiologists have the greatest amount of experience in interpreting. The lack of sensitivity and specificity of conventional imaging for identifying prostate cancer metastases has also been well documented [9]. For instance, in the POPSTAR trial, even with [^{18}F]-NaF PET bone scans, there was considerable understaging as distant progression-free survival (PFS) was about 40% at 2 years compared to 89–100% continued remission at the sites of irradiation [6]. This indicates that smaller deposits of disease had not been visible when metastasis-directed therapy was administered.

PSMA PET tracers significantly improve sensitivity [3]. However, due to limited access and difficulty in interpretation, the research community has largely opted to continue basing eligibility and response assessment on conventional imaging. In the ORIOLE trial, for example, conventional imaging formed the basis for treatment. The protocol specified that [^{18}F]DCFPyL-PET images were evaluated and compared to bone scans, but additional sites of suspected metastatic disease from the PET scan were not considered for treatment by SBRT nor required to undergo further evaluation [2]. This design resulted in the ability to analyze outcomes in patients whose PET-detected disease was fully treated (i.e., PET scan did not detect additional sites of disease beyond what was visible on conventional imaging) compared to those in whom some metastatic disease was left untreated, and it was noted that the former group had greater progression-free survival. Thus, future oligometastatic protocols are likely to rely on PSMA PET imaging. However, false positives will continue to be an important consideration since benign conditions such as Paget's disease have been reported to result in false-positive PSMA radiotracer uptake [10], and interpretation can be challenging for this relatively newer imaging modality.

Access to PET scans for prostate cancer patients remains a major limitation in community practice and in academic centers. A recent publication found that in a tertiary medical center, there were disparities in PET scanning, with African American prostate cancer patients less likely to undergo PSMA PET scan compared to non-Hispanic white patients [11]. In the community oncology setting, differences in health insurance coverage and imaging facility capabilities may exacerbate the lack of equitable access to PET imaging. Community oncologists may also have decreased access to multidisciplinary care via participation in tumor boards. In one survey of community practices, 53.8% of physicians reported participating in tumor boards weekly, while 42% participated less than once per week, with less attendance from medical oncologists compared to radiation oncologists [12]. In our experience, tumor board review was key in gaining confidence for inclusion of patients and treatment of oligometastatic sites, even if PET imaging was not available, and lack of access to optimal imaging should not preclude patients from accessing the potential benefit of metastasis-directed therapy.

While a major focus has been placed on determining the number of metastatic lesions that define the oligometastatic disease state, it seems that controversy surrounds whether a conventional imaging modality can adequately establish a patient as having the specified number of metastases. Bone scanning has been suggested to yield inconclusive results in about 16% of cases [13]. MRI has often been used to enhance detection of osseous metastases or clarify inconclusive bone scan findings. However, in a study of findings from pelvic MRI performed in 3765 patients for evaluation of presumed localized prostate cancer, 74% of patients had bone abnormalities, which were only rarely confirmed to be metastases [14]. This calls into question the use of MRI to adjudicate findings from conventional imaging. Bone biopsies may be helpful in confirming a bone metastasis, but even in experienced centers with specific protocols designed to maximize the yield, the detection rates from bone biopsy performed to obtain cancer tissue have been reported to be less than 80% [15], and in most community centers without expertise the yield will be lower. Therefore, bone biopsy may not have a high enough sensitivity to be used to exclude the presence of metastatic cancer. Some clinical characteristics may be helpful in selecting metastases for greater yield, including the size of the lesion in the bone, presence of a soft tissue component, intensity of scintigraphic uptake, or a newly apparent area of disease involvement, but without communication between oncologists and radiologists, optimal target selection is less likely to occur.

Overall, our experience raises concern about a potential lack of uniformity in the population of patients who are subject to protocols for oligometastatic prostate cancer, and how the results can subsequently be translated into clinical practice. There is no defined algorithm for how to confirm oligometastases identified in a prostate cancer patient using conventional imaging. We found that simple clinical factors may help guide clinicians as to when additional scrutiny is warranted. In this research, subjects with PSA over 10 were less likely to be deemed oligometastatic after further imaging or imaging review, although there were four subjects who were found eligible with PSA ranging from 10 to 45. Similarly, subjects were less likely to have metastases at all when PSA was less than 1, though four subjects were deemed to have oligometastatic disease at this PSA level, only one of whose disease was detected on a PET scan. While clearly not enough to define an oligometastatic state, higher or lower PSA should at least raise clinical suspicion and trigger additional imaging, consultation with radiology, or potentially a biopsy to better clarify the extent of the disease. Where PSMA PET scans are not available, any indeterminate bone findings, or a discordance between bone scanning and CT, or between the volume of disease and PSA, may warrant additional imaging and/or biopsy before metastasis-directed therapy is undertaken.

5. Conclusions

Defining the oligometastatic state depends on accurate interpretation of imaging. In our experience, prostate cancer patients initially thought to be oligometastatic were frequently reclassified when additional imaging was ordered to clarify indeterminate findings. Physicians should be encouraged to thoroughly review imaging, utilizing tumor boards or additional imaging modalities when appropriate, prior to applying metastasis-directed strategies for their patients who appear to have oligometastatic prostate cancer.

Author Contributions: Conceptualization: T.B.D. and S.D., Methodology: T.B.D. and S.K., Data Curation: T.B.D. and S.K., Writing—Original Draft Preparation: T.B.D. and S.K., Writing—Review and Editing: T.B.D., S.K., N.G., S.L., S.K.P., J.W. and S.D. All authors have read and agreed to the published version of the manuscript.

Funding: Conduct of the clinical trial was supported by research funds from Bayer Pharmaceuticals to City of Hope for an investigator sponsored trial.

Institutional Review Board Statement: This study was conducted with approval by the City of Hope IRB (protocol #17085), and was conducted according to the guidelines of the Declaration of Helsinki.

Informed Consent Statement: All subjects included and treated in the study provided informed consent; IRB exemption was granted to include information from patients who failed screening and, therefore, did not sign informed consent to enroll onto the study.

Data Availability Statement: Data sharing not applicable. This study is still accruing patients. Data sharing is not applicable to this article since full data from the clinical trial will be shared upon completion of the study.

Conflicts of Interest: T.D. has received consulting income from Astellas, AstraZeneca, Bayer, Exelixis, Janssen, Pfizer, and Sanofi. S.P. has received institutional research funding from Eisai, Genentech, Roche, Exelixis, Pfizer, Crispr, and Allogene. He has also received travel expenses from Crispr and Roche.

References

1. Gundem, G.; Van Loo, P.; Kremeyer, B.; Alexandrov, L.B.; Tubio, J.M.; Papaemmanuil, E.; Brewer, D.S.; Kallio, H.M.; Högnäs, G.; Annala, M. The evolutionary history of lethal metastatic prostate cancer. *Nature* **2015**, *520*, 353–357. [CrossRef] [PubMed]
2. Phillips, R.; Shi, W.Y.; Deek, M.; Radwan, N.; Lim, S.J.; Antonarakis, E.S.; Rowe, S.P.; Ross, A.E.; Gorin, M.A.; Deville, C.; et al. Outcomes of observation vs stereotactc ablative radiation for oligometastatic prostate cancer: The ORIOLE phase 2 randomized clinical trial. *JAMA Oncol.* **2020**, *6*, 650–659. [CrossRef] [PubMed]
3. Fendler, W.P.; Calais, J.; Eiber, M.; Flavell, R.R.; Mishoe, A.; Feng, F.Y.; Nguyen, H.G.; Reiter, R.E.; Rettig, M.B.; Okamoto, S.; et al. Assessment of 68Ga-PSMA-11 PET accuracy in localizing recurrent prostate cancer: A prospective single-arm clinical trial. *JAMA Oncol.* **2019**, *5*, 856–863. [CrossRef] [PubMed]
4. Morris, M.J.; Rowe, S.P.; Gorin, M.A.; Saperstein, L.; Pouliot, F.; Josephson, D.; Wong, J.Y.; Pantel, A.R.; Cho, S.Y.; Gage, K.L.; et al. Diagnostic performance of 18F-DCFPyL-PET/CT in men with biochemically recurrent prostate cancer: Results from the CONDOR phase III multicenter study. *Clin. Cancer Res.* **2021**, *27*, 3674–3682. [CrossRef] [PubMed]
5. Sobol, I.; Zaid, H.B.; Haloi, R.; Mynderse, L.A.; Froemming, A.T.; Lowe, V.J.; Davis, B.J.; Kwon, E.D.; Karnes, R.J. Contemporary mapping of post-prsotatectomy prostate cancer relapse with 11C-choline positron emission tomorgraphy and multiparametric magnetic resonance imaging. *J. Urol.* **2017**, *197*, 129–134. [CrossRef] [PubMed]
6. Siva, S.; Bressel, M.; Murphy, D.C.; Shaw, M.; Chander, S.; Violet, J.; Tai, K.H.; Udovicich, C.; Lim, A.; Selbie, L. Steroetactic ablative body radiotherapy (SABR) for oligometastatic prostate cancer: A prospective clinical trial. *Eur. Urol.* **2018**, *74*, 455–462. [CrossRef] [PubMed]
7. Ost, P.; Reynders, D.; Decaestecker, K.; Fonteyne, V.; Lumen, N.; De Bruycker, A.; Lambert, B.; Delrue, L.; Bultijnck, R.; Claeys, T.; et al. Surveillance or metastasis-directed therapy for oligometastatic prostate cancer: A prospective randomized multicenter phase II trial. *J. Clin. Oncol.* **2018**, *36*, 446–453. [CrossRef] [PubMed]
8. Karmakar, A.; Kumtakar, A.; Sehgal, H.; Kumar, S.; Kalyanpur, A. Interobserver Variation in Response Evaluation Criteria in Solid Tumors 1.1. *Acad Radiol.* **2019**, *26*, 489–501. [CrossRef] [PubMed]
9. Johnstone, P.A.; Tarman, G.H.; Riffenburgh, R.; Rohde, D.C.; Puckett, M.L.; Kane, C.J. Yield of imaging and scintigraphy assessing biochemical failure in prostate cancer patients. *Urol. Oncol.* **1997**, *3*, 108–112. [CrossRef] [PubMed]
10. Sasikumar, A.; Joy, A.; Nanabala, R.; Pillai, M.R.A.; Hari, T.A. 68Ga-PSMA PET/CT false-positive tracer uptake in paget disease. *Clin. Nucl. Med.* **2016**, *41*, e454–e455. [CrossRef] [PubMed]
11. Bucknor, M.D.; Lichtensztajn, D.Y.; Lin, T.K.; Borno, H.T.; Gomez, S.L.; Hope, T.A. Disparities in PET imaging for prostate cancer at a Tertiary Academic Medical Center. *J. Nucl. Med.* **2021**, *62*, 695–699. [CrossRef] [PubMed]
12. Kehl, K.L.; Landrum, M.B.; Kahn, K.L.; Gray, S.W.; Chen, A.B.; Keating, N.L. Tumor board participation among physicians caring for patients with lung or colorectal cancer. *J. Oncol. Pract.* **2015**, *11*, e267–e278. [CrossRef] [PubMed]
13. Wondergem, M.; van der Zant, F.M.; Knol, R.J.J.; Burgers, A.M.G.; Bos, S.D.; DeJong, I.J.; Pruim, J. 99mTc-HDP bone scintigraphy and 18F-sodium fluoride PET/CT in primary staging of patients with prostate cancer. *World J. Urol.* **2018**, *36*, 27–34. [CrossRef] [PubMed]
14. Vargas, H.A.; Schor-Bardach, R.; Long, N.; Kirzner, A.N.; Cunningham, J.D.; Goldman, D.A.; Moskowitz, C.S.; Sosa, R.E.; Sala, E.; Panicek, D.M.; et al. Prostate cancer bone metastases on staging prostate MRI: Prevalence and clinical features associated with their diagnosis. *Abdom. Radiol.* **2017**, *42*, 271–277. [CrossRef] [PubMed]
15. McKay, R.R.; Zukotynski, K.A.; Werner, L.; Voznesensky, O.; Wu, J.S.; Smith, S.E.; Jiang, Z.; Melnick, K.; Yuan, X.; Kantoff, P.W.; et al. Imaging, procedural and clinical variables associated with tumor yield on bone biopsy in metastatic castration resistant prostate cancer. *Prostate Cancer Prost. Dis* **2014**, *17*, 325–331. [CrossRef] [PubMed]

Journal of
Clinical Medicine

MDPI

Perspective

The Gut Microbiome and Metastatic Renal Cell Carcinoma

Luis Meza [1], Matthew Feng [1], Kyle Lee [1], Rubens Sperandio [2] and Sumanta Kumar Pal [1,*]

1 Department of Medical Oncology & Experimental Therapeutics, City of Hope Comprehensive Cancer Center, Duarte, CA 91010, USA
2 Hospital Israelita Albert Einstein, São Paulo 05652-900, Brazil
* Correspondence: spal@coh.org; Tel.: +1-(626)-256-4673; Fax: +1-(626)-301-8233

Abstract: The introduction of targeted therapy (TT) and immuno-oncology (IO) agents have revolutionized the treatment of metastatic renal cell carcinoma (mRCC). However, despite the significant improvements in survival and clinical response yielded by these agents, a significant percentage of patients still experience progressive disease. Evidence now suggests that microorganisms living in the gut (i.e., the gut microbiome) could be used as a biomarker for response and may also have utility in increasing response to these treatments. In this review, we present an overview of the role of the gut microbiome in cancer and its potential implications in the treatment of mRCC.

Keywords: renal cell carcinoma; gut microbiome; translational research

Citation: Meza, L.; Feng, M.; Lee, K.; Sperandio, R.; Pal, S.K. The Gut Microbiome and Metastatic Renal Cell Carcinoma. *J. Clin. Med.* **2023**, *12*, 1502. https://doi.org/10.3390/jcm12041502

Academic Editor: Hiroshi Tanaka

Received: 15 January 2023
Revised: 6 February 2023
Accepted: 11 February 2023
Published: 14 February 2023

1. Introduction

Approximately 82,000 new cases of kidney cancer will be diagnosed in the United States during 2023, with varying rates of progression to metastatic disease [1]. While treatment options for localized disease have remained largely unchanged, significant advances have occurred in the treatment landscape of metastatic renal cell carcinoma (mRCC). The last couple of decades have seen an explosion in the number of U.S. Food and Drug Administration (FDA) approvals for this disease setting with multiple targeted-therapy agents (TT), and immune checkpoint inhibitors (ICIs) being now available for this patient population. TT agents can be divided into (a) inhibitors of vascular endothelial growth factor (VEGF) signaling, which include drugs such as sunitinib, pazopanib, axitinib, cabozantinib and levantinib, and (b) inhibitors of the mammalian target of rapamycin (mTOR), represented by everolimus and temsirolimus [2–7]. In contrast, ICIs block coinhibitory molecules such as programmed death-1 (PD-1), the programmed death ligand-1 (PD-L1) and the cytotoxic T-lymphocyte activating protein-4 (CTLA-4) [8].

Despite the numerous available options for patients with mRCC with the use of the previously mentioned agents, either as monotherapy or in combination, response to these regimens remains heterogeneous, with some patients achieving a complete response (CR) while others experience progressive disease (PD). Moreover, the 5-year survival rate for patients in this stage is only 15% [9]. Therefore, selecting the approach that will yield the most benefit for a given patient remains a significant challenge [10]. Despite multiple efforts to identify biomarkers predictive of response, such as the gene expression signatures from the IMmotion 151 trial, tumor mutational burden (TMB), and PD-L1 expression, the International mRCC Database Consortium (IMDC) risk model remains the only predictive biomarker to be prospectively validated in a phase 3 trial to date. There is, therefore, a need to increase our understanding of the biological processes underlying the development and evolution of RCC to develop novel biomarkers of response that will allow for treatment selection in an individualized manner.

In recent years, fueled by the advent of next-generation sequencing technologies, there has been an increased interest in the evaluation of the gut microbiome and its role in cancer. Multiple studies now show that certain bacterial species might be associated with the development of certain cancers such as lung, melanoma and colon, as well as with treatment response to currently available regimens [11]. In the setting of RCC, there have also been efforts to characterize the role of the gut microbiome. Here, we provide an overview of the role of the gut microbiome in cancer with a special focus on RCC. In addition, we highlight the ongoing trials in the field and discuss the importance of intra–inter-institutional collaboration for creating a solid working framework for microbiome studies in the future.

2. Gut Microbiome

It is estimated that the human body is composed of around 3.7 × 1013 human cells [12]. In addition to these cells, the healthy human body also comprises a plethora of microbes including bacteria, viruses and fungi which are collectively known as the microbiome. Revised estimates suggest that these organisms amount to at least 3.8 × 1013 cells, accounting for approximately half of the total number of cells present in the body, and are intrinsically involved in the regulation and maintenance of human health [13,14]. However, although these organisms can be found in multiple tissues throughout the human body, such as in the skin, oral mucosa, and gastrointestinal tract, it is this last one, particularly the colon, that hosts the highest number of bacteria, exceeding all other organs by two orders of magnitude [13]. It is well established that the gut microbiome plays an integral role in a number of physiologic functions that include the metabolism and uptake of nutrients, the preservation of the intestinal barrier, and modulation of the immune system [14]. Indeed, it is now known that there is a complex interplay between the gut microbiota and the immune system of the host that impacts both local immunity and peripheral white blood cell dynamics [15–22].

It has been hypothesized that intestinal microbes confer many metabolic capabilities needed for the preservation of the host's immune homeostasis and that alterations of the gut microbiome composition (dysbiosis) could lead to immune alterations contributing to the development of a number of systemic disorders [23,24] Notably, numerous studies have shown its association with a number of inflammatory and autoimmune conditions such as inflammatory bowel disease and lupus nephritis, while a number of persuasive interventional studies have further demonstrated that microbiome modulating strategies, such as fecal microbiota transplantation (FMT), can induce remission of some of these conditions and modulate treatment response [25–29].

3. Gut Microbiome and Cancer

It is therefore not surprising that given the successes in establishing associations between the gut microbiome and several diseases, subsequent studies have sought to determine its influence in the context of cancer. Interestingly, and despite the increased interest in examining the role of the microbiome in cancer seen in recent years, there are historical reports dating back to 1868 suggesting a link between the presence of certain microbes and oncogenesis [11,30,31]. Among the microbes reported to have a role in carcinogenesis are viruses such as the Epstein–Barr, human papilloma, and hepatitis viruses and bacteria such as *Helicobacter pylori* [32,33]. Nevertheless, the path to characterization of other microbiome–cancer associations has been largely truncated by technical challenges of the time. Encouragingly, the advent of new laboratory techniques and technologies such as next-generation genomic sequencing is helping us to deepen our understanding of the contribution of bacteria present in the gut to the development of cancer and their influence in response to anti-cancer systemic therapies and their associated toxicities [34].

It is through the incorporation of these new technologies that pivotal investigations have been able to show the presence of distinct microbial profiles in the gut of cancer patients compared with their cancer-free counterparts [34,35]. Moreover, the preponderance of preclinical and clinical evidence now suggests that gut dysbiosis plays key role in the natural history of a number of malignancies including colorectal cancer, hepatocellular carcinoma, melanoma and breast cancer [36–41]. Furthermore, the influence of the gut microbiome has been investigated in the setting of different systemic therapy approaches, such as chemotherapy, stem cell transplantation and immunotherapy, where it has been shown to modulate toxicity and treatment response [31,42–46]. Particularly, significant efforts have been dedicated to investigating the association between the gut microbiome and immune-related adverse events (irAEs). Evidence now suggests that differences in gut microbiome profiles exist between patients who experience irAEs and those who do not [47–49]. This finding could potentially be used to develop biomarkers to predict their occurrence prior to initiation of therapy, as well as devising interventions to abrogate these events once they ensue [49].

Notably, associations between certain bacterial species and response to immune checkpoint blockade (anti-CTLA-4 and anti-PD-1) have also been demonstrated across different cancer types, suggesting the presence of "responder" and "non-responder" gut microbiome profiles [50–53] Indeed, there have been several efforts to recapitulate these favorable profiles through interventions such as FMT or bacterial supplementation that have shown some success in enhancing therapeutic response and overcoming resistance [50,54–57]. Likewise, dietary changes such as a higher fiber intake have also been associated with an increased benefit from ICIs in preclinical and clinical models [58]. All of the compounding evidence has resulted in the inclusion of "polymorphic microbes" as a new emerging hallmark of cancer [59,60]. However, despite these encouraging data, the cellular and molecular underpinnings that critically regulate these interactions are yet to be completely elucidated.

Although not fully understood, it is thought that the gut microbiome influences host immunity and carcinogenesis through positive and negative interaction with other recognized hallmarks of cancer [59]. This is mediated by a number of mechanisms including (1) direct DNA damage and the disruption of systems that aim to maintain genomic integrity, (2) production of ligand mimetics that stimulate epithelial proliferation, (3) secretion of gut hormones, (4) elicitation of immune responses through cross-reactive microbial and tumor-associated antigens and (5) shifts in the gut ecosystem causing changes in the levels of microbial metabolites [34,61–66]. Whereas it is certainly challenging to ascertain which of these factors has the biggest influence in the context of cancer, there is an increasing body of evidence suggesting that microbial metabolites and secondary metabolites not only play a key role in the onset and development of numerous malignancies, but could also be drivers of response of systemic treatment, namely immunotherapy [48,50,51,54,67]. One such group of metabolites are short-chain fatty acids (SCFAs), such as butyrate and propionate, which originate from the bacterial fermentation of non-digestible carbohydrates, and have been implicated in the reduction of inflammation and regulation of CD4+ and CD8+ T cells [66,68–73]. Moreover, butyrate has also been shown to have a role in tumor suppression through the up- and down-regulation of genes involved in carcinogenesis [66,74,75]. Indeed, this SCFA seems to induce a pro-apoptotic effect through the increased expression of genes such as *Bax* and *Bak*, and has been proposed to have an additional tumor suppressing effect by regulating the *Wnt*/β-catenin signaling pathway and by reducing the expression of anti-apoptotic genes such as *Bcl-2* [73,74,76–78].

4. Gut Microbiome and Renal Cell Carcinoma

The treatment of metastatic renal cell carcinoma has changed dramatically over the past decades with the introduction of targeted treatment strategies with tyrosine kinase inhibitors such as sunitinib, pazopanib and cabozantinib and more recently with the approval of ICIs that target inhibitory molecules such as PD-1, PD-L1 and CTDLA-4 [8]. The use of this latter treatment modality, either alone or in combination with TT, has further improved the outcome of patients with mRCC and is currently the standard of care for first-line treatment of this disease. However, unlike other malignancies such as non-small cell lung cancer and melanoma, where the use of ICIs can be guided by PD-L1 tumor expression or tumor mutational burden, there are currently no validated biomarkers to predict response in patients with mRCC receiving ICIs [79–82]. Moreover, despite the improvements in efficacy seen with current treatment approaches, up to 60% of patients receiving these regimens fail to respond [83]. Hence, there is increasing need for both biomarkers of response that will allow us to identify the group of patients that will benefit the most from these treatments, and interventions that can allow us to maximize the benefit conferred by these approaches.

Given this context, as well as the large body of evidence linking the gut microbiome with the host's immune system and treatment response to ICIs in other malignancies, the role of the gut microbiome in mRCC and its potential as a biomarker of response and an intervention to improve treatment effectiveness are also being studied. Initial observations from several studies, the majority of which were retrospective in nature, have sought to indirectly determine the impact of gut dysbiosis in treatment response to ICIs by assessing for changes in the context of antibiotic treatment. Overall, the resulting evidence indicates that treatment with antibiotics is associated with decreased overall survival (OS), progression-free survival (PFS) and objective response rate (ORR) in patients with mRCC treated with standard-of-care ICIs [84]. Moreover, a study by De Rosa and colleagues further suggested that antibiotic treatment was associated with an alteration in the composition of the intestinal microbiota and the taxonomic beta diversity. Namely, this study noted an over-representation of bacteria, such as *Erysipelotrichaceae bacterium* and *Clostridium hathewayi*, suggesting that akin to the observations made for other cancer types, gut dysbiosis could also affect treatment response in RCC [85].

Additional studies further extended this line of inquiry and aimed to delineate this effect by assessing the impact of baseline gut microbiome profiles in patients receiving ICIs. This was performed by collecting stool specimens prior to the initiation of treatment and looking for the relative abundance of different bacteria using whole genome sequencing (WGS). These studies found that an increase in microbial diversity, as well as in relative abundance of certain bacterial species such as *Akkermansia muciniphila* and *Bifidobacterium spp.*, was associated with response to ICIs [50,85,86]. In contrast, data published by Park and colleagues who evaluated a cohort of NSCLC and RCC patients showed that a lack of treatment response was associated with an over-representation of the *Enterocloster* genus [85]. Further work presented by Alves during the 2022 ESMO symposium supported these findings, noting that not only was the baseline overrepresentation of the *Enterocloster* genus linked with a lack of treatment response but that those patients who do respond to ICIs exhibited a decrease in the *Enterocloster* genus representation after treatment [87].

Preclinical models have been in turn devised to evaluate the impact of gut microbiome interventions in treatment response and have shown that the direct administration of bacterial species associated with response in previous studies, such as *Bifidobacterium* and *Akkermancia muciniphila*, could delay tumor progression and restore treatment efficacy in mice treated with an immune checkpoint blockade [50,88]. Interestingly, it has also been shown that bacterial supplementation with *Clostridium butyricum MIYAIRI 588 (CBM 588)*, a probiotic bacterium, could lead to an increase in relative abundance of previously identified "beneficial bacteria" such as *Bifidobacterium* and *Lactobacillus* in mice, while also enhancing the intestinal barrier function [89].

Current studies in humans have intended to harness this effect to achieve an increased response to treatment and a reduction in treatment-related side effects using several strategies including (1) bacterial supplementation, (2) fecal microbiota transplantation and (3) diet modulation.

The first randomized clinical trial in this space was conducted by Dizman and colleagues. In the study, twenty patients with mRCC who were initiating VEGF-TKIs in any line of therapy were randomized to a probiotic-supplemented arm receiving a Bifidobacterium-containing yogurt, or a probiotic-restricted arm. Notably, all patients enrolled to the intervention arm reached detectable levels of *Bifidobacterium animalis*. Although no difference in clinical benefit was seen between these arms, whole metagenome sequencing identified that *Barnesiella intestinihominis* and *Akkermansia muciniphila* were significantly more abundant in patients achieving clinical benefit [90].

Another study was then carried out by the same group evaluating the effect of live bacterial supplementation with CBM588 in treatment naïve mRCC patients receiving ipilimumab with nivolumab for first-line treatment [91]. A total of 30 patients were randomized in a 2:1 fashion to the probiotic-containing and probiotic restricted arms, respectively. Despite the robust preclinical and clinical rationale behind its' primary endpoint of characterizing the effect of CBM588 on the relative abundance of *Bifidobacterium* spp., this endpoint was not met [54,91]. However, a significant advantage in PFS was seen in those receiving live bacterial supplementation over those receiving ipilimumab with nivolumab alone (12.7 vs. 2.5 months, hazard ratio 0.15, 95% CI 0.05–0.47, $p < 0.001$). Additionally, a comparable safety profile was seen among the two groups, with grade 3 and 4 adverse events being reported in 50% and 52% of patients in the control and intervention arms, respectively.

Despite this encouraging PFS signal, a remaining question is whether the effects of CBM588 will also be relevant in the context of newer combination strategies combining ICIs and TKIs. This is especially true in light of our growing understanding of the effect of TKIs in immune responses with several pieces of evidence suggesting that common TKI-driven effects such as VEGF blockage or more specific activity such as inhibition of MET and the TAM kinases, as seen with cabozantinib, could play an immunomodulatory role [92–95]. To answer this question, and given the encouraging safety profile seen in the aforementioned trial, a currently ongoing study will evaluate the effect of CBM588 in treatment-naïve patients receiving treatment with a combination of cabozantinib plus nivolumab as first-line therapy for mRCC [96] (Figure 1).

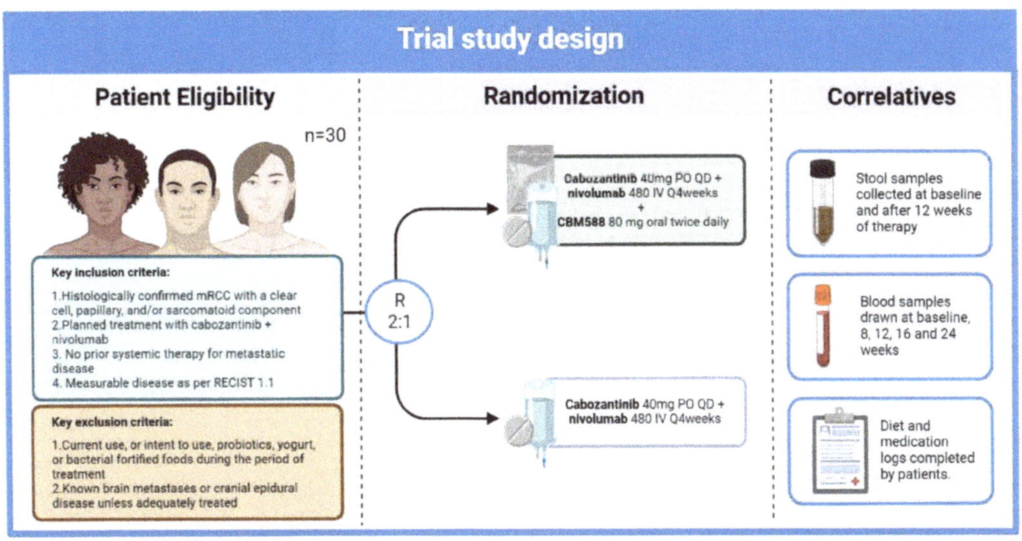

Figure 1. Study design for the phase I cabozantinib plus nivolumab +/− CBM588 trial [97].

FMT represents another microbiome-directed intervention with increasing momentum in the treatment of mRCC. Although there are still limited published data regarding the effect of this approach in this disease, current evidence suggests that FMT could improve mucosa-associated invariant T (MAIT) cell function in this patient population and boost immune surveillance against opportunistic pathogens that might be of relevance in the setting of cancer-mediated immunosuppression [98]. Furthermore, this intervention is also being evaluated as a way to reduce treatment-related toxicity. In a study conducted by Ianiro et al., FMT was employed to reduce TKI-induced diarrhea in mRCC patients. In his study, patients treated with donor-FMT showed a significant clinical improvement in TKI-induced diarrhea symptoms when compared to those receiving placebo [99]. Another interesting study is the currently ongoing PERFORM trial, one that will evaluate the prevention of treatment toxicity with immunotherapy utilizing this approach [100].

Beyond this, FMT is being evaluated as a tool to improve and induce response to ICIs in the TACITO and MITRIC trials, respectively. The TACITO trial is a randomized control trial of 50 mRCC patients to receive FMT or placebo and will evaluate the number of participants free of tumor progression [101]. In contrast, the MITRIC trial will enroll patients with solid tumors (including RCC) that have failed to respond to treatment. This is a single-arm, open-label study that will enroll 20 patients who will receive FMT from ICI-responders after experiencing progressive disease while on therapy with PD1/PD-L1 blockers and/or CTLA4-blockers [102]. The rationale behind these trials derives from pre-clinical evidence showing that FMT from patients responding to ICIs can successfully rescue primary resistance in RCC tumor-bearing mice [85]. Moreover, similar concepts have already been successfully implemented in cohorts of immunotherapy-refractory patients with melanoma [56,57].

Finally, dietary interventions are also underway in the KETOREIN trial. This is a non-randomized four-arm design that aims to evaluate a ketogenic diet used concomitantly with nivolumab plus ipilimumab in mRCC patients. This trial will evaluate objective response rate as its primary outcome and will enroll a total of 60 patients to one of four arms detailed in Figure 2 [103]. Results from this trial will build upon previously published pre-clinical data from Ferrere et al. suggesting that a ketogenic diet shifts the balance of the gut microbiota from tolerogenic to immunogenic bacteria (e.g., *Akkermancia muciniphila*) and induces an antineoplastic effect mediated by 3-hydroxybutyrate [104].

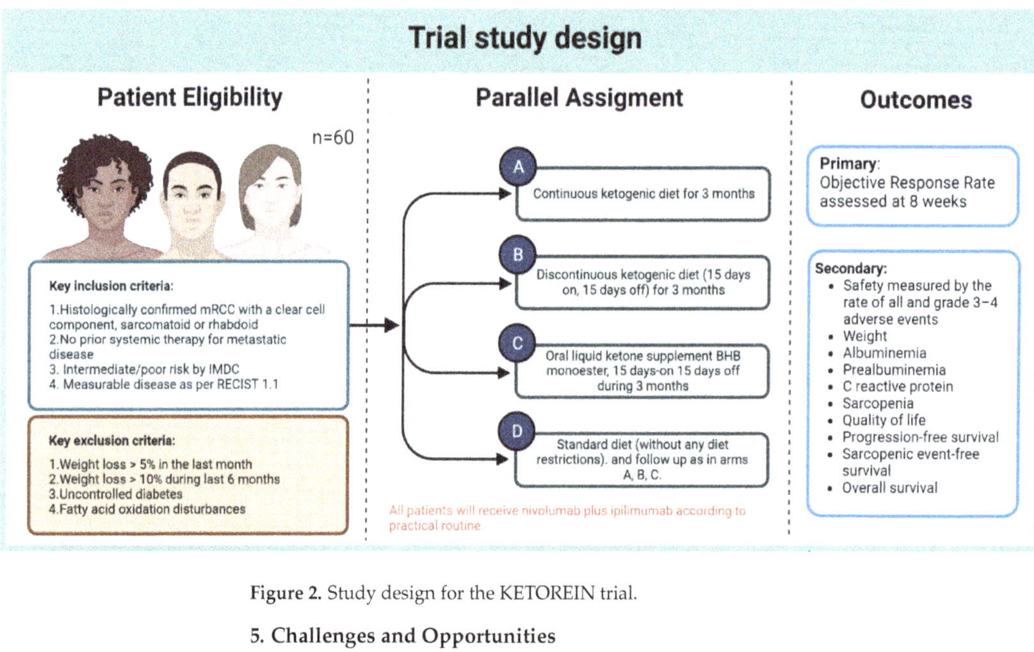

Figure 2. Study design for the KETOREIN trial.

5. Challenges and Opportunities

Historically, challenges related to the characterization of the microbiome were mostly attributable to technical limitations, especially considering that not all regular bacterial species are amenable to culture processes, and that cultivating viruses and fungi can be even more challenging. Moreover, body environments other than the gut are less colonized and have yielded disappointing results. It was only more recently, with the advent of advanced molecular techniques such as DNA sequencing and fluorescence in-situ hybridization of stool, blood and saliva samples, as well as intra-tumoral analysis, that a broader characterization of the human microbiome became independent from culture methods [105,106]. The most utilized tool as a strategy to surpass the challenge of obtaining reliable and high-quality samples is sequencing the 16S rRNA gene, which is present only in prokaryotic cells, with the drawback of identifying only bacteria [106]. Notwithstanding, even when high-throughput sequencing technologies are increasingly available, up to 50% of functional diversity remains unknown, a fact that is further complicated when including non-reference populations [105].

Furthermore, although evolutionary advances in next-generation sequencing technology have ushered in a new understanding of the interplay between the gut microbiome, immunity and cancer, several challenges are notable and represent barriers for its incorporation in routine clinical practice. Among these challenges are the lack of uniformity across the methodologies used for microbiome analysis (e.g., stool collection kits, probiotic restriction in the control arms, etc.), an issue that could explain the modest overlap in gut-microbiome profiles associated with response across studies (Table 1). Hence, the development and validation of a reference framework would be a promising approach to be incorporated in microbiome research that could facilitate collaboration and the comparison of results.

Table 1. Studies evaluating gut microbiome composition and treatment response in mRCC patients.

Study	Patient Population	Microorganism Associated with Response/Clinical Benefit	Treatment
Routy et al, Science (2018) [50]	Patients with metastatic RCC or NSCLC	*Akkermansia muciniphila, Ruminococcus, Alistipes*, and *Eubacterium*	Anti PD-L1
Derosa et al, European Urology, (2020) [85]	Patients with advanced RCC	*Akkermansia muciniphila, Bacteroides salyersiae*, and *Eubacterium siraeum*	Nivolumab
Salgia European Urology (2020) [86]	Patients with metastatic RCC	*Akkermansia municiphila, Prevotella copri, Feacalibacterium rumino Bifidobacterium adolescentis*, and *Barnesiella intestinihominis*	Nivolumab or nivolumab with ipilimumab
Dizman et al, Cancer Medicine (2021) [90]	Patients with metastatic RCC	*Akkermansia muciniphila, Barnesiella intestinihominis and Bacteroides caccae*	VEGF-TKI therapy
Dizman et al, Nature Medicine (2022) [91]	Patients with metastatic RCC	*Bifidobacterium* spp.	Nivolumab with ipilimumab +/− CBM588

Another important challenge is the limited sample size of most studies. Considering that microbiome profiling can be influenced by factors such as age, diet, socioeconomic status, geography and ethnicity, large sets of data are needed to identify and fully capture this heterogeneity [107]. Joint efforts analogous to The Cancer Genome Atlas (TCGA) could prove beneficial in better understanding the immune–microbiome interface. Initiated in 2006, TCGA consisted of a collaboration across multiple institutions and with the labor of a myriad of multidisciplinary specialists to collect and analyze data from over 20,000 samples across 33 different cancer types to elucidate genomic aspects of cancer. A similar approach would be an important step in microbiome research, with the collection of information from multiple centers, including academic and community sites, able to create a more robust database and provide the foundation for insights into different microbiome compositions. Ongoing population-wide initiatives such as The Human Microbiome Project (HMP) in the United States, the Metagenomes of the Human Intestinal Tract (MetaHIT) in Europe, and a diabetes cohort in China have already managed to survey around 2000 individuals [108–110].

The City of Hope is one medical group primed to help in this collaborative effort. With over 30 different locations across Southern California, this network is well positioned to conduct studies that collect samples representative of a broad population. Not only accounting for the diverse ethnic backgrounds present in the state of California, but also socioeconomic and cultural factors, can help broaden the resident microbiota. Additionally, the institution recently broadened its area of influence and cancer care beyond its original regional borders by acquiring the Cancer Treatment Centers of America group, which has a well-established presence in Georgia, Illinois and Arizona. This will hopefully allow for nation-wide studies that will provide a wider look at the composition of what constitutes a normal microbiome and will help better determine the changes seen during treatment and survey differences across various patient groups.

Admittedly, this collaborative endeavor would require the contribution of experts in many areas of biomedical research. Physicians and patient care personnel would identify eligible candidates to provide samples. Basic science researchers would identify strains, elucidate molecular pathways, and understand the gut microbiome's modulatory effects. In turn, bioinformaticians and data scientists would play a role in identifying correlations and scrutinizing data. With recent studies relating the gut microbiome to cancer treatment response and toxicity, it is particularly important for basic scientists to use animal models to understand the mechanism behind these findings. Communication across all levels of the chain of care is required to streamline such an effort and translate findings to patient care and the clinical setting.

With broad patient samples, physicians from multiple sites, and basic science labs working together, we can broaden our understanding of the mechanisms driving microbiome modulatory effects and use this knowledge to provide more personalized treatment options for patients. For example, in a certain cancer population, if malnutrition or a poor microbiota diversity is identified, we might be able to correct the course of treatment and increase the odds of response and perhaps extend survival by administering live bacterial products, as early-phase data have suggested, with larger confirmatory trials underway [111].

6. Conclusions and Future Directions

In summary, the gut microbiome represents an area of emerging interest in oncology. Difficulties faced during initial efforts for the characterization of the vast array of microorganisms that reside in the human body have now been largely addressed by the introduction and use of next-generation sequencing technologies. It is now well accepted that the microorganisms living in the gut have an impact across many disease settings, including cancer, and studies have further implicated the gut microbiome as a potential biomarker for response in many cancer types including mRCC.

Furthermore, randomized clinical trials in the mRCC space have produced encouraging results supporting the use of microbiome-based interventions to increase the effectiveness of systemic therapy and reduce toxicity. Ongoing clinical trials are seeking to validate these findings in larger cohorts, as well as address other clinically relevant questions, including the effect of dietary interventions in treatment outcomes. However, much work remains to be done before microbiome-based interventions can have a tangible impact in routine clinical practice.

Namely, there is an unmet need for longitudinal microbiome data at the individual and population level that can provide insights into the heterogeneity of the gut microbiome across different patient populations. Hence, future research efforts should aim to include diverse patient populations, as well as carefully annotated correlatives including genomic, epigenomic and metabolomic data, all of which will help us elucidate the factors driving differences between patient cohorts. Admittedly, such projects will necessitate large intra- and inter-institutional collaborations which remain, to date, a largely unfulfilled opportunity.

Author Contributions: Conceptualization, L.M., M.F., K.L., R.S. and S.K.P.; Data Curation, L.M., M.F., K.L., R.S. and S.K.P.; Writing—Original Draft Preparation L.M., M.F., K.L., R.S. and S.K.P.; Writing—Review & Editing, L.M., M.F., K.L., R.S. and S.K.P.; Visualization, L.M., M.F., K.L., R.S. and S.K.P.; Supervision, S.K.P. All authors have read and agreed to the published version of the manuscript.

Funding: No funding was received for the preparation of this manuscript.

Conflicts of Interest: Luis Meza, Matthew Feng, Kyle Lee, Rubens Sperandio and Sumanta K. Pal have no conflict of interest that might be relevant to the contents of this manuscript.

References

1. Key Statistics about Kidney Cancer. Available online: https://www.cancer.org/cancer/kidney-cancer/about/key-statistics.html (accessed on 27 May 2022).
2. Motzer, R.J.; Hutson, T.E.; Tomczak, P.; Michaelson, M.D.; Bukowski, R.M.; Rixe, O.; Oudard, S.; Negrier, S.; Szczylik, C.; Kim, S.T.; et al. Sunitinib versus Interferon Alfa in Metastatic Renal-Cell Carcinoma. *N. Engl. J. Med.* **2007**, *356*, 115–124. [CrossRef]
3. Sternberg, C.N.; Davis, I.D.; Mardiak, J.; Szczylik, C.; Lee, E.; Wagstaff, J.; Barrios, C.H.; Salman, P.; Gladkov, O.A.; Kavina, A.; et al. Pazopanib in Locally Advanced or Metastatic Renal Cell Carcinoma: Results of a Randomized Phase III Trial. *J. Clin. Oncol.* **2010**, *28*, 1061–1068. [CrossRef]
4. Rini, B.I.; Escudier, B.; Tomczak, P.; Kaprin, A.; Szczylik, C.; Hutson, T.E.; Michaelson, M.D.; Gorbunova, V.A.; Gore, M.E.; Rusakov, I.G.; et al. Comparative effectiveness of axitinib versus sorafenib in advanced renal cell carcinoma (AXIS): A randomised phase 3 trial. *Lancet* **2011**, *378*, 1931–1939. [CrossRef]
5. Choueiri, T.K.; Halabi, S.; Sanford, B.L.; Hahn, O.; Michaelson, M.D.; Walsh, M.K.; Feldman, D.; Olencki, T.; Picus, J.; Small, E.J.; et al. Cabozantinib Versus Sunitinib As Initial Targeted Therapy for Patients with Metastatic Renal Cell Carcinoma of Poor or Intermediate Risk: The Alliance A031203 CABOSUN Trial. *J. Clin. Oncol.* **2017**, *35*, 591–597. [CrossRef]
6. Motzer, R.J.; Hutson, T.E.; Glen, H.; Michaelson, M.D.; Molina, A.; Eisen, T.; Jassem, J.; Zolnierek, J.; Maroto, J.P.; Mellado, B.; et al. Lenvatinib, everolimus, and the combination in patients with metastatic renal cell carcinoma: A randomised, phase 2, open-label, multicentre trial. *Lancet Oncol.* **2015**, *16*, 1473–1482. [CrossRef]
7. Hudes, G.; Carducci, M.; Tomczak, P.; Dutcher, J.; Figlin, R.; Kapoor, A.; Staroslawska, E.; Sosman, J.; McDermott, D.; Bodrogi, I.; et al. Temsirolimus, Interferon Alfa, or Both for Advanced Renal-Cell Carcinoma. *N. Engl. J. Med.* **2007**, *356*, 2271–2281. [CrossRef]
8. Meza, L.; Malhotra, J.; Favorito, C.; Pal, S.K. Cabozantinib plus immunotherapy combinations in metastatic renal cell and urothelial carcinoma. *Futur. Oncol.* **2022**, *18*, 21–33. [CrossRef]
9. SEER. Cancer of the Kidney and Renal Pelvis—Cancer Stat Facts. Available online: https://seer.cancer.gov/statfacts/html/kidrp.html (accessed on 4 February 2023).
10. Dudani, S.; Savard, M.-F.; Heng, D.Y. An Update on Predictive Biomarkers in Metastatic Renal Cell Carcinoma. *Eur. Urol. Focus* **2020**, *6*, 34–36. [CrossRef]
11. Sepich-Poore, G.D.; Zitvogel, L.; Straussman, R.; Hasty, J.; Wargo, J.A.; Knight, R. The microbiome and human cancer. *Science* **2021**, *371*, eabc4552. [CrossRef]
12. Bianconi, E.; Piovesan, A.; Facchin, F.; Beraudi, A.; Casadei, R.; Frabetti, F.; Vitale, L.; Pelleri, M.C.; Tassani, S.; Piva, F.; et al. An estimation of the number of cells in the human body. *Ann. Hum. Biol.* **2013**, *40*, 463–471. [CrossRef]
13. Sender, R.; Fuchs, S.; Milo, R. Revised Estimates for the Number of Human and Bacteria Cells in the Body. *PLOS Biol.* **2016**, *14*, e1002533. [CrossRef]
14. Thomas, S.; Izard, J.; Walsh, E.; Batich, K.; Chongsathidkiet, P.; Clarke, G.; Sela, D.A.; Muller, A.J.; Mullin, J.M.; Albert, K.; et al. The Host Microbiome Regulates and Maintains Human Health: A Primer and Perspective for Non-Microbiologists. *Cancer Res.* **2017**, *77*, 1783–1812. [CrossRef]
15. Schluter, J.; Peled, J.; Taylor, B.P.; Markey, K.A.; Smith, J.A.; Taur, Y.; Niehus, R.; Staffas, A.; Dai, A.; Fontana, E.; et al. The gut microbiota is associated with immune cell dynamics in humans. *Nature* **2020**, *588*, 303–307. [CrossRef]
16. Mazmanian, S.K.; Liu, C.H.; Tzianabos, A.O.; Kasper, D.L. An Immunomodulatory Molecule of Symbiotic Bacteria Directs Maturation of the Host Immune System. *Cell* **2005**, *122*, 107–118. [CrossRef]
17. Gomez de Agüero, M.; Ganal-Vonarburg, S.C.; Fuhrer, T.; Rupp, S.; Uchimura, Y.; Li, H.; Steinert, A.; Heikenwalder, M.; Hapfelmeier, S.; Sauer, U.; et al. The maternal microbiota drives early postnatal innate immune development. *Science* **2016**, *351*, 1296–1302. [CrossRef]
18. Olin, A.; Henckel, E.; Chen, Y.; Lakshmikanth, T.; Pou, C.; Mikes, J.; Gustafsson, A.; Bernhardsson, A.K.; Zhang, C.; Bohlin, K.; et al. Stereotypic Immune System Development in Newborn Children. *Cell* **2018**, *174*, 1277–1292.e14. [CrossRef]
19. Tan, T.G.; Sefik, E.; Geva-Zatorsky, N.; Kua, L.; Naskar, D.; Teng, F.; Pasman, L.; Ortiz-Lopez, A.; Jupp, R.; Wu, H.-J.J.; et al. Identifying species of symbiont bacteria from the human gut that, alone, can induce intestinal Th17 cells in mice. *Proc. Natl. Acad. Sci. USA* **2016**, *113*, E8141–E8150. [CrossRef]
20. Deshmukh, H.S.; Liu, Y.; Menkiti, O.R.; Mei, J.; Dai, N.; O'Leary, C.E.; Oliver, P.M.; Kolls, J.K.; Weiser, J.N.; Worthen, G.S. The microbiota regulates neutrophil homeostasis and host resistance to Escherichia coli K1 sepsis in neonatal mice. *Nat. Med.* **2014**, *20*, 524–530. [CrossRef]
21. Ivanov, I.I.; de Llanos Frutos, R.; Manel, N.; Yoshinaga, K.; Rifkin, D.B.; Sartor, R.B.; Finlay, B.B.; Littman, D.R. Specific Microbiota Direct the Differentiation of IL-17-Producing T-Helper Cells in the Mucosa of the Small Intestine. *Cell Host Microbe* **2008**, *4*, 337–349. [CrossRef]
22. Geva-Zatorsky, N.; Sefik, E.; Kua, L.; Pasman, L.; Tan, T.G.; Ortiz-Lopez, A.; Yanortsang, T.B.; Yang, L.; Jupp, R.; Mathis, D.; et al. Mining the Human Gut Microbiota for Immunomodulatory Organisms. *Cell* **2017**, *168*, 928–943.e11. [CrossRef]
23. Zaneveld, J.; Turnbaugh, P.; Lozupone, C.; Ley, R.; Hamady, M.; Gordon, J.I.; Knight, R. Host-bacterial coevolution and the search for new drug targets. *Curr. Opin. Chem. Biol.* **2008**, *12*, 109–114. [CrossRef] [PubMed]
24. Khan, I.; Khan, I.; Usman, M.; Wei, Z.X.; Ping, X.; Khan, S.; Khan, F.; Jianye, Z.; Zhiqiang, L.; Lizhe, A. Circulating microbiota and metabolites: Insights into cardiovascular diseases. *J. Clin. Lab. Anal.* **2022**, *36*, e24779. [CrossRef]
25. Lloyd-Price, J.; Arze, C.; Ananthakrishnan, A.N.; Schirmer, M.; Avila-Pacheco, J.; Poon, T.W.; Andrews, E.; Ajami, N.J.; Bonham, K.S.; Brislawn, C.J.; et al. Multi-omics of the gut microbial ecosystem in inflammatory bowel diseases. *Nature* **2019**, *569*, 655–662. [CrossRef]

26. Azzouz, D.; Omarbekova, A.; Heguy, A.; Schwudke, D.; Gisch, N.; Rovin, B.H.; Caricchio, R.; Buyon, J.P.; Alekseyenko, A.V.; Silverman, G.J. Lupus nephritis is linked to disease-activity associated expansions and immunity to a gut commensal. *Ann. Rheum. Dis.* **2019**, *78*, 947–956. [CrossRef]
27. Ianiro, G.; Maida, M.; Burisch, J.; Simonelli, C.; Hold, G.; Ventimiglia, M.; Gasbarrini, A.; Cammarota, G. Efficacy of different faecal microbiota transplantation protocols for Clostridium difficile infection: A systematic review and meta-analysis. *United Eur. Gastroenterol. J.* **2018**, *6*, 1232–1244. [CrossRef] [PubMed]
28. Kong, L.; Lloyd-Price, J.; Vatanen, T.; Seksik, P.; Beaugerie, L.; Simon, T.; Vlamakis, H.; Sokol, H.; Xavier, R.J. Linking Strain Engraftment in Fecal Microbiota Transplantation with Maintenance of Remission in Crohn's Disease. *Gastroenterology* **2020**, *159*, 2193–2202.e5. [CrossRef] [PubMed]
29. Koote, R.S.; Levin, E.; Salojärvi, J.; Smits, L.P.; Hartstra, A.V.; Udayappan, S.D.; Hermes, G.; Bouter, K.E.; Koopen, A.M.; Holst, J.J.; et al. Improvement of Insulin Sensitivity after Lean Donor Feces in Metabolic Syndrome Is Driven by Baseline Intestinal Microbiota Composition. *Cell Metab.* **2017**, *26*, 611–619.e6. [CrossRef]
30. Rous, P. A Sarcoma of the Fowl Transmissible by an Agent Separable from the Tumor Cells. *J. Exp. Med.* **1911**, *13*, 397–411. [CrossRef]
31. Yang, S.; Zhao, S.; Ye, Y.; Jia, L.; Lou, Y. Global research trends on the links between gut microbiota and cancer immunotherapy: A bibliometric analysis (2012–2021). *Front. Immunol.* **2022**, *13*, 952546. [CrossRef]
32. White, M.K.; Pagano, J.S.; Khalili, K. Viruses and Human Cancers: A Long Road of Discovery of Molecular Paradigms. *Clin. Microbiol. Rev.* **2014**, *27*, 463–481. [CrossRef]
33. Knippel, R.J.; Drewes, J.L.; Sears, C.L. The Cancer Microbiome: Recent Highlights and Knowledge Gaps. *Cancer Discov.* **2021**, *11*, 2378–2395. [CrossRef] [PubMed]
34. Park, E.M.; Chelvanambi, M.; Bhutiani, N.; Kroemer, G.; Zitvogel, L.; Wargo, J.A. Targeting the gut and tumor microbiota in cancer. *Nat. Med.* **2022**, *28*, 690–703. [CrossRef] [PubMed]
35. Yonekura, S.; Terrisse, S.; Silva, C.A.C.; Lafarge, A.; Iebba, V.; Ferrere, G.; Goubet, A.-G.; Fahrner, J.-E.; Lahmar, I.; Ueda, K.; et al. Cancer Induces a Stress Ileopathy Depending on β-Adrenergic Receptors and Promoting Dysbiosis that Contributes to Carcinogenesis. *Cancer Discov.* **2021**, *12*, 1128–1151. [CrossRef] [PubMed]
36. Brennan, C.A.; Garrett, W.S. Gut Microbiota, Inflammation, and Colorectal Cancer. *Annu. Rev. Microbiol.* **2016**, *70*, 395–411. [CrossRef] [PubMed]
37. Yang, Y.; Jobin, C. Novel insights into microbiome in colitis and colorectal cancer. *Curr. Opin. Gastroenterol.* **2017**, *33*, 422–427. [CrossRef]
38. Mei, S.; Deng, Z.; Chen, Y.; Ning, D.; Guo, Y.; Fan, X.; Wang, R.; Meng, Y.; Zhou, Q.; Tian, X. Dysbiosis: The first hit for digestive system cancer. *Front. Physiol.* **2022**, *13*, 1040991. [CrossRef]
39. Mima, K.; Nakagawa, S.; Sawayama, H.; Ishimoto, T.; Imai, K.; Iwatsuki, M.; Hashimoto, D.; Baba, Y.; Yamashita, Y.-I.; Yoshida, N.; et al. The microbiome and hepatobiliary-pancreatic cancers. *Cancer Lett.* **2017**, *402*, 9–15. [CrossRef]
40. Wu, R.; Yu, I.; Tokumaru, Y.; Asaoka, M.; Oshi, M.; Yan, L.; Okuda, S.; Ishikawa, T.; Takabe, K. Elevated bile acid metabolism and microbiome are associated with suppressed cell proliferation and better survival in breast cancer. *Am. J. Cancer Res.* **2022**, *12*, 5271–5285.
41. Mekadim, C.; Skalnikova, H.K.; Cizkova, J.; Cizkova, V.; Palanova, A.; Horak, V.; Mrazek, J. Dysbiosis of skin microbiome and gut microbiome in melanoma progression. *BMC Microbiol.* **2022**, *22*, 63. [CrossRef]
42. Viaud, S.; Saccheri, F.; Mignot, G.; Yamazaki, T.; Daillère, R.; Hannani, D.; Enot, D.P.; Pfirschke, C.; Engblom, C.; Pittet, M.J.; et al. The Intestinal Microbiota Modulates the Anticancer Immune Effects of Cyclophosphamide. *Science* **2013**, *342*, 971–976. [CrossRef]
43. Iida, N.; Dzutsev, A.; Stewart, C.A.; Smith, L.; Bouladoux, N.; Weingarten, R.A.; Molina, D.A.; Salcedo, R.; Back, T.; Cramer, S.; et al. Commensal Bacteria Control Cancer Response to Therapy by Modulating the Tumor Microenvironment. *Science* **2019**, *342*, 967–970. [CrossRef] [PubMed]
44. Weber, D.; Oefner, P.J.; Hiergeist, A.; Koestler, J.; Gessner, A.; Weber, M.; Hahn, J.; Wolff, D.; Stämmler, F.; Spang, R.; et al. Low urinary indoxyl sulfate levels early after transplantation reflect a disrupted microbiome and are associated with poor outcome. *Blood* **2015**, *126*, 1723–1728. [CrossRef]
45. Jenq, R.R.; Ubeda, C.; Taur, Y.; Menezes, C.C.; Khanin, R.; Dudakov, J.A.; Liu, C.; West, M.L.; Singer, N.V.; Equinda, M.J.; et al. Regulation of intestinal inflammation by microbiota following allogeneic bone marrow transplantation. *J. Exp. Med.* **2012**, *209*, 903–911. [CrossRef] [PubMed]
46. Simpson, R.C.; Shanahan, E.R.; Batten, M.; Reijers, I.L.M.; Read, M.; Silva, I.P.; Versluis, J.M.; Ribeiro, R.; Angelatos, A.S.; Tan, J.; et al. Diet-driven microbial ecology underpins associations between cancer immunotherapy outcomes and the gut microbiome. *Nat. Med.* **2022**, *28*, 2344–2352. [CrossRef]
47. Liu, W.; Ma, F.; Sun, B.; Liu, Y.; Tang, H.; Luo, J.; Chen, H.; Luo, Z. Intestinal Microbiome Associated with Immune-Related Adverse Events for Patients Treated with Anti-PD-1 Inhibitors, a Real-World Study. *Front. Immunol.* **2021**, *12*, 5334. Available online: https://www.frontiersin.org/articles/10.3389/fimmu.2021.756872 (accessed on 5 February 2023). [CrossRef]
48. Andrews, M.C.; Duong, C.P.M.; Gopalakrishnan, V.; Iebba, V.; Chen, W.-S.; Derosa, L.; Khan, A.W.; Cogdill, A.P.; White, M.G.; Wong, M.C.; et al. Gut microbiota signatures are associated with toxicity to combined CTLA-4 and PD-1 blockade. *Nat. Med.* **2021**, *27*, 1432–1441. [CrossRef] [PubMed]

49. Wang, Y.; Wiesnoski, D.H.; Helmink, B.A.; Gopalakrishnan, V.; Choi, K.; DuPont, H.L.; Jiang, Z.-D.; Abu-Sbeih, H.; Sanchez, C.A.; Chang, C.-C.; et al. Fecal microbiota transplantation for refractory immune checkpoint inhibitor-associated colitis. *Nat. Med.* **2018**, *24*, 1804–1808. [CrossRef]
50. Routy, B.; le Chatelier, E.; DeRosa, L.; Duong, C.P.M.; Alou, M.T.; Daillère, R.; Fluckiger, A.; Messaoudene, M.; Rauber, C.; Roberti, M.P.; et al. Gut microbiome influences efficacy of PD-1–based immunotherapy against epithelial tumors. *Science* **2018**, *359*, 91–97. [CrossRef]
51. Gopalakrishnan, V.; Spencer, C.N.; Nezi, L.; Reuben, A.; Andrews, M.C.; Karpinets, T.V.; Prieto, P.A.; Vicente, D.; Hoffman, K.; Wei, S.C.; et al. Gut microbiome modulates response to anti–PD-1 immunotherapy in melanoma patients. *Science* **2018**, *359*, 97–103. [CrossRef]
52. Frankel, A.E.; Coughlin, L.A.; Kim, J.; Froehlich, T.W.; Xie, Y.; Frenkel, E.P.; Koh, A.Y. Metagenomic Shotgun Sequencing and Unbiased Metabolomic Profiling Identify Specific Human Gut Microbiota and Metabolites Associated with Immune Checkpoint Therapy Efficacy in Melanoma Patients. *Neoplasia* **2017**, *19*, 848–855. [CrossRef]
53. Fernandes, M.R.; Aggarwal, P.; Costa, R.G.F.; Cole, A.M.; Trinchieri, G. Targeting the gut microbiota for cancer therapy. *Nat. Rev. Cancer* **2022**, *22*, 703–722. [CrossRef] [PubMed]
54. Tomita, Y.; Ikeda, T.; Sakata, S.; Saruwatari, K.; Sato, R.; Iyama, S.; Jodai, T.; Akaike, K.; Ishizuka, S.; Saeki, S.; et al. Association of Probiotic Clostridium butyricum Therapy with Survival and Response to Immune Checkpoint Blockade in Patients with Lung Cancer. *Cancer Immunol. Res.* **2020**, *8*, 1236–1242. [CrossRef]
55. Matson, V.; Fessler, J.; Bao, R.; Chongsuwat, T.; Zha, Y.; Alegre, M.-L.; Luke, J.J.; Gajewski, T.F. The commensal microbiome is associated with anti–PD-1 efficacy in metastatic melanoma patients. *Science* **2018**, *359*, 104–108. [CrossRef]
56. Davar, D.; Dzutsev, A.K.; McCulloch, J.A.; Rodrigues, R.R.; Chauvin, J.-M.; Morrison, R.M.; Deblasio, R.N.; Menna, C.; Ding, Q.; Pagliano, O.; et al. Fecal microbiota transplant overcomes resistance to anti–PD-1 therapy in melanoma patients. *Science* **2021**, *371*, 595–602. [CrossRef] [PubMed]
57. Baruch, E.N.; Youngster, I.; Ben-Betzalel, G.; Ortenberg, R.; Lahat, A.; Katz, L.; Adler, K.; Dick-Necula, D.; Raskin, S.; Bloch, N.; et al. Fecal microbiota transplant promotes response in immunotherapy-refractory melanoma patients. *Science* **2020**, *371*, 602–609. [CrossRef]
58. Spencer, C.N.; McQuade, J.L.; Gopalakrishnan, V.; McCulloch, J.A.; Vetizou, M.; Cogdill, A.P.; Khan, A.W.; Zhang, X.; White, M.G.; Peterson, C.B.; et al. Dietary fiber and probiotics influence the gut microbiome and melanoma immunotherapy response. *Science* **2021**, *374*, 1632–1640. [CrossRef]
59. Lythgoe, M.P.; Mullish, B.H.; Frampton, A.E.; Krell, J. Polymorphic microbes: A new emerging hallmark of cancer. *Trends Microbiol.* **2022**, *30*, 1131–1134. [CrossRef]
60. Hanahan, D. Hallmarks of Cancer: New Dimensions. *Cancer Discov.* **2022**, *12*, 31–46. [CrossRef]
61. Pleguezuelos-Manzano, C.; Puschhof, J.; Rosendahl Huber, A.; Van Hoeck, A.; Wood, H.M.; Nomburg, J.; Gurjao, C.; Manders, F.; Dalmasso, G.; Stege, P.B.; et al. Mutational signature in colorectal cancer caused by genotoxic pks+ E coli. *Nature* **2020**, *580*, 269–273. [CrossRef]
62. Helmink, B.A.; Khan, M.A.W.; Hermann, A.; Gopalakrishnan, V.; Wargo, J.A. The microbiome, cancer, and cancer therapy. *Nat. Med.* **2019**, *25*, 377–388. [CrossRef] [PubMed]
63. Yoon, H.S.; Cho, C.H.; Yun, M.S.; Jang, S.J.; You, H.J.; Kim, J.-H.; Han, D.; Cha, K.H.; Moon, S.H.; Lee, K.; et al. Akkermansia muciniphila secretes a glucagon-like peptide-1-inducing protein that improves glucose homeostasis and ameliorates metabolic disease in mice. *Nat. Microbiol.* **2021**, *6*, 563–573. [CrossRef] [PubMed]
64. Derosa, L.; Routy, B.; Desilets, A.; Daillère, R.; Terrisse, S.; Kroemer, G.; Zitvogel, L. Microbiota-Centered Interventions: The Next Breakthrough in Immuno-Oncology? *Cancer Discov.* **2021**, *11*, 2396–2412. [CrossRef] [PubMed]
65. Okumura, S.; Konishi, Y.; Narukawa, M.; Sugiura, Y.; Yoshimoto, S.; Arai, Y.; Sato, S.; Yoshida, Y.; Tsuji, S.; Uemura, K.; et al. Gut bacteria identified in colorectal cancer patients promote tumourigenesis via butyrate secretion. *Nat. Commun.* **2021**, *12*, 5674. [CrossRef]
66. Rossi, T.; Vergara, D.; Fanini, F.; Maffia, M.; Bravaccini, S.; Pirini, F. Microbiota-Derived Metabolites in Tumor Progression and Metastasis. *Int. J. Mol. Sci.* **2020**, *21*, 5786. [CrossRef] [PubMed]
67. Blake, S.J.; James, J.; Ryan, F.J.; Caparros-Martin, J.; Eden, G.L.; Tee, Y.C.; Salamon, J.R.; Benson, S.C.; Tumes, D.J.; Sribnaia, A.; et al. The immunotoxicity, but not anti-tumor efficacy, of anti-CD40 and anti-CD137 immunotherapies is dependent on the gut microbiota. *Cell Rep. Med.* **2021**, *2*, 100464. [CrossRef]
68. Furusawa, Y.; Obata, Y.; Fukuda, S.; Endo, T.A.; Nakato, G.; Takahashi, D.; Nakanishi, Y.; Uetake, C.; Kato, K.; Kato, T.; et al. Commensal microbe-derived butyrate induces the differentiation of colonic regulatory T cells. *Nature* **2013**, *504*, 446–450. [CrossRef]
69. Makki, K.; Deehan, E.C.; Walter, J.; Bäckhed, F. The Impact of Dietary Fiber on Gut Microbiota in Host Health and Disease. *Cell Host Microbe* **2018**, *23*, 705–715. [CrossRef]
70. Morrison, D.J.; Preston, T. Formation of short chain fatty acids by the gut microbiota and their impact on human metabolism. *Gut Microbes* **2016**, *7*, 189–200. [CrossRef]
71. Kespohl, M.; Vachharajani, N.; Luu, M.; Harb, H.; Pautz, S.; Wolff, S.; Sillner, N.; Walker, A.; Schmitt-Kopplin, P.; Boettger, T.; et al. The Microbial Metabolite Butyrate Induces Expression of Th1-Associated Factors in CD4+ T Cells. *Front. Immunol.* **2017**, *8*, 1036. [CrossRef]

72. Luu, M.; Weigand, K.; Wedi, F.; Breidenbend, C.; Leister, H.; Pautz, S.; Adhikary, T.; Visekruna, A. Regulation of the effector function of CD8+ T cells by gut microbiota-derived metabolite butyrate. *Sci. Rep.* **2018**, *8*, 14430. [CrossRef]
73. Salvi, P.S.; Cowles, R.A. Butyrate and the Intestinal Epithelium: Modulation of Proliferation and Inflammation in Homeostasis and Disease. *Cells* **2021**, *10*, 1775. [CrossRef]
74. Bordonaro, M.; Lazarova, D.L.; Augenlicht, L.H.; Sartorelli, A.C. Cell type- and promoter-dependent modulation of the Wnt signaling pathway by sodium butyrate. *Int. J. Cancer* **2001**, *97*, 42–51. [CrossRef] [PubMed]
75. Bordonaro, M.; Lazarova, D.L.; Sartorelli, A.C. Role of Tcf-DNA binding and the chromatin remodeling factor Brg-1 in the modulation of Wnt activity by butyrate. *Cell Cycle* **2008**, *7*, 3472–3473. [CrossRef] [PubMed]
76. Mandal, M.; Olson, D.J.; Sharma, T.; Vadlamudi, R.K.; Kumar, R. Butyric acid induces apoptosis by up-regulating Bax expression via stimulation of the c-Jun N-terminal kinase/activation protein-1 pathway in human colon cancer cells. *Gastroenterology* **2001**, *120*, 71–78. [CrossRef] [PubMed]
77. Thangaraju, M.; Cresci, G.A.; Liu, K.; Ananth, S.; Gnanaprakasam, J.P.; Browning, D.D.; Mellinger, J.D.; Smith, S.B.; Digby, G.J.; Lambert, N.A.; et al. GPR109A Is a G-protein–Coupled Receptor for the Bacterial Fermentation Product Butyrate and Functions as a Tumor Suppressor in Colon. *Cancer Res.* **2009**, *69*, 2826–2832. [CrossRef]
78. Chen, D.; Jin, D.; Huang, S.; Wu, J.; Xu, M.; Liu, T.; Dong, W.; Liu, X.; Wang, S.; Zhong, W.; et al. Clostridium butyricum, a butyrate-producing probiotic, inhibits intestinal tumor development through modulating Wnt signaling and gut microbiota. *Cancer Lett.* **2019**, *469*, 456–467. [CrossRef]
79. Miao, D.; Margolis, C.A.; Vokes, N.I.; Liu, D.; Taylor-Weiner, A.; Wankowicz, S.M.; Adeegbe, D.; Keliher, D.; Schilling, B.; Tracy, A.; et al. Genomic correlates of response to immune checkpoint blockade in microsatellite-stable solid tumors. *Nat. Genet.* **2018**, *50*, 1271–1281. [CrossRef]
80. McDermott, D.F.; Huseni, M.A.; Atkins, M.B.; Motzer, R.J.; Rini, B.I.; Escudier, B.; Fong, L.; Joseph, R.W.; Pal, S.K.; Reeves, J.A.; et al. Clinical activity and molecular correlates of response to atezolizumab alone or in combination with bevacizumab versus sunitinib in renal cell carcinoma. *Nat. Med.* **2018**, *24*, 749–757. [CrossRef]
81. Garon, E.B.; Rizvi, N.A.; Hui, R.; Leighl, N.; Balmanoukian, A.S.; Eder, J.P.; Patnaik, A.; Aggarwal, C.; Gubens, M.; Horn, L.; et al. Pembrolizumab for the Treatment of Non–Small-Cell Lung Cancer. *N. Engl. J. Med.* **2015**, *372*, 2018–2028. [CrossRef]
82. Tumeh, P.C.; Harview, C.L.; Yearley, J.H.; Shintaku, I.P.; Taylor, E.J.M.; Robert, L.; Chmielowski, B.; Spasic, M.; Henry, G.; Ciobanu, V.; et al. PD-1 blockade induces responses by inhibiting adaptive immune resistance. *Nature* **2014**, *515*, 568–571. [CrossRef]
83. Albiges, L.; Tannir, N.M.; Burotto, M.; McDermott, D.; Plimack, E.R.; Barthélémy, P.; Porta, C.; Powles, T.; Donskov, F.; George, S.; et al. Nivolumab plus ipilimumab versus sunitinib for first-line treatment of advanced renal cell carcinoma: Extended 4-year follow-up of the phase III CheckMate 214 trial. *ESMO Open* **2020**, *5*, e001079. [CrossRef]
84. Derosa, L.; Hellmann, M.D.; Spaziano, M.; Halpenny, D.; Fidelle, M.; Rizvi, H.; Long, N.; Plodkowski, A.J.; Arbour, K.C.; Chaft, J.E.; et al. Negative association of antibiotics on clinical activity of immune checkpoint inhibitors in patients with advanced renal cell and non-small-cell lung cancer. *Ann. Oncol.* **2018**, *29*, 1437–1444. [CrossRef]
85. Derosa, L.; Routy, B.; Fidelle, M.; Iebba, V.; Alla, L.; Pasolli, E.; Segata, N.; Desnoyer, A.; Pietrantonio, F.; Ferrere, G.; et al. Gut Bacteria Composition Drives Primary Resistance to Cancer Immunotherapy in Renal Cell Carcinoma Patients. *Eur. Urol.* **2020**, *78*, 195–206. [CrossRef] [PubMed]
86. Salgia, N.J.; Bergerot, P.G.; Maia, M.C.; Dizman, N.; Hsu, J.; Gillece, J.D.; Folkerts, M.; Reining, L.; Trent, J.; Highlander, S.K.; et al. Stool Microbiome Profiling of Patients with Metastatic Renal Cell Carcinoma Receiving Anti–PD-1 Immune Checkpoint Inhibitors. *Eur. Urol.* **2020**, *78*, 498–502. [CrossRef]
87. Silva, C.A.C. Longitudinal analysis reveals gut microbiota shift during standard therapies in metastatic renal cell carcinoma (mRCC). 2022, 33 (Suppl. 7), pp. S660–S680. Available online: https://oncologypro.esmo.org/meeting-resources/esmo-congress/longitudinal-analysis-reveals-gut-microbiota-shift-during-standard-therapies-in-metastatic-renal-cell-carcinoma-mrcc (accessed on 10 January 2023).
88. Sivan, A.; Corrales, L.; Hubert, N.; Williams, J.B.; Aquino-Michaels, K.; Earley, Z.M.; Benyamin, F.W.; Lei, Y.M.; Jabri, B.; Alegre, M.-L.; et al. Commensal Bifidobacterium promotes antitumor immunity and facilitates anti-PD-L1 efficacy. *Science* **2015**, *350*, 1084–1089. [CrossRef] [PubMed]
89. Hagihara, M.; Kuroki, Y.; Ariyoshi, T.; Higashi, S.; Fukuda, K.; Yamashita, R.; Matsumoto, A.; Mori, T.; Mimura, K.; Yamaguchi, N.; et al. Clostridium butyricum Modulates the Microbiome to Protect Intestinal Barrier Function in Mice with Antibiotic-Induced Dysbiosis. *Iscience* **2019**, *23*, 100772. [CrossRef] [PubMed]
90. Dizman, N.; Hsu, J.; Bergerot, P.G.; Gillece, J.D.; Folkerts, M.; Reining, L.; Trent, J.; Highlander, S.K.; Pal, S.K. Randomized trial assessing impact of probiotic supplementation on gut microbiome and clinical outcome from targeted therapy in metastatic renal cell carcinoma. *Cancer Med.* **2020**, *10*, 79–86. [CrossRef]
91. Dizman, N.; Meza, L.; Bergerot, P.; Alcantara, M.; Dorff, T.; Lyou, Y.; Frankel, P.; Cui, Y.; Mira, V.; Llamas, M.; et al. Nivolumab plus ipilimumab with or without live bacterial supplementation in metastatic renal cell carcinoma: A randomized phase 1 trial. *Nat. Med.* **2022**, *28*, 704–712. [CrossRef]
92. Balan, M.; Teran, E.M.Y.; Waaga-Gasser, A.M.; Gasser, M.; Choueiri, T.K.; Freeman, G.; Pal, S. Novel Roles of c-Met in the Survival of Renal Cancer Cells through the Regulation of HO-1 and PD-L1 Expression. *J. Biol. Chem.* **2015**, *290*, 8110–8120. [CrossRef]
93. Paolino, M.; Penninger, J.M. The Role of TAM Family Receptors in Immune Cell Function: Implications for Cancer Therapy. *Cancers* **2016**, *8*, 97. [CrossRef]

94. Voron, T.; Marcheteau, E.; Pernot, S.; Colussi, O.; Tartour, E.; Taieb, J.; Terme, M. Control of the Immune Response by Pro-Angiogenic Factors. *Front. Oncol.* **2014**, *4*, 70. [CrossRef] [PubMed]
95. Hughes, P.E.; Caenepeel, S.; Wu, L.C. Targeted Therapy and Checkpoint Immunotherapy Combinations for the Treatment of Cancer. *Trends Immunol.* **2016**, *37*, 462–476. [CrossRef] [PubMed]
96. Meza, L.A.; Malhotra, J.; Zengin, Z.B.; Dizman, N.; Hsu, J.; Chawla, N.S.; Chehrazi-Raffle, A.; Muddasani, R.; Govindarajan, A.; Castro, D.V.; et al. A phase I trial to evaluate the biologic effect of CBM588 (Clostridium butyricum) in combination with cabozantinib plus nivolumab for patients with metastatic renal cell carcinoma (mRCC). *J. Clin. Oncol.* **2022**, *40*, TPS4606. Available online: https://ascopubs.org/doi/abs/10.1200/JCO.2022.40.16_suppl.TPS4606 (accessed on 24 August 2022). [CrossRef]
97. Meza, L.; Govindarajan, A.; Feng, M.; Pal, S.K. Live bacterial supplementation for improving treatment response in metastatic renal cell carcinoma. *Clin. Transl. Med.* **2022**, *12*, e948. [CrossRef] [PubMed]
98. Ninkov, M.; Schmerk, C.L.; Moradizadeh, M.; Parvathy, S.N.; Figueredo, R.; Burton, J.P.; Silverman, M.S.; Fernandes, R.; Vareki, S.M.; Haeryfar, S.M.M. Improved MAIT cell functions following fecal microbiota transplantation for metastatic renal cell carcinoma. *Cancer Immunol. Immunother.* **2022**, 1–14. [CrossRef]
99. Ianiro, G.; Rossi, E.; Thomas, A.M.; Schinzari, G.; Masucci, L.; Quaranta, G.; Settanni, C.R.; Lopetuso, L.R.; Armanini, F.; Blanco-Miguez, A.; et al. Faecal microbiota transplantation for the treatment of diarrhoea induced by tyrosine-kinase inhibitors in patients with metastatic renal cell carcinoma. *Nat. Commun.* **2020**, *11*, 4333. [CrossRef]
100. Lawson Health Research Institute. Preventing Immune-Related Adverse Events in Renal Cell Carcinoma Patients Treated with Combination Immunotherapy Using Fecal Microbiota Transplantation. 2022. Available online: https://clinicaltrials.gov/ct2/show/NCT04163289 (accessed on 19 December 2022).
101. GIANLUCA, I. Targeting Gut Microbiota to Improve Efficacy of Immune Checkpoint Inhibitors in Patients with Advanced Renal Cell Carcinoma. 2022. Available online: https://clinicaltrials.gov/ct2/show/NCT04758507 (accessed on 19 December 2022).
102. Kyte, J.A. MITRIC: Microbiota Transplant to Cancer Patients Who Have Failed Immunotherapy Using Faeces from Clincal Responders. 2022. Available online: https://clinicaltrials.gov/ct2/show/NCT05286294 (accessed on 19 December 2022).
103. Roussy, G.; Campus, C.; Paris, G. A Pilot Study Evaluating a Ketogenic Diet Concomitant to Nivolumab and Ipilimumab in Patients with Metastatic Renal Cell Carcinoma. 2021. Available online: https://clinicaltrials.gov/ct2/show/NCT05119010 (accessed on 19 December 2022).
104. Ferrere, G.; Alou, M.T.; Liu, P.; Goubet, A.-G.; Fidelle, M.; Kepp, O.; Durand, S.; Iebba, V.; Fluckiger, A.; Daillère, R.; et al. Ketogenic diet and ketone bodies enhance the anticancer effects of PD-1 blockade. *J. Clin. Investig.* **2021**, *6*, e145207. [CrossRef]
105. Lloyd-Price, J.; Abu-Ali, G.; Huttenhower, C. The healthy human microbiome. *Genome Med.* **2016**, *8*, 51. [CrossRef]
106. Cogdill, A.P.; Gaudreau, P.O.; Arora, R.; Gopalakrishnan, V.; Wargo, J.A. The Impact of Intratumoral and Gastrointestinal Microbiota on Systemic Cancer Therapy. *Trends Immunol.* **2018**, *39*, 900–920. [CrossRef]
107. Brooks, A.W.; Priya, S.; Blekhman, R.; Bordenstein, S.R. Gut microbiota diversity across ethnicities in the United States. *PLOS Biol.* **2018**, *16*, e2006842. [CrossRef]
108. The Integrative HMP (iHMP) Research Network Consortium. The Integrative Human Microbiome Project. *Nature* **2019**, *569*, 641–648. [CrossRef] [PubMed]
109. Qin, J.; Li, R.; Raes, J.; Arumugam, M.; Burgdorf, K.S.; Manichanh, C.; Nielsen, T.; Pons, N.; Levenez, F.; Yamada, T.; et al. A human gut microbial gene catalogue established by metagenomic sequencing. *Nature* **2010**, *464*, 59–65. [CrossRef] [PubMed]
110. Qin, J.; Li, Y.; Cai, Z.; Li, S.; Zhu, J.; Zhang, F.; Liang, S.; Zhang, W.; Guan, Y.; Shen, D.; et al. A metagenome-wide association study of gut microbiota in type 2 diabetes. *Nature* **2012**, *490*, 55–60. [CrossRef]
111. Li, X.; Zhang, S.; Guo, G.; Han, J.; Yu, J. Gut microbiome in modulating immune checkpoint inhibitors. *Ebiomedicine* **2022**, *82*, 104163. [CrossRef]

Journal of
Clinical Medicine

MDPI

Opinion

Proposed Implementation of a Patient-Centered Self-Assessment Tool for Patients with Neuroendocrine Tumors among Academic and Community Practice Sites: The City of Hope Model

Christiana Joy Crook [1], Lisa Yen [2], Kathleen Ta [1], Misagh Karimi [3], Danny Nguyen [4], Richard T. Lee [5] and Daneng Li [1,*]

1 Department of Medical Oncology & Therapeutics Research, City of Hope National Medical Center, Duarte, CA 91010, USA
2 Learn Advocate Connect Neuroendocrine Tumor Society, Denver, CO 80237, USA
3 Department of Medical Oncology & Therapeutics Research, City of Hope Newport Beach Fashion Island, Newport Beach, CA 92660, USA
4 Department of Medical Oncology & Therapeutics Research, City of Hope Irvine Sand Canyon, Irvine, CA 92618, USA
5 Department of Supportive & Integrative Medicine, City of Hope Orange County Lennar Foundation Cancer Center, Irvine, CA 92618, USA
* Correspondence: danli@coh.org; Tel.: +1-626-471-9200

Abstract: Neuroendocrine tumors are a rare type of cancer found in hormone-producing cells throughout the body. Research on disease-specific patient education assessments in this population is lacking. We previously demonstrated the feasibility and validity of NET VITALS, a patient-centered self-assessment designed to improve patients' knowledge of their neuroendocrine tumor diagnosis/treatment and facilitate communication with their physician. In this report, we provide a brief overview of patient assessments that have been used for patients with neuroendocrine tumors. We summarize NET VITALS and present a proposed infrastructure for its implementation into standard clinical care in both academic and community practice settings at City of Hope. Incorporating NET VITALS into standard of care treatment for patients with neuroendocrine tumors may improve patients' overall clinical care experience.

Keywords: neuroendocrine tumors; self-assessment; NET VITALS; implementation science

Citation: Crook, C.J.; Yen, L.; Ta, K.; Karimi, M.; Nguyen, D.; Lee, R.T.; Li, D. Proposed Implementation of a Patient-Centered Self-Assessment Tool for Patients with Neuroendocrine Tumors among Academic and Community Practice Sites: The City of Hope Model. *J. Clin. Med.* **2023**, *12*, 1229. https://doi.org/10.3390/jcm12031229

Academic Editor: Bruno Annibale

Received: 15 December 2022
Revised: 30 January 2023
Accepted: 1 February 2023
Published: 3 February 2023

1. Introduction

Neuroendocrine tumors (NETs) are hormone-producing tumors that develop from endocrine cells throughout the body [1]. Although the incidence of NETs is rising, NETs are a rare diagnosis, with 8.3 cases per 100,000 individuals diagnosed in the United States in 2018 [2,3]. Given that NETs often present with nonspecific symptoms, such as diarrhea, bloating, abdominal cramping, and flushing [4,5], delays in diagnosis are common, with patients reporting a median of 9.2 years between the development of symptoms and final diagnosis [6]. As a result, patients often present with advanced or metastatic disease at diagnosis [7,8].

The relatively small number of patients with NETs may be a contributing factor to the lack of research regarding education and treatment experiences of this patient population. Patients with NETs often report poor clinical experiences, with many patients expressing frustration with the lack of information provided and poor communication with their treating physician [7–9]. Educational tools for patients with breast, prostate, and liver cancers have demonstrated improvements in patient understanding and satisfaction with the information received from their treating physician [10,11]. These positive results suggest that patients with NETs could also benefit from patient-centric educational tools.

In this report, we provide a brief overview of patient-centered self-assessments, with an emphasis on tools specifically designed for patients with NETs. We describe NET VITALS, a patient-centered self-assessment tool created by NET patient advocates and physicians, and propose a strategy for its implementation at City of Hope. Our goal is to establish a robust clinical infrastructure for the implementation of NET VITALS that could improve the treatment experience of patients with NETs and contribute to an increase in patients' overall well-being.

2. Patient-Centered Assessments

Patient self-assessments have been developed for a variety of situations. Here, we describe disease-agnostic and disease-specific self-assessments that have been used among patients with NETs.

2.1. Disease-Agnostic Patient Assessments: Quality of Life

Many disease-agnostic patient self-assessments are designed to assess quality of life. A well-known example is the European Organization for Research and Treatment of Cancer (EORTC) Quality of Life Questionnaire-Core 30 (QLQ-C30) [12]. This 30-item questionnaire has been used among patients with a variety of diagnoses [13–15]. Other examples of patient-centered quality of life tools include the Health Outcomes Tool, Attitude Scale, Now vs Later tool, Prognosis and Treatment Perceptions Questionnaire, and Short Form 36 health survey questionnaire [16–21]. These tools allow patients to clearly identify treatment goals and current health status and can provide physicians with clarity regarding patients' preferences, which can potentially improve patient–physician communication. Moreover, these tools have successfully been used in NET patient populations in the contexts of clinical trials and observational studies, emphasizing their utility [22–29]. However, patients' knowledge of their disease and treatment is not analyzed with these tools, leaving a gap in patient education. While these tools can increase patient awareness and potentially improve patient–physician communication, they lack disease-specific questions that may provide greater insight for both the patient and physician.

2.2. NET-Specific Patient Assessments

Although there are numerous examples of disease-specific patient educational tools and assessments [10,11,30], there are few examples of assessments designed specifically for patients with NETs. The International Neuroendocrine Cancer Alliance (INCA) conducted a global survey of patients with NETs in 2014 that investigated patients' knowledge of their disease and their perspective on disease burden and treatment experience [7,8,31]. While this anonymous survey provided the NET research community with additional details regarding patient perspectives on their NET disease burden, it did not allow for any follow-up with patients, nor did it provide patients with a way to improve communication with their physician. In 2017, INCA conducted another international survey of patients/families, healthcare providers, and patient advocates and reported that patients with NETs continue to struggle with a lack of information [32]. Only 30% of patients stated that they were provided with sufficient information from their healthcare provider at diagnosis. However, 59% of healthcare providers surveyed believed that they provided patients with sufficient information, highlighting a lack of communication between patients with NETs and their healthcare providers.

The EORTC created a quality of life questionnaire for patients with gastrointestinal NETs (EORTC QLQ-GINET21); this 21-item questionnaire has been validated in patients with liver, pancreatic, and other gastrointestinal NETs [33,34]. This questionnaire is often used in conjunction with the EORTC QLQ-C30 to obtain a more comprehensive picture of a NET patient's health-related quality of life [23,26,27,35]. While the EORTC QLQ-GINET21 has allowed patients with gastrointestinal NETs to provide their physicians with information about their current quality of life, some of the questions may not be relevant for patients with non-functional tumors, and patients with non-gastrointestinal NETs are not included in the target patient population. Spolverato et al. designed a quality of life questionnaire specifically for patients with NET liver metastases that incorporated elements from the EORTC QLQ-C30, EORTC QLQ-GINET21, and Norfolk Quality of Life tool for NETs [36,37]. While this questionnaire is useful for this patient population, it does not assess patient knowledge or perception of information received.

To address the issue of patient satisfaction with information received, Bouma et al. developed a web-based information system designed to improve patient satisfaction with the amount and quality of information they were able to access about their diagnosis [38]. Although an initial feasibility study suggested that patients experienced an improvement in quality of life and were satisfied with the application, a randomized controlled trial comparing the web-based application to standard of care treatment found no difference in perception of information received or satisfaction with information received [39]. However, a 26-week multidisciplinary educational intervention for patients recently diagnosed with NETs reported improvements in patients' general self-efficacy and health-related quality of life [28], suggesting that educational interventions in this patient population require further optimization to maximize their benefit to patients.

The NET Cancer Health Storylines mobile application was developed to allow patients with NETs to track the frequency and severity of their symptoms and monitor additional health outcomes such as nutritional concerns, medications, and sleep [40]. Adams et al. investigated the health-related quality of life of patients using the application who were receiving somatostatin analog treatment; they observed a decrease in reported physical symptoms on the EORTC QLQ-C30 and EORTC QLQ-GINET21 over time, suggesting that the act of tracking symptoms may improve patients' perception of changes in symptoms [41].

In terms of treatment planning, Wagner et al. designed a multicriteria decision analysis framework for NET patients and physicians to use together when deciding on treatment plans [42]. Although this framework is limited in its treatment options, it provides an example of cooperative decision making, which has the potential to provide NET patients with increased autonomy and feelings of improved communication with their physician.

In summary, there is a lack of NET-specific tools that have the goal of increasing patients' self-knowledge of their diagnosis and treatment journey.

3. NET VITALS

In light of complaints from NET patients that they did not have enough information about their diagnosis and treatment from their physician [7–9], we decided to create a tool to allow patients to address these issues. Patient advocates from the Learn Advocate Connect Neuroendocrine Tumor Society (LACNETS) collaborated with NET physicians from City of Hope to create NET VITALS, a patient-centered self-assessment tool [43,44]. NET VITALS comprises six sections: diagnosis information, pathology/functional status/symptoms, imaging and diagnostic information, laboratory test results, surgery and treatments received, and additional information (genetic testing information, level of social support) (Figure 1 and Figure S1) [43]. The goal of NET VITALS is to give patients a sense of autonomy and control as they navigate their NET diagnosis and treatment odyssey.

In 2019, we introduced NET VITALS to patients attending the Los Angeles NET Education Conference [43]. Patients were invited to complete NET VITALS after attending a seminar that explained how to fill it out. The feasibility of NET VITALS was demonstrated, with an 88.3% response rate (68 out of 77 patients) and a median of 85.7% of items completed. NET patients were satisfied with NET VITALS as a potential tool to use with their physicians, with 74.6% of patients agreeing that NET VITALS was a useful communication tool and 76.3% of patients indicating that they would recommend NET VITALS to someone else. In terms of disease and treatment knowledge, NET VITALS highlighted areas where NET patients may not have as much knowledge about their diagnosis or treatment, including tumor Ki-67 index, grade, functional status, differentiation status, and receipt of liver-directed therapy. These gaps in knowledge suggest that NET VITALS could be used to spur communication between patients and physicians to increase NET patients' knowledge of their disease.

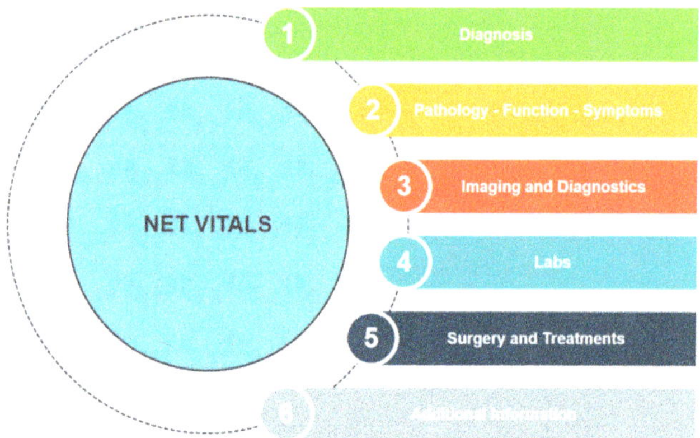

Figure 1. Sections of NET VITALS.

Given the feasibility and high level of patient satisfaction with NET VITALS in our preliminary cross-sectional survey study, we present a suggested infrastructure for implementing NET VITALS in the clinic.

4. Implementation of NET VITALS: The City of Hope Model

NET patient advocates, physicians experienced in treating patients with NETs across the City of Hope enterprise, and healthcare providers with experience integrating patient assessments into clinical care were approached to determine the best way to implement NET VITALS into clinical practice. A proposed model for NET VITALS integration is shown in Figure 2.

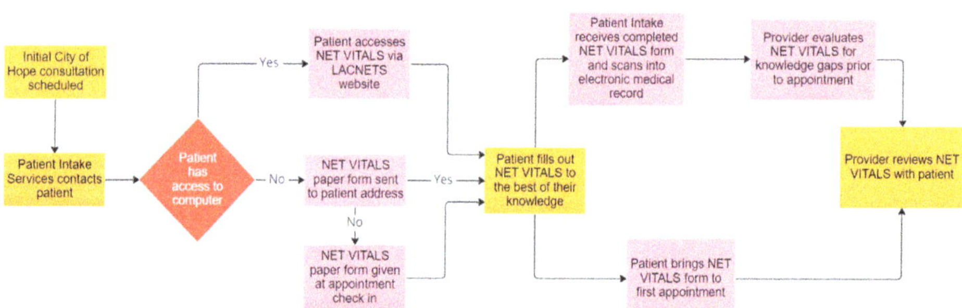

Figure 2. Proposed implementation of NET VITALS.

In this model, NET VITALS will be built into the list of intake tasks for patients with NETs seeking consultation at City of Hope. Intake coordinators will direct patients to the LACNETS website to download and complete NET VITALS. Once completed, the patient will send a copy of NET VITALS to the intake coordinator to scan and upload into the patient's medical record for physician review prior to the consultation appointment.

For patients unable to access the LACNETS website, the intake coordinator will mail a paper copy of NET VITALS to the patient for completion prior to their appointment. If completion of NET VITALS prior to the consultation appointment is not feasible, patients will be given a paper copy of NET VITALS upon check-in on the day of their appointment. Paper copies of NET VITALS will be scanned and uploaded into the patient's medical record on the day of their initial consultation at City of Hope.

5. Opportunities and Challenges of NET VITALS Integration within the City of Hope Enterprise

The implementation model for NET VITALS will be piloted initially on the City of Hope main Duarte campus and select community practice sites throughout southern California. NET VITALS will also continue to be promoted through the LACNETS website and outreach platform. Once the feasibility of this model has been demonstrated, expansion to all City of Hope locations across the enterprise will be performed.

Strengths of the proposed infrastructure include the strong relationships between the academic and community centers of City of Hope. With a primary center in Duarte, the City of Hope Orange County Lennar Foundation Cancer Center in Irvine, over two dozen community practice sites across southern California, and three Cancer Treatment Centers of America locations in Phoenix, Chicago, and Atlanta, the potential for collaboration is enormous. The connectivity of the main campus at Duarte and various community practice locations has been previously demonstrated [45]. This interconnectivity allows physicians from satellite clinics that may not see many patients with NETs on a routine basis to have access to specialists at other locations to better understand their patient's diagnosis and develop an optimal treatment plan. City of Hope's new relationship with Cancer Treatment Centers of America furthers this connectivity, allowing patients at all locations to have the opportunity to potentially benefit from NET VITALS.

A potential challenge of this implementation strategy is its reliance on electronic medical records. To complete NET VITALS before their appointment, patients are expected to have access to their online patient portal and the LACNETS website (which also implies Internet access). We have suggested alternative pathways to complete NET VITALS that are not dependent on Internet access to ensure that all patients have an opportunity to complete this assessment. Additionally, City of Hope is in the process of promoting access to the online patient portal, which includes converting Cancer Treatment Centers of America to the electronic medical record system used by City of Hope to ensure uniform access across the enterprise. The integration of NET VITALS into the electronic medical record could allow care teams to easily compare patients' responses to pre-existing data and track patients' care over time.

Another perceived limitation may be the inability of patients to complete all sections of NET VITALS before their appointment. However, patients are not expected to be familiar with everything covered in NET VITALS [44]. Identifying areas where knowledge is lacking allows patients to have a more guided discussion with their physician during their appointment, potentially improving overall patient–physician communication.

6. Conclusions

Patient-centered self-assessments, such as NET VITALS, may help increase patients' knowledge about their NET diagnosis/treatment and promote dialogue with their physician and healthcare providers. Identifying and implementing a strategy for the incorporation of NET VITALS into clinical care can be significant to care teams that strive to provide the best personalized care possible for patients with NETs.

Supplementary Materials: The following supporting information can be downloaded at: https://www.mdpi.com/article/10.3390/jcm12031229/s1, Figure S1: NET VITALS.

Author Contributions: Conceptualization, C.J.C., L.Y., K.T., R.T.L. and D.L.; investigation, C.J.C., L.Y., K.T., M.K., D.N., R.T.L. and D.L.; writing—original draft preparation, C.J.C., K.T. and D.L.; writing—review and editing, C.J.C., L.Y., K.T., M.K., D.N., R.T.L. and D.L.; supervision, D.L. All authors have read and agreed to the published version of the manuscript.

Funding: This research received no external funding.

Institutional Review Board Statement: Not applicable.

Informed Consent Statement: Not applicable.

Data Availability Statement: Not applicable.

Acknowledgments: The authors would like to acknowledge Giovanna Joyce Imbesi for her role in the development of NET VITALS and the patients and families who participated in the 2019 feasibility/satisfaction study of NET VITALS.

Conflicts of Interest: Daneng Li reports research funding to his institution from AstraZeneca and Brooklyn ImmunoTherapeutics. He serves as a consultant and has received honoraria from Adagene, Advanced Accelerator Applications, Coherus, Delcath, Eisai, Exelixis, Genentech, Ipsen Biopharmaceuticals, Lexicon, Merck, MiNA Therapeutics, QED, Servier, Sun Pharma, and TerSera Therapeutics, all outside the submitted work. Lisa Yen serves as a consultant and has received honoraria from Ipsen Biopharmaceuticals and Sun Pharma. The remaining authors declare no conflict of interest.

References

1. Yao, J.C.; Hassan, M.; Phan, A.; Dagohoy, C.; Leary, C.; Mares, J.E.; Abdalla, E.K.; Fleming, J.B.; Vauthey, J.-N.; Rashid, A.; et al. One hundred years after "carcinoid": Epidemiology of and prognostic factors for neuroendocrine tumors in 35,825 cases in the United States. *J. Clin. Oncol.* **2008**, *26*, 3063–3072. [CrossRef] [PubMed]
2. Dasari, A.; Shen, C.; Halperin, D.; Zhao, B.; Zhou, S.; Xu, Y.; Shih, T.; Yao, J.C. Trends in the Incidence, Prevalence, and Survival Outcomes in Patients with Neuroendocrine Tumors in the United States. *JAMA Oncol.* **2017**, *3*, 1335–1342. [CrossRef] [PubMed]
3. Surveillance, Epidemiology, and End Results (SEER) Program SEER*Stat Database. Available online: https://seer.cancer.gov/ (accessed on 13 July 2021).
4. Kulke, M.H.; Hörsch, D.; Caplin, M.E.; Anthony, L.B.; Bergsland, E.; Öberg, K.; Welin, S.; Warner, R.R.P.; Lombard-Bohas, C.; Kunz, P.L.; et al. Telotristat Ethyl, a Tryptophan Hydroxylase Inhibitor for the Treatment of Carcinoid Syndrome. *J. Clin. Oncol.* **2017**, *35*, 14–23. [CrossRef] [PubMed]
5. Kulke, M.H.; Kennecke, H.F.; Murali, K.; Joish, V.N. Changes in Carcinoid Syndrome Symptoms Among Patients Receiving Telotristat Ethyl in US Clinical Practice: Findings from the TELEPRO-II Real-World Study. *Cancer Manag. Res.* **2021**, *13*, 7439–7446. [CrossRef] [PubMed]
6. Vinik, A.I.; Woltering, E.A.; Warner, R.R.P.; Caplin, M.; O'Dorisio, T.M.; Wiseman, G.A.; Coppola, D.; Go, V.L.W. NANETS Consensus Guidelines for the Diagnosis of Neuroendocrine Tumor. *Pancreas* **2010**, *39*, 713–734. [CrossRef]
7. Singh, S.; Granberg, D.; Wolin, E.; Warner, R.; Sissons, M.; Kolarova, T.; Goldstein, G.; Pavel, M.; Öberg, K.; Leyden, J. Patient-Reported Burden of a Neuroendocrine Tumor (NET) Diagnosis: Results from the First Global Survey of Patients with NETs. *J. Glob. Oncol.* **2017**, *3*, 43–53. [CrossRef]
8. Wolin, E.M.; Leyden, J.; Goldstein, G.; Kolarova, T.; Hollander, R.; Warner, R.R.P. Patient-Reported Experience of Diagnosis, Management, and Burden of Neuroendocrine Tumors: Results from a Large Patient Survey in the United States. *Pancreas* **2017**, *46*, 639–647. [CrossRef]
9. Feinberg, Y.; Law, C.; Singh, S.; Wright, F.C. Patient experiences of having a neuroendocrine tumour: A qualitative study. *Eur. J. Oncol. Nurs.* **2013**, *17*, 541–545. [CrossRef]
10. Berger, O.; Grønberg, B.H.; Loge, J.H.; Kaasa, S.; Sand, K. Cancer patients' knowledge about their disease and treatment before, during and after treatment: A prospective, longitudinal study. *BMC Cancer* **2018**, *18*, 381. [CrossRef]
11. Yang, T.; Tan, T.; Yang, J.; Pan, J.; Hu, C.; Li, J.; Zou, Y. The impact of using three-dimensional printed liver models for patient education. *J. Int. Med Res.* **2018**, *46*, 1570–1578. [CrossRef]
12. Aaronson, N.K.; Ahmedzai, S.; Bergman, B.; Bullinger, M.; Cull, A.; Duez, N.J.; Filiberti, A.; Flechtner, H.; Fleishman, S.B.; De Haes, J.C.J.M.; et al. The European Organization for Research and Treatment of Cancer QLQ-C30: A Quality-of-Life Instrument for Use in International Clinical Trials in Oncology. *J. Natl. Cancer Inst.* **1993**, *85*, 365–376. [CrossRef] [PubMed]
13. Galle, P.R.; Finn, R.S.; Qin, S.; Ikeda, M.; Zhu, A.X.; Kim, T.-Y.; Kudo, M.; Breder, V.; Merle, P.; Kaseb, A.; et al. Patient-reported outcomes with atezolizumab plus bevacizumab versus sorafenib in patients with unresectable hepatocellular carcinoma (IMbrave150): An open-label, randomised, phase 3 trial. *Lancet Oncol.* **2021**, *22*, 991–1001. [CrossRef] [PubMed]
14. Harbeck, N.; Iyer, S.; Turner, N.; Cristofanilli, M.; Ro, J.; André, F.; Loi, S.; Verma, S.; Iwata, H.; Bhattacharyya, H.; et al. Quality of life with palbociclib plus fulvestrant in previously treated hormone receptor-positive, HER2-negative metastatic breast cancer: Patient-reported outcomes from the PALOMA-3 trial. *Ann. Oncol.* **2016**, *27*, 1047–1054. [CrossRef] [PubMed]
15. Hui, R.; Özgüroğlu, M.; Villegas, A.; Daniel, D.; Vicente, D.; Murakami, S.; Yokoi, T.; Chiappori, A.; Lee, K.H.; de Wit, M.; et al. Patient-reported outcomes with durvalumab after chemoradiotherapy in stage III, unresectable non-small-cell lung cancer (PACIFIC): A randomised, controlled, phase 3 study. *Lancet Oncol.* **2019**, *20*, 1670–1680. [CrossRef] [PubMed]
16. Case, S.M.; Fried, T.R.; O'Leary, J. How to ask: Older adults' preferred tools in health outcome prioritization. *Patient Educ. Couns.* **2013**, *91*, 29–36. [CrossRef]
17. Fried, T.R.; Tinetti, M.; Agostini, J.; Iannone, L.; Towle, V. Health outcome prioritization to elicit preferences of older persons with multiple health conditions. *Patient Educ. Couns.* **2011**, *83*, 278–282. [CrossRef] [PubMed]
18. Case, S.M.; Towle, V.R.; Fried, T.R. Considering the Balance: Development of a Scale to Assess Patient Views on Trade-Offs in Competing Health Outcomes. *J. Am. Geriatr. Soc.* **2013**, *61*, 1331–1336. [CrossRef]

19. El-Jawahri, A.; Traeger, L.; Park, E.R.; Greer, J.A.; Pirl, W.; Lennes, I.T.; Jackson, V.A.; Gallagher, E.R.; Temel, J.S. Associations among prognostic understanding, quality of life, and mood in patients with advanced cancer. *Cancer* **2014**, *120*, 278–285. [CrossRef]
20. Nipp, R.D.; Greer, J.A.; El-Jawahri, A.; Moran, S.M.; Traeger, L.; Jacobs, J.M.; Jacobsen, J.C.; Gallagher, E.R.; Park, E.R.; Ryan, D.P.; et al. Coping and Prognostic Awareness in Patients with Advanced Cancer. *J. Clin. Oncol.* **2017**, *35*, 2551–2557. [CrossRef]
21. Brazier, J.E.; Harper, R.; Jones, N.M.; O'Cathain, A.; Thomas, K.J.; Usherwood, T.; Westlake, L. Validating the SF-36 health survey questionnaire: New outcome measure for primary care. *BMJ* **1992**, *305*, 160–164. [CrossRef]
22. Vinik, A.; Bottomley, A.; Korytowsky, B.; Bang, Y.-J.; Raoul, J.-L.; Valle, J.W.; Metrakos, P.; Hörsch, D.; Mundayat, R.; Reisman, A.; et al. Patient-Reported Outcomes and Quality of Life with Sunitinib Versus Placebo for Pancreatic Neuroendocrine Tumors: Results from an International Phase III Trial. *Target. Oncol.* **2016**, *11*, 815–824. [CrossRef] [PubMed]
23. Strosberg, J.; Wolin, E.; Chasen, B.; Kulke, M.; Bushnell, D.; Caplin, M.; Baum, R.P.; Kunz, P.; Hobday, T.; Hendifar, A.; et al. Health-Related Quality of Life in Patients with Progressive Midgut Neuroendocrine Tumors Treated With ^{177}Lu-Dotatate in the Phase III NETTER-1 Trial. *J. Clin. Oncol.* **2018**, *36*, 2578–2584. [CrossRef] [PubMed]
24. Martini, C.; Buxbaum, S.; Rodrigues, M.; Nilica, B.; Scarpa, L.; Holzner, B.; Virgolini, I.; Gamper, E.-M. Quality of Life in Patients with Metastatic Gastroenteropancreatic Neuroendocrine Tumors Receiving Peptide Receptor Radionuclide Therapy: Information from a Monitoring Program in Clinical Routine. *J. Nucl. Med.* **2018**, *59*, 1566–1573. [CrossRef] [PubMed]
25. Chen, L.; Navalkissoor, S.; Quigley, A.-M.; Gnanasegaran, G.; Mandair, D.; Toumpanakis, C.; Caplin, M.E.; Hayes, A.R. 177Lu-DOTATATE in older patients with metastatic neuroendocrine tumours: Safety, efficacy and health-related quality of life. *Eur. J. Nucl. Med.* **2021**, *48*, 3582–3594. [CrossRef]
26. Sorbye, H.; Meyer, L.S.; Mordal, K.E.; Myhre, S.; Thiis-Evensen, E. Patient reported symptoms, coping and quality of life during somatostatin analogue treatment for metastatic small- intestinal neuroendocrine tumours. *Health Qual. Life Outcomes* **2020**, *18*, 188. [CrossRef]
27. Scandurra, C.; Modica, R.; Maldonato, N.M.; Dolce, P.; Dipietrangelo, G.G.; Centello, R.; Di Vito, V.; Bottiglieri, F.; de Cicco, F.; Giannetta, E.; et al. Quality of Life in Patients with Neuroendocrine Neoplasms: The Role of Severity, Clinical Heterogeneity, and Resilience. *J. Clin. Endocrinol. Metab.* **2020**, *106*, e316–e327. [CrossRef]
28. Haugland, T.; Veenstra, M.; Vatn, M.H.; Wahl, A.K. Improvement in Stress, General Self-Efficacy, and Health Related Quality of Life following Patient Education for Patients with Neuroendocrine Tumors: A Pilot Study. *Nurs. Res. Pract.* **2013**, *2013*, 695820. [CrossRef]
29. Li, D.; Sun, C.-L.; Kim, H.; Crook, C.; Zhang, Y.-H.; Allen, R.; Ballena, R.; Hyder, S.; Koczywas, M.; Chung, V.; et al. Patient-Defined Goals and Preferences Among Adults with Advanced Neuroendocrine Tumors. *J. Natl. Compr. Canc. Netw.* **2022**, *20*, 1330–1337.e1333. [CrossRef]
30. Coughlin, C.C.; Pérez, M.; Kumar, M.G.; Jeffe, D.B.; Bayliss, S.J.; Sternhell-Blackwell, K. Skin cancer risk education in pediatric solid organ transplant patients: An evaluation of knowledge, behavior, and perceptions over time. *Pediatr. Transplant.* **2017**, *21*, e12817. [CrossRef]
31. Leyden, J.; Pavlakis, N.; Chan, D.; Michael, M.; Clarke, S.; Khasraw, M.; Price, T. Patient-reported experience of the impact and burden of neuroendocrine tumors: Oceania patient results from a large global survey. *Asia-Pacific J. Clin. Oncol.* **2018**, *14*, 256–263. [CrossRef]
32. Leyden, S.; Kolarova, T.; Bouvier, C.; Caplin, M.; Conroy, S.; Davies, P.; Dureja, S.; Falconi, M.; Ferolla, P.; Fisher, G.; et al. Unmet needs in the international neuroendocrine tumor (NET) community: Assessment of major gaps from the perspective of patients, patient advocates and NET health care professionals. *Int. J. Cancer* **2020**, *146*, 1316–1323. [CrossRef] [PubMed]
33. Davies, A.H.G.; Larsson, G.; Ardill, J.; Friend, E.; Jones, L.; Falconi, M.; Bettini, R.; Koller, M.; Sezer, O.; Fleissner, C.; et al. Development of a disease-specific quality of life questionnaire module for patients with gastrointestinal neuroendocrine tumours. *Eur. J. Cancer* **2006**, *42*, 477–484. [CrossRef]
34. Yadegarfar, G.; Friend, L.; Jones, L.; Plum, L.M.; Ardill, J.; Taal, B.; Larsson, G.; Jeziorski, K.; Kwekkeboom, D.; Ramage, J.K.; et al. Validation of the EORTC QLQ-GINET21 questionnaire for assessing quality of life of patients with gastrointestinal neuroendocrine tumours. *Br. J. Cancer* **2013**, *108*, 301–310. [CrossRef] [PubMed]
35. Milanetto, A.C.; Nordenström, E.; Sundlöv, A.; Almquist, M. Health-Related Quality of Life After Surgery for Small Intestinal Neuroendocrine Tumours. *World J. Surg.* **2018**, *42*, 3231–3239. [CrossRef] [PubMed]
36. Spolverato, G.; Bagante, F.; Wagner, D.; Buettner, S.; Gupta, R.; Kim, Y.; Maqsood, H.; Pawlik, T.M. Quality of life after treatment of neuroendocrine liver metastasis. *J. Surg. Res.* **2015**, *198*, 155–164. [CrossRef] [PubMed]
37. Vinik, E.; Carlton, C.A.; Silva, M.P.; Vinik, A.I. Development of the Norfolk Quality of Life Tool for Assessing Patients with Neuroendocrine Tumors. *Pancreas* **2009**, *38*, e87–e95. [CrossRef]
38. Bouma, G.; de Hosson, L.D.; van Woerkom, C.E.; van Essen, H.; de Bock, G.H.; Admiraal, J.M.; Reyners, A.K.L.; Walenkamp, A.M.E. Web-based information and support for patients with a newly diagnosed neuroendocrine tumor: A feasibility study. *Support. Care Cancer* **2017**, *25*, 2075–2083. [CrossRef]
39. De Hosson, L.D.; Bouma, G.; Stelwagen, J.; Van Essen, H.; De Bock, G.H.; De Groot, D.J.A.; De Vries, E.G.E.; Walenkamp, A.M.E. Web-based personalised information and support for patients with a neuroendocrine tumour: Randomised controlled trial. *Orphanet J. Rare Dis.* **2019**, *14*, 60. [CrossRef]

40. NET Cancer Health Storylines: All of Your Tools for Managing NET Cancer in One Place. Available online: https://www.healthstorylines.com/net-cancer-healthstorylines (accessed on 21 October 2022).
41. Adams, J.R.; Ray, D.; Willmon, R.; Pulgar, S.; Dasari, A. Living with Neuroendocrine Tumors: Assessment of Quality of Life Through a Mobile Application. *JCO Clin. Cancer Informatics* **2019**, *3*, 1–10. [CrossRef]
42. Wagner, M.; Samaha, D.; Khoury, H.; O'Neil, W.M.; Lavoie, L.; Bennetts, L.; Badgley, D.; Gabriel, S.; Berthon, A.; Dolan, J.; et al. Development of a Framework Based on Reflective MCDA to Support Patient–Clinician Shared Decision-Making: The Case of the Management of Gastroenteropancreatic Neuroendocrine Tumors (GEP-NET) in the United States. *Adv. Ther.* **2017**, *35*, 81–99. [CrossRef]
43. Li, D.; Imbesi, G.J.; Yen, L.; Kim, H.; Sun, C.-L.; Crook, C.J.; Ballena, R.; Zhang, Y.-H.; Allen, R.; Sedrak, M.; et al. Feasibility and Satisfaction of Using NET VITALS Self-assessment Tool Among Patients with Neuroendocrine Tumors. *Pancreas* **2022**, *51*, 319–324. [CrossRef] [PubMed]
44. Learn Advocate Connect Neuroendocrine Tumor Society. NET VITALS-Your NET Communication Tool. Available online: https://www.lacnets.org/netvitals (accessed on 13 July 2021).
45. Salgia, R.; Kulkarni, P. Integrating Clinical and Translational Research Networks—Building Team Medicine. *J. Clin. Med.* **2020**, *9*, 2975. [CrossRef] [PubMed]

Journal of
Clinical Medicine

MDPI

Perspective

A Systems Biology Approach for Addressing Cisplatin Resistance in Non-Small Cell Lung Cancer

Sravani Ramisetty [1], Prakash Kulkarni [1,2], Supriyo Bhattacharya [3], Arin Nam [4], Sharad S. Singhal [1], Linlin Guo [1], Tamara Mirzapoiazova [1], Bolot Mambetsariev [1], Sandeep Mittan [5], Jyoti Malhotra [6], Evan Pisick [7], Shanmuga Subbiah [8], Swapnil Rajurkar [9], Erminia Massarelli [1], Ravi Salgia [1] and Atish Mohanty [1,*]

1 Department of Medical Oncology and Therapeutics Research, City of Hope National Medical Center, Duarte, CA 91010, USA
2 Department of Systems Biology, City of Hope National Medical Center, Duarte, CA 91010, USA
3 Translational Bioinformatics, Center for Informatics, Department of Computational and Quantitative Medicine, City of Hope National Medical Center, 1500 Duarte Rd, Duarte, CA 91010, USA
4 Department of Pathology, University of California, La Jolla, San Diego, CA 92093, USA
5 Montefiore Medical Center, The University Hospital for Albert Einstein College of Medicine, Bronx, NY 10467, USA
6 Department of Medical Oncology and Therapeutics Research, City of Hope National Medical Center, 1000 FivePoint, Irvine, CA 92618, USA
7 Cancer Treatment Centers of America (CTCA) Chicago, 2520 Elisha Avenue, Zion, IL 60099, USA
8 Department of Medical Oncology and Therapeutics Research, City of Hope National Medical Center, 1250 S. Sunset Ave., Suite 303, West Covina, CA 91790, USA
9 Department of Medical Oncology and Therapeutics Research, City of Hope National Medical Center, 1100 San Bernardino Road, Suite 1100, Upland, CA 91786, USA
* Correspondence: amohanty@coh.org

Abstract: Translational research in medicine, defined as the transfer of knowledge and discovery from the basic sciences to the clinic, is typically achieved through interactions between members across scientific disciplines to overcome the traditional silos within the community. Thus, translational medicine underscores 'Team Medicine', the partnership between basic science researchers and clinicians focused on addressing a specific goal in medicine. Here, we highlight this concept from a City of Hope perspective. Using cisplatin resistance in non-small cell lung cancer (NSCLC) as a paradigm, we describe how basic research scientists, clinical research scientists, and medical oncologists, in true 'Team Science' spirit, addressed cisplatin resistance in NSCLC and identified a previously approved compound that is able to alleviate cisplatin resistance in NSCLC. Furthermore, we discuss how a 'Team Medicine' approach can help to elucidate the mechanisms of innate and acquired resistance in NSCLC and develop alternative strategies to overcome drug resistance.

Keywords: drug resistance; cisplatin; non-small cell lung cancer; group behavior; IDPs; phenotype switching; mathematical modeling

Citation: Ramisetty, S.; Kulkarni, P.; Bhattacharya, S.; Nam, A.; Singhal, S.S.; Guo, L.; Mirzapoiazova, T.; Mambetsariev, B.; Mittan, S.; Malhotra, J.; et al. A Systems Biology Approach for Addressing Cisplatin Resistance in Non-Small Cell Lung Cancer. *J. Clin. Med.* **2023**, *12*, 599. https://doi.org/10.3390/jcm12020599

Academic Editor: Paola Concolino

Received: 6 December 2022
Revised: 6 January 2023
Accepted: 10 January 2023
Published: 11 January 2023

1. Introduction

Cancer is one of the major contributors to global mortality. According to Cancer Statistics, which is published every year [1], 1.9 million new cancer cases and 609,360 deaths from cancer are estimated in 2022 in the US alone, which is about 1670 deaths a day. Among all the cancer types, lung cancer (both small and non-small cell) is the most prevalent, and it is estimated that a total of 236,740 people will be diagnosed with lung cancer in 2022, which is 1 in 16 people in the US alone [2–5]. Lung cancer is generally a disease of middle-aged and elderly smokers, usually with several comorbid smoking-related conditions, such as emphysema, chronic bronchitis, widespread atherosclerosis, and degenerative disorders of the central nervous system (CNS) and other organs [6–8]. Despite significant developments in preventing, screening, and treating lung cancer over the past decade, innate and acquired resistance to chemotherapeutic agents and radiation remains a vexing problem, and success in increasing the life expectancy of patients is limited.

A vast majority (~85%) of lung cancer patients have a group of histological subtypes collectively known as non-small cell lung cancer (NSCLC). Among the various subtypes, lung adenocarcinoma (LUAD) and lung squamous cell carcinoma (LSCC) are the most common subtypes [9]. LUAD is mostly driven by driver oncogenes, such as EGFR, KRAS ($G^{12}C$), MET, ALK, etc., against which targeted therapies are available. On the other hand, no known targetable driver oncogenes have been identified for LSCC; thus, the therapeutic options for LSSC patients are limited.

NSCLC patients are offered a broad range of genotoxic drugs, such as cisplatin or carboplatin alone or in combination with immunotherapy [10], and often respond to treatment initially. However, most patients develop drug resistance, and numerous mechanisms underlying drug resistance have been identified, mostly in preclinical models of the disease [11–13]. Unfortunately, many of these mechanisms do not always hold in vivo and, very often, are not effective in the clinic even if they appear promising in the in vivo models. Therefore, a collaborative effort integrating the preclinical studies with clinical outcomes could help to better understand the mechanism of drug resistance.

Here, we summarize the concept of 'Integrating Clinical and Translational Research Networks—Building Team Medicine' from a City of Hope perspective (Figure 1). Using cisplatin resistance in NSCLC as a paradigm, we describe how basic research scientists with expertise in fields as varied as cancer biology, cell and molecular biology, biochemistry, biophysics, structural biology, and mathematical and computational biology; clinical research scientists; and medical oncologists working together with a true 'Team Medicine' spirit, uncovered a non-genetic mechanism underlying cisplatin resistance in NSCLC and identified carfilzomib (CFZ), a previously approved proteasome inhibitor, to alleviate resistance. The team was led by the Department Chair, a thoracic oncologist who helped coordinate the team's efforts, much like a tumor board that comprises clinical specialists, nurses, and care providers does in a hospital setting. Furthermore, we discuss how this 'Team Medicine' approach also helped explore novel treatment strategies that could potentially preclude, attenuate, or at least delay, the onset of cisplatin resistance in these patients.

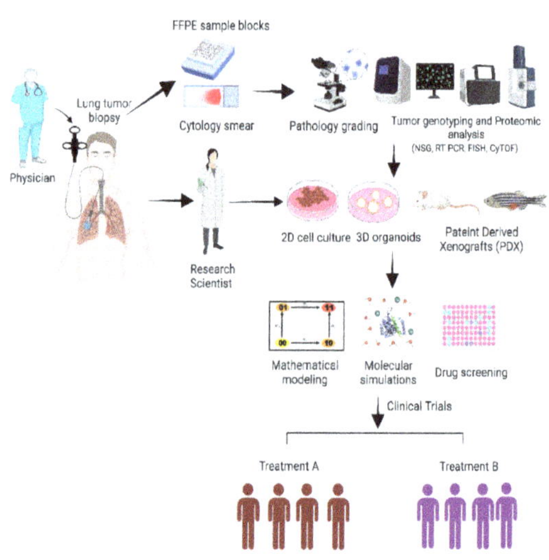

Figure 1. Schematic representing the 'Team Medicine' approach.

2. Mechanisms of Drug Resistance

Drug resistance is the major obstacle to long-term patient survival [14]. Cancer cells can escape therapy and exhibit drug resistance by different routes and many of these routes remain unpredictable and difficult to characterize [15]. A better understanding of the molecular mechanisms that help in tumor progression and drug resistance is essential in designing cancer subtype specific treatments. Drug resistance is defined as the inheritable ability of the cells to survive clinically relevant drug concentrations [16]. Drug resistance can either pre-exist before the start of the treatment, referred to as innate resistance, or develop in response to the treatment and is referred to as acquired resistance [15–19]. Innate drug resistance is typically thought to involve genetic mutations, while acquired resistance is generally believed to be due to both genetic and non-genetic/epigenetic changes. In either case, resistance to therapy is associated with metastatic disease and poor survival rates in patients [16,20–22]. Regardless, however, numerous mechanisms [23,24] that promote drug resistance have been reported in the literature, adding to the resistance conundrum (Figure 2).

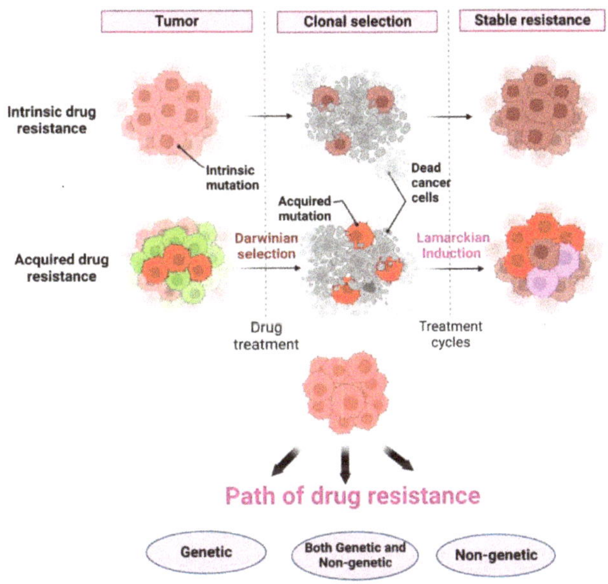

Figure 2. Schematic representation of acquired resistance. Acquired resistance can arise through the Darwinian section or Lamarckian induction. Path of drug resistance where a tumor cell can become resistant purely due to genetic changes through non-genetic alteration in a particular genotype or through initial non-genetic changes combined with genetic mutations.

Cancer drug resistance is multi-factorial and not solely driven by genetic mechanisms [25]. In fact, an increasing body of evidence shows that non-genetic mechanisms, such as lineage plasticity [26] (change in cell identity), epigenetic factors that regulate gene expression, and phenotype plasticity, contribute to cancer drug resistance [27]. Cancer cells escape the drug assault by two phenomena; 'tolerance', which is the ability of the cell to survive transient exposure to high drug concentration, and 'persistence', which is the ability of a subpopulation of a clonal population to survive exposure to high concentrations of a drug [27]. Drug-tolerant persisters (DTP) remain major factors in cancer relapse and in developing drug resistance [28]. Persistence is observed in low frequency in tumor cells with reduced proliferation rate and metabolism that helps them in tolerating drug insult. The genetic makeup of the DTPs is indistinguishable from the bulk tumor population and the resistance exhibited by them reverts to the sensitive state upon drug removal [29–34].

However, Shaffer et al. [35] showed that persisters cells exhibit significant variability at the single-cell level. Furthermore, these variabilities eventually decide the fate of the cell regarding whether to irreversibly become resistant to drug treatment.

Two phenomena that determine whether a given cancer cell population will undergo non-genetic evolution of drug resistance are epigenetic heterogeneity and epigenetic plasticity [27,36–38]. Epigenetic heterogeneity refers to the overall variability in the epigenetic landscape across a given cell population, which is influenced by both cell-intrinsic and cell-extrinsic stimuli. Epigenetic plasticity, on the other hand, is the capacity of a cell to alter its epigenetic state in response to either internal or external stimuli [27,36–38]. It is crucial to understand that both epigenetic heterogeneity and epigenetic plasticity are not completely independent variables; for example, various cancer cell types exhibit heterogeneity because the epigenetic state of the population is more plastic.

A stable mechanism of non-genetic resistance can result in the pre-existence of resistant clones in the subpopulation, in which case drug resistance simply emerges through Darwinian selection and is completely dependent on epigenetic heterogeneity. Alternatively, the stable origin of a non-genetic resistance can also be a result of gradual Darwinian or Lamarckian induction (Figure 2).

Recent evidence indicates that the genetic and non-genetic mechanisms of drug resistance are not mutually exclusive but indeed co-exist (meaning that both evolutionary phenomena of Darwinian selection and Lamarckian induction may be active) within a given cancer type and drive the drug resistance that eventually led to therapy failure. The genetic/nongenetic duality as described in the review is believed to be a major contributor to the complexity of drug resistance [39]. Designing drugs that target only the genetic mutations is like playing a whack-a-mole game where the player has zero chance of winning, even after multiple attempts. Thus, it is important to gain a deeper understanding of the relative contributions of genetic and non-genetic mechanisms especially, by understanding how, why, and when these non-genetic alterations occur so that one can hit the desired target consistently.

In addition to genetic mutations/epigenetic changes, protein interaction networks (PIN) also contribute to drug resistance [40–42]. PIN dynamics are orchestrated by the hub proteins, which are typically intrinsically disordered proteins (IDPs). IDPs lack 3D structure but exist as conformational ensembles. Indeed, ~80% of cancer-associated proteins, for example, p53, cyclins, MYC, SOX2, paxillin, etc., are IDPs [43]. This article focuses on acquired resistance to cisplatin resistance in NSCLC.

3. Cisplatin Resistance in NSCLC

Cisplatin is one of the platinum-based frontline chemotherapeutic agents used to treat solid tumors in a wide spectrum of cancers, including lung, ovarian, colorectal, head and neck, and testicular [44–46]. Cisplatin delivers its attack by entering cancer cells and binding to DNA, thus forming DNA adducts. These adducts block transcription and DNA synthesis, which activates the intracellular signal transduction that helps to eliminate the tumor lesions [43]. Patients usually have a good initial response to cisplatin-based chemotherapy but relapse later, because the development of acquired or innate resistance markedly reduces its clinical effectiveness [46–50]. Various molecular mechanisms, such as altered DNA repair and the cellular accumulation of the drug, as well as the cytoplasmic inactivation of the drug, are a few of many pathways through which patients usually develop resistance to cisplatin. The goal of personalized medicine is to develop better responses to the drug in the clinic. Here in this review, we will comprehensively discuss the non-genetic mechanisms of cisplatin resistance in NSCLC [51,52]. Cellular resistance to cisplatin may conceivably be based upon the overexpression or inactivation of certain oncogenes both in genetic and epigenetic pathways [47,49,53]. One such epigenetic mechanism involves focal adhesion complex (FA) and the components that contribute to cisplatin-resistance in NSCLC.

3.1. Focal Adhesion and Cisplatin Resistance

Focal adhesions (FAs) are contact points for a cell that interact with the extracellular matrix (ECM) and regulate diverse cellular processes, such as apoptosis, proliferation, migration, and differentiation. The principal components of FAs are integrin, paxillin, focal adhesion kinase (FAK), SRC (Src Oncogene), talin, tensin, vinculin, and actin [54]. Integrins function as transmembrane receptors for extracellular ligands and transduce biochemical signals into the cell. Integrins, when bound to ligands, are shown to be involved in a variety of signaling pathways, such as the cell cycle; the organization of the intracellular cytoskeleton; and in mediating the translocation of new receptors to the cell membranes with an α and a β subunit [55,56]. In mammals, there are 24α and 9β integrins among which integrin β4 (ITGB4) is believed to be unique due to its >1000 amino acid cytoplasmic domain when compared to other β-forms that typically have cytoplasmic domains of ~59 amino acids [57]. Furthermore, the unique property of ITGB4 is that it heterodimerizes with ITGα6 as well as ITGα7 [52,58]. However, ITGβ4 and its role in cisplatin resistance remained poorly understood until recently.

Another important component of the FA complex is paxillin (PXN). Human PXN is a 68-kDa (591 amino acids) protein [59]. The LUAD upregulation of PXN is associated with tumor progression and metastasis [60,61]. The phosphorylation of PXN leads to the activation of the downstream pathways of MAPF/ERK, resulting in cisplatin resistance [50]. PXN contains an N-terminus proline-rich region that anchors SH3-containing proteins along with five leucine-rich residues (LD domains 1–5) with a consensus sequence of LDXLLXXL [62,63]. The LD2-LD4 region includes sequences for the recruitment of signaling and structural molecules, such as FAK, vinculin, and Crk [62,64,65]. This region has also been reported to interact with integrin; more specifically, integrin α4 (ITGA4). Interestingly, PXN is an IDP [66]. The C-terminal region of PXN is believed to be involved in anchoring PXN to the plasma membrane and targeting to FA complex. The C-terminal of the FA complex harbor Cysteine-Histidine-enriched Lin11/Isl1/Mec3 (LIM) domains that form zinc fingers, suggesting that PXN could bind DNA and act as a transcription factor [67]. In addition to LD domains, LIM domains contain the SH3 domain and SH2 domain that forms a docking site for many tyrosine and threonine kinases and recruit additional enzymes into the complex, eventually leading to the activation of canonical signaling through the Ras-mitogen-activated protein kinase (MAPK), phosphoinositide-3-kinase (PI3K)-Akt, and phospholipase C-gamma (PLC-γ) pathways.

Our recent work showed that NSCLC tumor tissue has the heterogenous expression of PXN/ITGB4, and patients with the increased expression of both these genes have poor overall survival [50]. Furthermore, the cell lines that were identified to be cisplatin-resistant were also observed to have elevated levels of ITGB4/PXN. The knocking down of ITGB4 and PXN attenuated cell growth and enhanced apoptosis in 2D and 3D cultures. Interestingly, the double knockdown affected the expression of several genes, including USP1/VDAC1. Chromatin immunoprecipitation revealed a reduced binding of acetylated H3K27 at the promoter region of USP1 on the knocking down of ITGB4/PXN, highlighting the epigenetic regulation of various genes by these two proteins. Further, the knocking down of USP1 and VDAC1 generated a similar phenotype as the knockdown of ITGB4/PXN (Figure 3). The suppression of VDAC1 resulted in increased mitochondrial respiration and the generation of reactive oxygen species, leading to DNA damage, whereas the suppression of USP1 affected the DNA repair caused by adduct formation induced by cisplatin. Thus, these results highlighted the important role of the FA-associated complex-associated genes in cisplatin resistance and suggested that disrupting the interactions between the key components could potentially alleviate cisplatin resistance.

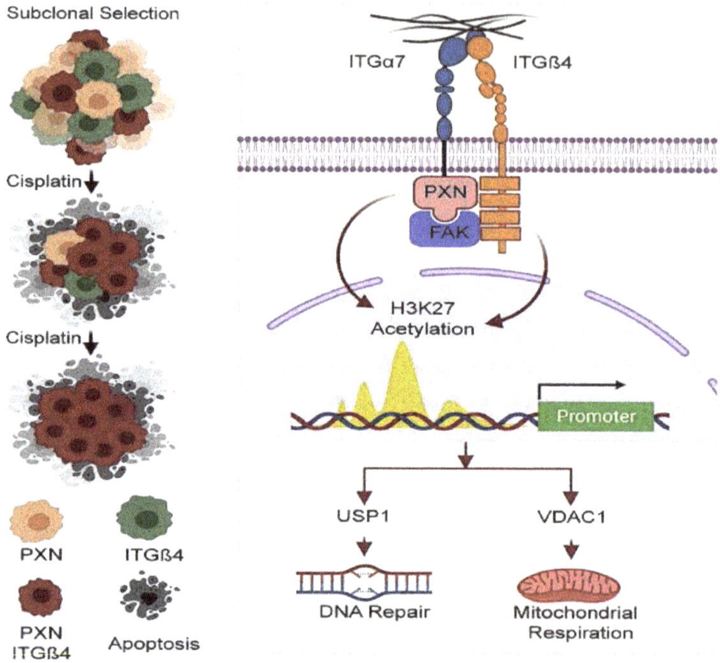

Figure 3. Schematic depicting the interaction between ITGB4 and PXN regulating downstream proteins USP1 and VDAC1 at the transcriptional level to coordinate cisplatin resistance; taken from Reprinted/adapted with permission from Ref. [51].

3.2. *Mathematical Modeling Suggests Bistability Drives Phenotypic Switching*

Our work also highlighted the role of ITGB4 in defining tumor heterogeneity. An immunohistochemistry analysis of patient samples and NSCLC cell lines confirmed the differential expression of ITGB4 [52]. Therefore, ITGB4 was used as a marker to sort low and high ITGB4-expressing NSCLC cells. Interestingly, the low ITGB4-expressing cells, after a few days in culture, were able to recreate the heterogeneous population of ITGB4-expressing cells, but the cells sorted for high ITGB4 failed to recreate the heterogeneous expression of ITGB4. These results were suggestive that low ITGB4-expressing cells have more plasticity to recreate the heterogeneity compared to high ITGB4-expressing cells. Further, RNA seq analysis carried out on the ITGB4 knockdown cells suggested a bistable relation between the microRNA 1-3p and ITGB4. A mathematical model developed based on the expression of these two genes indicated bistability; in a mixture of heterogeneous cells, some cells express high ITGB4/low microRNA 1-3p, low ITGB4/high microRNA 1-3p, or an equal expression of both ITGB4 and miRNA 1-3p (intermediary cells). Intermediary cells can shift in either direction depending on the environmental cues, for example, they could increase ITGB4 expression to tolerate cisplatin toxicity, and in absence of a drug, they could return to normal or low expressing subtypes [52].

3.3. *Novel Alternatives to Alleviate Cisplatin Resistance*

To identify small molecules that could potentially disrupt the interaction between PXN and FAK and, hence, perturb the focal adhesion complex and its downstream signaling, we used an in silico screening approach to screen a library of FDA-approved compounds. The screen identified several compounds that were found to sensitize the platinum-resistant NSCLC cells. Of these, carfilzomib (CFZ) was the most efficacious (IC50 in the low nanomolar range) and was able to induce DNA damage and apoptosis in NSCLC cells [68]. Furthermore, CFZ was also found to significantly inhibit migration, wound-healing, and ITGB4 expression at sublethal doses. Altogether, the data revealed an alternative and more efficient approach to treating lung cancer patients with cisplatin resistance.

3.4. Group Behavior and Phenotypic Switching Enable NSCLC Cells to Evade Chemotherapy

Phenotypic plasticity is critical for cancer cells to adapt themselves and survive [69]. Because of phenotypic plasticity, cancer cells are adept at switching their phenotypes in response to either intrinsic or extrinsic (environmental) cues. Thus, phenotypic plasticity enables cancer cells to undergo epithelial-to-mesenchymal transition (EMT) in order to facilitate distant metastasis; switch from being drug-sensitive to becoming drug-tolerant and, eventually, -resistant; or acquire stem cell-like characteristics. These phenomena help cancer cells to adapt to the fitness landscape and withstand drug treatment. Emerging evidence also indicates that both genetic and non-genetic mechanisms play crucial roles in the adaptability or cooperation between cancer cells to withstand stressful conditions. Hata et al. [30] provided clinical evidence showing that drug-resistant cells can both pre-exist and can evolve from drug-tolerant cells. If so, how does the co-existence of drug-sensitive and drug-tolerant/resistant clones impact their ability to cooperate or compete (group behavior) to evade drug toxicity [70,71]?

To address this question, we employed an approach that embodied both experimental methods and mathematical modeling, again using the cisplatin treatment of NSCLC cell lines [53]. Cisplatin-sensitive H23 and cisplatin-tolerant H2009 NSCLC cells were co-cultured and monitored in real-time in order to discern differences in their behavior. The two cell line cultures were grown as either monotypic (grown by themselves) or as heterotypic cultures (co-culture of tolerant and sensitive cells) in different ratios 1:1, 2:1, 4:1, and 8:1 and their growth rates were monitored in real time using a live cell imaging system. The data revealed that the tolerant cell proliferation was suppressed in the presence of sensitive cells at a 1:1 ratio and the proliferation could be rescued by increasing the fraction of tolerant cells in the co-cultures (Figure 4). The experiment was also repeated for the alternative ratios where the sensitive cells were increased, keeping the tolerant cells constant and the same result was observed, i.e., tolerant cell growth was inhibited by sensitive cells; however, the addition of a drug or increase in the ratio of tolerant cells in the population favored the growth of tolerant cells.

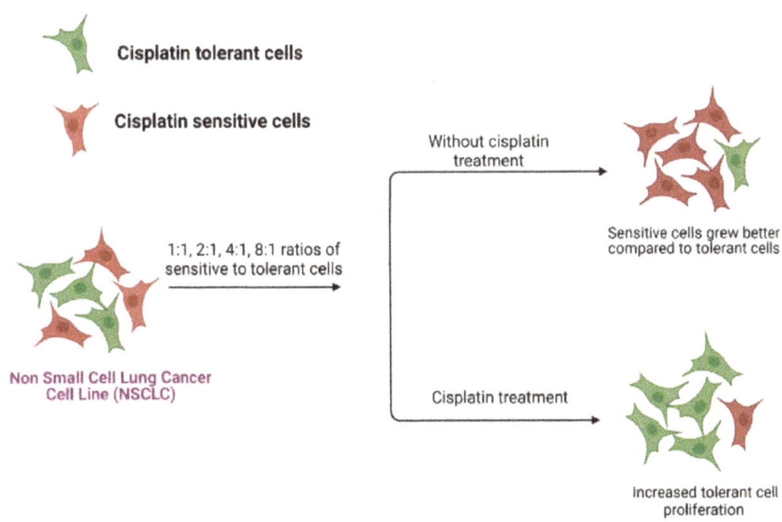

Figure 4. Behavior of cisplatin-sensitive and tolerant NSCLC cells in the 2D co-culture.

Considering the key observations from these in vitro studies, such as (i) sensitive cells inhibiting tolerant cell proliferation in a co-culture in the absence of cisplatin, (ii) the suppressive effect being stronger upon longer incubation compared to shorter incubation, and (iii) the competition by the sensitive cells being eliminated in the presence of cisplatin, a new mathematical approach called the Phenotypic Switch Model with Stress Response (PSMSR) was developed to fit the observed growth curves, the model conglomerate concepts from chemical reaction kinetics, and the cooperative behavior of drug-tolerant phenotypes in the community. A distinguishing feature of the PSMSR model is that it considers the ability of cancer cells to switch phenotypes. In addition to several testable predictions, the most important takeaway from the modeling exercise is that a small population of the tolerant cells may help the drug-sensitive cells to sustain proliferation. However, high levels of or continuous drug treatment, such stress removal mechanisms, may be insufficient to sustaining sensitive cell viability. Thus, it follows that it is essential to turn off phenotypic switching in such situations and allow the sensitive cells to become extinct and the tolerant cells to proliferate, which is the fundamental basis of 'adaptive' therapy or intermittent therapy strategy [72,73].

4. Conclusions and Future Perspective

In the cancer world, drug attrition rates are notorious—several drugs are effective in preclinical studies but only a few are approved for clinical use [74]. Furthermore, while most approved cancer drugs do help in improving the life expectancy of the patients, cancer cells often develop resistance against these therapies and relapse as resistant and metastatic diseases. Moreover, underlying mechanisms remain poorly understood.

The prescribed strategy that a physician follows—administering the maximal dose continuously in the shortest possible time—can lead to counterproductive and potentially adverse outcomes, such as drug resistance through genetic and non-genetic mechanisms, as discussed above. This led to exploring the emergence of alternative therapies and approaches [75–78], for example, adaptive/intermittent therapies [70,79]. The basic principle of intermittent therapy is to administer a lower therapeutic dose of the drug than the maximally tolerated dose to maintain a stable disease. The major advantage of intermittent therapy is an improved quality of life for the patient due to low drug dosage and, thus, fewer side effects. By keeping the drug doses low and intermittent (with drug holidays in between), the proliferation of resistant subclones can be delayed. Some of the success stories of intermittent therapy have been seen in rectal, pediatric sarcoma, prostate, and breast cancer [80–85].

Based on our observations and those of others from the literature, we believe that maintaining a stable disease may be more prudent. A good example is a study by Klotz et [78], where they treated 20 patients with advanced prostate cancer with intermittent endocrine therapy (diethylstilbesterol in 19 cases and flutamide in 1 case). These patients were treated until a clinical response was demonstrated, with a mean initial treatment duration of 10 months (range 2–70 months). The treatment was then stopped and re-started when tumors relapsed, with mean interval times of 8 months (range 1–24 months). All relapsed patients responded to the re-administration of the drug. Patients had a better quality of life during the drug holidays of the treatment. Indeed, subsequent studies [86], including a meta-analysis (Marlon et al.) [87], also concluded that intermittent androgen deprivation can be considered as an option for recurrent or metastatic prostate cancer.

Therefore, the data from our studies on cisplatin resistance in NSCLC not only lend further credence to the paradigm of intermittent strategy to maintain stable disease but also underscore the nuances and benefits of a 'Team Medicine' approach from a systems biology perspective.

Author Contributions: Conceptualization, S.R. (Sravani Ramisetty) and A.M; resources R.S.; writing—original draft preparation, S.R. (Sravani Ramisetty), P.K. and A.M.; writing—review and editing, visualization, P.K., S.B., A.N., S.S.S., L.G., T.M., B.M., S.M., J.M., E.P., S.S., S.R. (Swapnil Rajurkar), E.M., R.S. and A.M.; supervision, P.K., A.M. and R.S.; project administration, R.S. All authors have read and agreed to the published version of the manuscript.

Funding: This work was supported in part by grants from the United States Department of Defense (W81XWH-22-0331).

Institutional Review Board Statement: Not applicable for studies not involving humans or animals.

Informed Consent Statement: Not applicable for studies not involving humans.

Data Availability Statement: The data are available upon request from the corresponding author.

Conflicts of Interest: The authors declare no conflict of interest.

Abbreviations

CFZ, carfilzomib; DTP drug-tolerant persisters; FA, focal adhesion; FAK, focal adhesion kinase; IDP, intrinsically disordered proteins; ITGB4, integrin beta 4; LSCC, lung squamous cell carcinoma; LUAD, lung adenocarcinoma; MAPK, Ras-mitogen-activated protein kinase, NSCLC, non-small cell lung cancer; PI3K, Phosphoinositide 3-Kinase; PIN, protein interaction networks; PLC-γ, phospholipase C-gamma; PSMR, phenotypic switch model with stress response; PXN, paxillin.

References

1. Siegel, R.L.; Miller, K.D.; Fuchs, H.E.; Jemal, A. Cancer statistics, 2022. *CA A Cancer J. Clin.* **2022**, *72*, 7–33. [CrossRef] [PubMed]
2. Bray, F.; Ferlay, J.; Soerjomataram, I.; Siegel, R.L.; Torre, L.A.; Jemal, A. Global cancer statistics 2018: GLOBOCAN estimates of incidence and mortality worldwide for 36 cancers in 185 countries. *CA Cancer J. Clin.* **2018**, *68*, 394–424. [CrossRef] [PubMed]
3. Ettinger, D.S.; Akerley, W.; Bepler, G.; Blum, M.G.; Chang, A.; Cheney, R.T.; Chirieac, L.R.; D'Amico, T.A.; Demmy, T.L.; Ganti, A.K.P.; et al. Non-small cell lung cancer. *J. Natl. Compr. Canc Netw.* **2010**, *8*, 740–801. [CrossRef] [PubMed]
4. Salgia, R. Mutation testing for directing upfront targeted therapy and post-progression combination therapy strategies in lung adenocarcinoma. *Expert Rev. Mol. Diagn.* **2016**, *16*, 737–749. [CrossRef] [PubMed]
5. Herbst, R.S.; Morgensztern, D.; Boshoff, C. The biology and management of non-small cell lung cancer. *Nature* **2018**, *553*, 446–454. [CrossRef]
6. Walser, T.; Cui, X.; Yanagawa, J.; Lee, J.M.; Heinrich, E.; Lee, G.; Sharma, S.; Dubinett, S.M. Smoking and lung cancer: The role of inflammation. *Proc. Am. Thorac. Soc.* **2008**, *5*, 811–815. [CrossRef] [PubMed]
7. Proctor, R.N. Tobacco and the global lung cancer epidemic. *Nat. Rev. Cancer* **2001**, *1*, 82–86. [CrossRef]
8. Sasco, A.; Secretan, M.; Straif, K. Tobacco smoking and cancer: A brief review of recent epidemiological evidence. *Lung Cancer* **2004**, *45*, S3–S9. [CrossRef]
9. Travis, W.D. *Pathology & Genetics Tumors of the Lung, Pleura, Thymus, and Heart*; World Health Organization Classification of tumors; WHO: Geneva, Switzerland, 2004.
10. Non-small Cell Lung Cancer Collaborative Group. Chemotherapy in non-small cell lung cancer: A meta-analysis using updated data on individual patients from 52 randomized clinical trials. *BMJ* **1995**, *311*, 899–909. [CrossRef]
11. Wang, S.; Song, Y.; Yan, F.; Liu, D. Mechanisms of resistance to third-generation EGFR tyrosine kinase inhibitors. *Front. Med.* **2016**, *10*, 383–388. [CrossRef]
12. Trowe, T.; Boukouvala, S.; Calkins, K.; Cutler, R.E.; Fong, R.; Funke, R.; Gendreau, S.B.; Kim, Y.D.; Miller, N.; Woolfrey, J.R.; et al. EXEL-7647 inhibits mutant forms of ErbB2 associated with lapatinib resistance and neoplastic transformation. *Clin. Cancer Res.* **2008**, *14*, 2465–2475. [CrossRef] [PubMed]
13. Tetsu, O.; Hangauer, M.; Phuchareon, J.J.; Eisele, D.W.; McCormick, F. Drug Resistance to EGFR Inhibitors in Lung Cancer. *Chemotherapy* **2016**, *61*, 223–235. [CrossRef] [PubMed]
14. Lohitesh, K.; Chowdhury, R.; Mukherjee, S. Resistance a major hindrance to chemotherapy in hepatocellular carcinoma: An insight. *Cancer Cell Int.* **2018**, *18*, 44. [CrossRef] [PubMed]
15. Mansoori, B.; Mohammadi, A.; Davudian, S.; Shirjang, S.; Baradaran, B. The Different Mechanisms of Cancer Drug Resistance: A Brief Review. *Adv. Pharm. Bull.* **2017**, *7*, 339–348. [CrossRef] [PubMed]
16. Housman, G.; Byler, S.; Heerboth, S.; Lapinska, K.; Longacre, M.; Snyder, N.; Sarkar, S. Drug resistance in cancer: An overview. *Cancers* **2014**, *6*, 1769–1792. [CrossRef]
17. Ward, R.A.; Fawell, S.; Floc'h, N.; Flemington, V.; McKerrecher, D.; Smith, P.D. Challenges and Opportunities in Cancer Drug Resistance. *Chem. Rev.* **2021**, *121*, 3297–3351. [CrossRef]

18. Holohan, C.; Van Schaeybroeck, S.; Longley, D.B.; Johnston, P.G. Cancer drug resistance: An evolving paradigm. *Nat. Rev. Cancer* **2013**, *13*, 714–726. [CrossRef]
19. Alfarouk, K.O.; Stock, C.-M.; Taylor, S.; Walsh, M.; Muddathir, A.K.; Verduzco, D.; Bashir, A.H.H.; Mohammed, O.Y.; O ElHassan, G.; Harguindey, S.; et al. Resistance to cancer chemotherapy: Failure in drug response from ADME to P-gp. *Cancer Cell Int.* **2015**, *15*, 71. [CrossRef]
20. Wang, X.; Zhang, H.; Chen, X. Drug resistance and combating drug resistance in cancer. *Cancer Drug Resist.* **2019**, *2*, 141–160. [CrossRef]
21. Vasan, N.; Baselga, J.; Hyman, D.M. A view on drug resistance in cancer. *Nature* **2019**, *575*, 299–309. [CrossRef]
22. Senthebane, D.A.; Rowe, A.; Thomford, N.E.; Shipanga, H.; Munro, D.; Al Mazeedi, M.A.M.; Almazyadi, H.A.M.; Kallmeyer, K.; Dandara, C.; Pepper, M.S.; et al. The Role of Tumor Microenvironment in Chemoresistance: To Survive, Keep Your Enemies Closer. *Int. J. Mol. Sci.* **2017**, *18*, 1586. [CrossRef] [PubMed]
23. Gonzalez Rajal, A.; Marzec, K.A.; McCloy, R.A.; Nobis, M.; Chin, V.; Hastings, J.F.; Lai, K.; Kennerson, M.; E Hughes, W.; Vaghjiani, V.; et al. A non-genetic, cell cycle-dependent mechanism of platinum resistance in lung adenocarcinoma. *eLife* **2021**, *10*, e65234. [CrossRef] [PubMed]
24. Stewart, D.J. Mechanisms of resistance to cisplatin and carboplatin. *Crit. Rev. Oncol./Hematol.* **2007**, *63*, 12–31. [CrossRef]
25. Konieczkowski, D.J.; Johannessen, C.M.; Garraway, L.A. A convergence-based framework for cancer drug resistance. *Cancer Cell* **2018**, *33*, 801–815. [CrossRef]
26. Le Magnen, C.; Shen, M.M.; Abate-Shen, C. Lineage plasticity in cancer progression and treatment. *Annu. Rev. Cancer Biol.* **2018**, *2*, 271–289. [CrossRef]
27. Bell, C.C.; Gilan, O. Principles and mechanisms of non-genetic resistance in cancer. *Br. J. Cancer* **2020**, *122*, 465–472. [CrossRef] [PubMed]
28. Dhanyamraju, P.K.; Schell, T.D.; Amin, S.; Robertson, G.P. Drug-Tolerant Persister Cells in Cancer Therapy Resistance. *Cancer Res.* **2022**, *82*, 2503–2514. [CrossRef]
29. Sharma, S.V.; Lee, D.Y.; Li, B.; Quinlan, M.P.; Takahashi, F.; Maheswaran, S.; McDermott, U.; Azizian, N.; Zou, L.; Fischbach, M.A.; et al. A chromatin-mediated reversible drug-tolerant state in cancer cell subpopulations. *Cell* **2010**, *141*, 69–80. [CrossRef]
30. Hata, A.N.; Niederst, M.J.; Archibald, H.L.; Gomez-Caraballo, M.; Siddiqui, F.M.; Mulvey, H.E.; Maruvka, Y.E.; Ji, F.; Bhang, H.-E.C.; Radhakrishna, V.K.; et al. Tumor cells can follow distinct evolutionary paths to become resistant to epidermal growth factor receptor inhibition. *Nat. Med.* **2016**, *22*, 262–269. [CrossRef]
31. Hangauer, M.J.; Viswanathan, V.S.; Ryan, M.J.; Bole, D.; Eaton, J.K.; Matov, A.; Galeas, J.; Dhruv, H.D.; Berens, M.E.; Schreiber, S.L.; et al. Drug-tolerant persister cancer cells are vulnerable to GPX4 inhibition. *Nature* **2017**, *551*, 247–250. [CrossRef]
32. Raha, D.; Wilson, T.R.; Peng, J.; Peterson, D.; Yue, P.; Evangelista, M.; Wilson, C.; Merchant, M.; Settleman, J. The Cancer Stem Cell Marker Aldehyde Dehydrogenase Is Required to Maintain a Drug-Tolerant Tumor Cell Subpopulation Aldehyde Dehydrogenase Contributes to Cancer Drug Resistance. *Cancer Res.* **2014**, *74*, 3579–3590. [CrossRef] [PubMed]
33. Menon, D.R.; Das, S.; Krepler, C.; Vultur, A.; Rinner, B.; Schauer, S.; Kashofer, K.; Wagner, K.; Zhang, G.; Rad, E.B.; et al. A stress-induced early innate response causes multidrug tolerance in melanoma. *Oncogene* **2015**, *34*, 4545. [CrossRef] [PubMed]
34. Liau, B.B.; Sievers, C.; Donohue, L.K.; Gillespie, S.M.; Flavahan, W.A.; Miller, T.E.; Venteicher, A.S.; Hebert, C.H.; Carey, C.D.; Rodig, S.J.; et al. Adaptive chromatin remodeling drives glioblastoma stem cell plasticity and drug tolerance. *Cell Stem Cell* **2017**, *20*, 233–246. [CrossRef] [PubMed]
35. Shaffer, S.M.; Dunagin, M.C.; Torborg, S.R.; Torre, E.A.; Emert, B.; Krepler, C.; Beqiri, M.; Sproesser, K.; Brafford, P.A.; Xiao, M.; et al. Rare cell variability and drug-induced reprogramming as a mode of cancer drug resistance. *Nature* **2017**, *546*, 431–435. [CrossRef]
36. Flavahan, W.A.; Gaskell, E.; Bernstein, B.E. Epigenetic plasticity and the hallmarks of cancer. *Science* **2017**, *357*, eaal2380. [CrossRef]
37. Paksa, A.; Rajagopal, J. The epigenetic basis of cellular plasticity. *Curr. Opin. Cell Biol.* **2017**, *49*, 116–122. [CrossRef]
38. Wainwright, E.N.; Scaffidi, P. Epigenetics and cancer stem cells: Unleashing, hijacking, and restricting cellular plasticity. *Trends Cancer* **2017**, *3*, 372–386. [CrossRef]
39. Salgia, R.; Kulkarni, P. The Genetic/Non-genetic Duality of Drug 'Resistance' in Cancer. *Trends Cancer* **2018**, *4*, 110–118. [CrossRef]
40. Kulkarni, V.; Kulkarni, P. Intrinsically disordered proteins and phenotypic switching: Implications in cancer. *Prog. Mol. Biol. Transl. Sci.* **2019**, *166*, 63–84.
41. Kulkarni, P.; Mohanty, A.; Bhattacharya, S.; Singhal, S.; Guo, L.; Ramisetty, S.; Mirzapoiazova, T.; Mambetsariev, B.; Mittan, S.; Malhotra, J.; et al. Addressing Drug Resistance in Cancer: A Team Medicine Approach. *J. Clin. Med.* **2022**, *11*, 5701. [CrossRef]
42. Mészáros, B.; Hajdu-Soltész, B.; Zeke, A.; Dosztányi, Z. Mutations of Intrinsically Disordered Protein Regions Can Drive Cancer but Lack Therapeutic Strategies. *Biomolecules* **2021**, *11*, 381. [CrossRef] [PubMed]
43. Brown, A.; Kumar, S.; Tchounwou, P.B. Cisplatin-Based Chemotherapy of Human Cancers. *J. Cancer Sci. Ther.* **2019**, *11*, 97. [PubMed]
44. Dasari, S.; Tchounwou, P.B. Cisplatin in cancer therapy: Molecular mechanisms of action. *Eur. J. Pharmacol.* **2014**, *740*, 364–378. [CrossRef] [PubMed]
45. Horwich, A.; Shipley, J.; Huddart, R. Testicular germ-cell cancer. *Lancet* **2006**, *367*, 754–765. [CrossRef]
46. Florea, A.-M.; Büsselberg, D. Cisplatin as an anti-tumor drug: Cellular mechanisms of activity, drug resistance, and induced side effects. *Cancers* **2011**, *3*, 1351–1371. [CrossRef] [PubMed]

47. Giaccone, G. Clinical perspectives on platinum resistance. *Drugs* **2000**, *59* (Suppl. S4), 9–17, discussion 37–38. [CrossRef]
48. Köberle, B.; Tomicic, M.T.; Usanova, S.; Kaina, B. Cisplatin resistance: Preclinical findings and clinical implications. *Biochim. Biophys. Acta (BBA) Rev. Cancer* **2010**, *1806*, 172–182. [CrossRef]
49. Dempke, W.; Voigt, W.; Grothey, A.; Hill, B.T.; Schmoll, H.-J. Cisplatin resistance and oncogenes—A review. *Anti-Cancer Drugs* **2000**, *11*, 225–236. [CrossRef]
50. Amable, L. Cisplatin resistance and opportunities for precision medicine. *Pharmacol. Res.* **2016**, *106*, 27–36. [CrossRef]
51. Galluzzi, L.; Senovilla, L.; Vitale, I.; Michels, J.; Martins, I.; Kepp, O.; Castedo, M.; Kroemer, G. Molecular mechanisms of cisplatin resistance. *Oncogene* **2012**, *31*, 1869–1883. [CrossRef]
52. Mohanty, A.; Nam, A.; Pozhitkov, A.; Yang, L.; Srivastava, S.; Nathan, A.; Wu, X.; Mambetsariev, I.; Nelson, M.; Subbalakshmi, A.R.; et al. A Non-genetic Mechanism Involving the Integrin β4/Paxillin Axis Contributes to Chemoresistance in Lung Cancer. *iScience* **2020**, *23*, 101496. [CrossRef] [PubMed]
53. Nam, A.; Mohanty, A.; Bhattacharya, S.; Kotnala, S.; Achuthan, S.; Hari, K.; Srivastava, S.; Guo, L.; Nathan, A.; Chatterjee, R.; et al. Dynamic Phenotypic Switching, and Group Behavior Help Non-Small Cell Lung Cancer Cells Evade Chemotherapy. *Biomolecules* **2022**, *12*, 8. [CrossRef] [PubMed]
54. Shen, D.-W.; Pouliot, L.M.; Hall, M.D.; Gottesman, M.M. Cisplatin resistance: A cellular self-defense mechanism resulting from multiple epigenetic and genetic changes. *Pharmacol. Rev.* **2012**, *64*, 706–721. [CrossRef] [PubMed]
55. Kutuk, O.; Arisan, E.D.; Tezil, T.; Hoshan, M.C.; Basaga, H. Cisplatin overcomes Bcl-2-mediated resistance to apoptosis via preferential engagement of Bak: Critical role of Noxa-mediated lipid peroxidation. *Carcinogenesis* **2009**, *30*, 1517–1527. [CrossRef]
56. Giancotti, F.G.; Ruoslahti, E. Integrin Signaling. *Science* **1999**, *285*, 1028–1033. [CrossRef]
57. Maziveyi, M.; Alahari, S.K. Cell matrix adhesions in cancer: The proteins that form the glue. *Oncotarget* **2017**, *8*, 48471–48487. [CrossRef]
58. Cary, L.A.; Guan, J.L. Focal adhesion kinase in integrin-mediated signaling. *Front. Biosci.* **1999**, *4*, D102–D113. [CrossRef]
59. Moreno-Layseca, P.; Streuli, C.H. Signalling pathways linking integrins with cell cycle progression. *Matrix Biol.* **2014**, *34*, 144–153. [CrossRef]
60. Alberts, B. *Molecular Biology of the Cell*, 5th ed.; Garland Science: New York, NY, USA, 2008; pp. 906–911.
61. Mainiero, F.; Murgia, C.; Wary, K.K.; Curatola, A.M.; Pepe, A.; Blumemberg, M.; Westwick, J.K.; Der, C.J.; Giancotti, F.G. The coupling of α6β4 integrin to Ras–MAP kinase pathways mediated by Shc controls keratinocyte proliferation. *EMBO J.* **1997**, *16*, 2365–2375. [CrossRef]
62. Salgia, R.; Li, J.-L.; Lo, S.H.; Brunkhorst, B.; Kansas, G.S.; Sobhany, E.S.; Sun, Y.; Pisick, E.; Hallek, M.; Ernst, T.; et al. Molecular Cloning of Human Paxillin, a Focal Adhesion Protein Phosphorylated by P210BCR/ABL(∗). *J. Biol. Chem.* **1995**, *270*, 5039–5047. [CrossRef]
63. Song, J.; Li, M.; Tretiakova, M.; Salgia, R.; Cagle, P.T.; Husain, A.N. Expression Patterns of PAX5, c-Met, and Paxillin in Neuroendocrine Tumors of the Lung. *Arch Pathol. Lab. Med.* **2010**, *134*, 1702–1705. [CrossRef] [PubMed]
64. Mackinnon, A.C.; Tretiakova, M.; Henderson, L.; Mehta, R.G.; Yan, B.C.; Joseph, L.; Krausz, T.; Husain, A.N.; E Reid, M.; Salgia, R. Paxillin expression and amplification in early lung lesions of high-risk patients, lung adenocarcinoma and metastatic disease. *J. Clin. Pathol.* **2011**, *64*, 16. [CrossRef]
65. Turner, C.E. Paxillin and focal adhesion signaling. *Nat. Cell Biol.* **2000**, *2*, E231–E236. [CrossRef] [PubMed]
66. Turner, C.E. Paxillin. *Int. J. Biochem. Cell Biol.* **1998**, *30*, 955–959. [CrossRef] [PubMed]
67. Dong, J.M.; Lau, L.S.; Ng, Y.W.; Lim, L.; Manser, E. Paxillin nuclear-cytoplasmic localization is regulated by phosphorylation of the LD4 motif: Evidence that nuclear paxillin promotes cell proliferation. *Biochem. J.* **2009**, *418*, 173–184. [CrossRef]
68. López-Colomé, A.M.; Lee-Rivera, I.; Benavides-Hidalgo, R.; López, E. Paxillin: A crossroad in pathological cell migration. *J. Hematol. Oncol.* **2017**, *10*, 50. [CrossRef]
69. Neerathilingam, M.; Bairy, S.G.; Mysore, S. Deciphering Mode of Action of Functionally Important Regions in the Intrinsically Disordered Paxillin (Residues 1-313) Using Its Interaction with FAT (Focal Adhesion Targeting Domain of Focal Adhesion Kinase). *PLoS ONE* **2016**, *11*, e0150153. [CrossRef] [PubMed]
70. Bertolucci, C.M.; Guibao, C.D.; Zheng, J. Structural features of the focal adhesion kinase-paxillin complex give insight into the dynamics of focal adhesion assembly. *Protein Sci. A Publ. Protein Soc.* **2005**, *14*, 644–652. [CrossRef]
71. Mohanty, A.; Nam, A.; Pozhitkov, A.; Bhattacharya, S.; Yang, L.; Nathan, A.; Wu, X.; Srivastava, S.; Mambetsariev, I.; Nelson, M.; et al. A Non-genetic Mechanism for Chemoresistance in Lung Cancer: The Role of Integrin β4/Paxillin Axis. *bioRxiv* **2019**. [CrossRef]
72. Gupta, P.B.; Pastushenko, I.; Skibinski, A.; Blanpain, C.; Kuperwasser, C. Phenotypic Plasticity: Driver of Cancer Initiation, Progression, and Therapy Resistance. *Cell Stem Cell* **2019**, *24*, 65–78. [CrossRef]
73. Bhattacharya, S.; Mohanty, A.; Achuthan, S.; Kotnala, S.; Jolly, M.K.; Kulkarni, P.; Salgia, R. Group Behavior and Emergence of Cancer Drug Resistance. *Trends Cancer* **2021**, *7*, 323–334. [CrossRef] [PubMed]
74. Emond, R.; Griffiths, J.I.; Grolmusz, V.K.; Sousa, R.S.; Bild, A.H.; Adler, F.R. Ecological interactions in breast cancer: Cell facilitation promotes growth and survival under drug pressure. *bioRxiv* **2021**. [CrossRef]
75. Stanková, K.; Brown, J.S.; Dalton, W.S.; Gatenby, R.A. Optimizing Cancer Treatment Using Game Theory: A Review. *JAMA Oncol.* **2019**, *5*, 96–103. [CrossRef] [PubMed]

76. Enriquez-Navas, P.M.; Wojtkowiak, J.W.; Gatenby, R.A. Application of Evolutionary Principles to Cancer Therapy. *Cancer Res.* **2015**, *75*, 4675–4680. [CrossRef]
77. Hutchinson, L.; Kirk, R. High drug attrition rates–where are we going wrong? *Nat Rev. Clin. Oncol.* **2011**, *8*, 189–190. [CrossRef] [PubMed]
78. Klotz, L.H.; Herr, H.W.; Morse, M.J.; Whitmore, W.F. Intermittent endocrine therapy for advanced prostate cancer. *Cancer* **1986**, *58*, 2546–2550. [CrossRef]
79. Goldenberg, S.L.; Bruchovsky, N.; Gleave, M.E.; Sullivan, L.D.; Akakura, K. Intermittent androgen suppression in the treatment of prostate cancer: A preliminary report. *Urology* **1995**, *45*, 839–844, discussion 844–845. [CrossRef]
80. Bruchovsky, N.; Klotz, L.H.; Sadar, M.; Crook, J.M.; Hoffart, D.; Godwin, L.; Warkentin, M.; E Gleave, M.; Goldenberg, S.L. Intermittent androgen suppression for prostate cancer: Canadian Prospective Trial and related observations. *Mol. Urol.* **2000**, *4*, 191–199, discussion 201.
81. Gatenby, R.A.; Silva, A.S.; Gillies, R.J.; Frieden, B.R. Adaptive therapy. *Cancer Res.* **2009**, *69*, 4894–4903. [CrossRef]
82. Kaznatcheev, A.; Vander Velde, R.; Scott, J.G.; Basanta, D. Cancer treatment scheduling and dynamic heterogeneity in social dilemmas of tumor acidity and vasculature. *Br. J. Cancer* **2017**, *116*, 785–792. [CrossRef]
83. Kavran, A.J.; Stuart, S.A.; Hayashi, K.R.; Basken, J.M.; Brandhuber, B.J.; Ahn, N.G. Intermittent treatment of BRAF(V600E) melanoma cells delays resistance by adaptive resensitization to drug rechallenge. *Proc. Natl. Acad. Sci. USA* **2022**, *119*, e2113535119. [CrossRef] [PubMed]
84. Felder, S.I.; Fleming, J.B.; Gatenby, R.A. Treatment-induced evolutionary dynamics in nonmetastatic locally advanced rectal adenocarcinoma. *Adv. Cancer Res.* **2021**, *151*, 39–67. [PubMed]
85. Reed, D.R.; Metts, J.; Pressley, M.; Fridley, B.L.; Hayashi, M.; Isakoff, M.S.; Loeb, D.M.; Makanji, R.; Roberts, R.D.; Trucco, M.; et al. An evolutionary framework for treating pediatric sarcomas. *Cancer* **2020**, *126*, 2577–2587. [CrossRef]
86. Zhang, J.; Cunningham, J.; Brown, J.; Gatenby, R. Evolution-based mathematical models significantly prolong response to abiraterone in metastatic castrate-resistant prostate cancer and identify strategies to further improve outcomes. *eLife* **2022**, *11*, e76284. [CrossRef]
87. Perera, M.; Roberts, M.J.; Klotz, L.; Higano, C.S.; Papa, N.; Sengupta, S.; Bolton, D.; Lawrentschuk, N. Intermittent versus continuous androgen deprivation therapy for advanced prostate cancer. *Nat. Rev. Urol.* **2020**, *17*, 469–481. [CrossRef] [PubMed]

Journal of
Clinical Medicine

MDPI

Perspective

Addressing Drug Resistance in Cancer: A Team Medicine Approach

Prakash Kulkarni [1,2], Atish Mohanty [1], Supriyo Bhattacharya [3], Sharad Singhal [1], Linlin Guo [1], Sravani Ramisetty [1], Tamara Mirzapoiazova [1], Bolot Mambetsariev [1], Sandeep Mittan [4], Jyoti Malhotra [5], Naveen Gupta [6], Pauline Kim [7], Razmig Babikian [1], Swapnil Rajurkar [6], Shanmuga Subbiah [8], Tingting Tan [9], Danny Nguyen [10], Amartej Merla [11], Sudarsan V. Kollimuttathuillam [12], Tanyanika Phillips [13], Peter Baik [14], Bradford Tan [14], Pankaj Vashi [14], Sagun Shrestha [15], Benjamin Leach [16], Ruchi Garg [17], Patricia L. Rich [17], F. Marc Stewart [1], Evan Pisick [14] and Ravi Salgia [1,*]

1 Department of Medical Oncology and Therapeutics Research, City of Hope National Medical Center, Duarte, CA 91010, USA
2 Department of Systems Biology, City of Hope National Medical Center, Duarte, CA 91010, USA
3 Integrative Genomics Core, City of Hope National Medical Center, Duarte, CA 91010, USA
4 Montefiore Medical Center, The University Hospital for Albert Einstein College of Medicine, Bronx, NY 10467, USA
5 Department of Medical Oncology and Therapeutics Research, City of Hope National Medical Center, 1000 FivePoint, Irvine, CA 92618, USA
6 Department of Medical Oncology and Therapeutics Research, City of Hope National Medical Center, 1100 San Bernardino Road, Suite 1100, Upland, CA 91786, USA
7 Department of Pharmacy, City of Hope National Medical Center, Duarte, CA 91010, USA
8 Department of Medical Oncology and Therapeutics Research, City of Hope National Medical Center, 1250 S. Sunset Ave., Suite 303, West Covina, CA 91790, USA
9 Department of Medical Oncology and Therapeutics Research, City of Hope National Medical Center, 1601 Avocado Ave., Newport Beach, CA 92660, USA
10 Department of Medical Oncology and Therapeutics Research, City of Hope National Medical Center, 19671 Beach Blvd. #315, Huntington Beach, CA 92648, USA
11 Department of Medical Oncology and Therapeutics Research, City of Hope National Medical Center, 38660 Medical Center Dr, Suite A380, Palmdale, CA 93551, USA
12 Department of Medical Oncology and Therapeutics Research, City of Hope National Medical Center, 16300 Sand Canyon Ave., Suite 207, Irvine, CA 92618, USA
13 Department of Medical Oncology and Therapeutics Research, City of Hope National Medical Center, 44151 15th St. West, Lancaster, CA 93534, USA
14 Cancer Treatment Centers of America, CTCA Chicago, 2520 Elisha Avenue, Zion, IL 60099, USA
15 Cancer Treatment Centers of America, CTCA Phoenix, 14200 West Celebrate Life Way, Goodyear, AZ 85338, USA
16 Department of Medical Oncology and Therapeutics Research, City of Hope National Medical Center, 15031 Rinaldi St., Suite 150, Mission Hills, CA 91345, USA
17 Cancer Treatment Centers of America, CTCA Atlanta, 600 Celebrate Life Parkway, Newnan, GA 30265, USA
* Correspondence: rsalgia@coh.org

Citation: Kulkarni, P.; Mohanty, A.; Bhattacharya, S.; Singhal, S.; Guo, L.; Ramisetty, S.; Mirzapoiazova, T.; Mambetsariev, B.; Mittan, S.; Malhotra, J.; et al. Addressing Drug Resistance in Cancer: A Team Medicine Approach. *J. Clin. Med.* **2022**, *11*, 5701. https://doi.org/10.3390/jcm11195701

Academic Editor: Claude Lambert

Received: 30 August 2022
Accepted: 23 September 2022
Published: 27 September 2022

Abstract: Drug resistance remains one of the major impediments to treating cancer. Although many patients respond well initially, resistance to therapy typically ensues. Several confounding factors appear to contribute to this challenge. Here, we first discuss some of the challenges associated with drug resistance. We then discuss how a 'Team Medicine' approach, involving an interdisciplinary team of basic scientists working together with clinicians, has uncovered new therapeutic strategies. These strategies, referred to as intermittent or 'adaptive' therapy, which are based on eco-evolutionary principles, have met with remarkable success in potentially precluding or delaying the emergence of drug resistance in several cancers. Incorporating such treatment strategies into clinical protocols could potentially enhance the precision of delivering personalized medicine to patients. Furthermore, reaching out to patients in the network of hospitals affiliated with leading academic centers could help them benefit from such innovative treatment options. Finally, lowering the dose of the drug and its frequency (because of intermittent rather than continuous therapy) can also have a significant impact on lowering the toxicity and undesirable side effects of the drugs while lowering the financial burden carried by the patient and insurance providers.

Keywords: drug resistance; drug tolerance; eco-evolutionary; intermittent therapy; adaptive therapy; continuous therapy; Team Medicine; intrinsically disordered proteins

1. Introduction

Cancer is a major healthcare crisis and one of the leading causes of death in the world. In 2020 alone, worldwide, there were ~18 million cancer cases, and cancer accounted for nearly 10 million deaths, or nearly one in six deaths, in the same year, Sung et al., 2020 [1]. In less than two decades from now (by 2040), the number of new cancer cases per year is expected to rise to >27 million and the number of cancer-related deaths to >16 million [2]. These whopping numbers underscore the magnitude of the health care crisis and the economic burden of this burgeoning problem across the globe. Although the disease presents several challenges depending on the cancer type and, in some cases, ethnicity, the emergence of drug resistance remains a major common concern in treating all cancer patients.

Several confounding factors appear to be contributing to this challenge. Here, we first discuss some of the challenges associated with drug resistance in cancer. We then discuss how a 'Team Medicine' approach [3], involving an interdisciplinary team of scientists with expertise in physics, biophysics, mathematics, evolutionary biology, bioinformatics, data science, computational biology, and cancer biology, working with clinicians, has provided new opportunities and new therapeutic strategies that could potentially preclude or delay the emergence of drug resistance in several cancers. We conclude by proposing a few innovations to our approach in treating cancer: (1) the novel therapeutic strategies such as intermittent drug treatment at moderate dosage as opposed to continuous treatment at high dosage and (2) leveraging precise knowledge of the tumor phenotypic landscape in designing personalized therapy through deeper consideration of genetic, epigenetic, and transcriptomic information.

2. Is Drug Resistance Genetic or, Are Non-Genetic Mechanisms Involved?

For well over a century, since Theodore Boveri's ground-breaking observations in the early 1900s (*Zur Frage der Entstehung maligner Tumouren*), cancer has been thought to be a genetic disease [4–7]. In fact, today, a genetic basis underlying the origin of cancer, its progression through distant metastasis, and the emergence of drug resistance is practically common knowledge. The following excerpt from an influential review in Nature Medicine by Vogelstein and Kinzler [8], titled "Cancer genes and the pathways they control", underscores the prevailing ethos: "*The revolution in cancer research can be summed up in a single sentence: cancer is, in essence, a genetic disease. In the last decade, many important genes responsible for the genesis of various cancers have been discovered, their mutations precisely identified, and the pathways through which they act characterized.*" (Our bold for emphasis). Pursuant to this landmark review, a decade later, Vogelstein and Kinzler [9] further emphasized the genetic nature of cancer in a perspective article, "The path to cancer—three strikes and you're out", in the leading medical journal, New England Journal of Medicine. The authors wrote "*Focusing on driver-gene mutations and the pathways they control has rendered complex cancer-genome landscapes intelligible. In solid tumors of adults, alterations in as few as three driver genes appear to suffice for a cell to evolve into an advanced cancer.*" (Our bold is for emphasis).

Highly influential articles like these and countless others have helped to firmly establish the genetic basis of cancer and, much like the modern synthesis in evolutionary theory [10], provide a conceptual framework to understand cancer and its link to Darwinian evolution. Furthermore, specific mutations have been leveraged as hallmarks for a conclusive diagnosis and the staging of specific cancer types, and highly potent drugs that specifically bind to the mutant target oncoproteins have been developed, adding to the precision with which individual patients are treated [11–13].

However, contrary to the prevailing wisdom, emerging evidence indicates that mechanisms such as epigenetic modifications and protein interaction network (PIN) rewiring—defined as changes in the interaction of individual proteins in signaling pathways in response to environmental changes—can also contribute to various aspects of cancer, including its origin, progression, and the emergence of drug resistance [14–20], underscoring the role of non-genetic mechanisms. It is now evident that both non-genetic and genetic mechanisms are involved, especially in acquired drug resistance and, furthermore, as discussed below, the irreversible resistance to a drug that can be acquired via an intermediate, reversible tolerant state via non-genetic mechanisms.

3. Discerning Drug Tolerance and Resistance

Perhaps one of the main reasons for our failure to overcome drug resistance in cancer, whether innate or acquired, may have to do with the difficulty in how we perceive the phenomenon. Unfortunately, and erroneously (even if inadvertently), it is assumed that drug resistance, tolerance, and persistence are synonymous or equivalent and hence, are used indiscriminately [21]. However, these are quite distinct and nuanced phenomena, as elegantly demonstrated in microbiology.

In bacteria for example, resistance is defined as the ability of an organism to grow at high concentrations within the presence of a drug. Resistance is typically due to mutations and is heritable transgenerationally. On the other hand, tolerance is more generally used to describe the ability, whether inherited or not, to survive transient exposure to high concentrations of a drug, and persistence is defined as the ability of a subpopulation to survive long-term exposure to high concentrations of a drug. Persistence is typically observed when the majority of the population is rapidly killed following drug treatment while a subpopulation persists for a much longer period of time [22,23] (Figure 1). Since rigorous definitions are lacking in the cancer field, the term 'resistance' remains ambiguous or confusing at best; thus, this further adds to the difficulty of defining a patient's response to a drug. For example, if a patient does not respond to a drug, is the patient's tumor tolerant and hence potentially reversible? or is it truly resistant and hence irreversible? Perhaps, publicly available databases dedicated to cancer drug resistance could help alleviate some of the confusion [24].

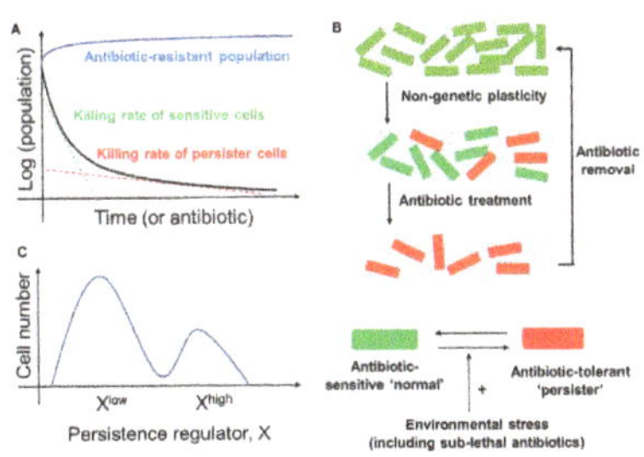

Figure 1. Bacterial Persistence. (**A**) Biphasic time-kill curve in bacterial populations exposed to antibiotics: faster killing rate of sensitive cell (green dotted line) followed by a slower killing rate (red dotted line) of the persisters. In contrast, the antibiotic-resistant population continues to grow in the presence of antibiotic (blue curve). (**B**) (top) An isogenic population of antibiotic sensitive cells can give rise to persisters via non-genetic/phenotypic plasticity. These slow cycling persisters survive in the antibiotic treatment and tend to resume growth and generate a new population identical to the original population upon antibiotic removal (bottom). Persisters and non-persisters can switch among one another; the switching rate can be influenced by external stress factors. (**C**) Non-genetic heterogeneity of a key regulator of persistence (say X) in an isogenic population may give rise to two (or more) subpopulations that may continue switching stochastically among themselves to maintain persistence [25].

4. Current Treatment Strategies May Be Counterproductive

To further complicate the issue, in addition to being taken for granted that cancer is a genetic disease, it is also believed that cancers arise by Darwinian evolution, involving a reiterative process of clonal expansion, genetic diversification, and selection within the adaptive landscapes of the tissue ecosystems they inhabit [26]. Therefore, while therapeutic intervention can destroy cancer clones and erode their habitats, the same intervention, especially when administered continuously, may also inadvertently provide a potent selective pressure for the expansion of drug-resistant phenotypes. However, recent developments in sequencing and omics technologies, coupled with theoretical advancements, have provided an expanded understanding of the cancer phenotypic landscape. They highlight non-genetic mechanisms that enable cancer cells to reversibly adapt to their environment, unlike genetic mechanisms that are irreversible, underscoring the dire need to reconcile the two mechanisms. Additionally, plastic phenotypes such as cancer stem cells (CSCs) are also well-recognized entities contributing to drug resistance in many cancers [27–29]. Therefore, the dynamic heterogeneity and the dynamic transitions between CSCs and non-CSCs and their significance in metastasis and drug resistance warrant a deeper understanding of the underlying mechanisms.

Indeed, a recent article by Sui Huang [30] in a Special Issue of *Trends in Cancer* entitled, 'Quantitative Cancer Biology', further emphasizes the need to reconcile non-genetic plasticity with somatic evolution in cancer: "Posttreatment progression of tumors is commonly explained by somatic Darwinian evolution (i.e., selection of cells carrying genetic mutations that create more aggressive cell traits). But cancer genome and transcriptome analyses now paint a picture far more complex, prompting us to see beyond the Darwinian scheme: non-genetic cell phenotype plasticity explained by alternative stable gene expression states ('attractors'), may also produce aggressive phenotypes that can be selected for, without mutations. Worse, treatment may even induce cell state transitions into more malignant attractors." (Our bold for emphasis).

5. Emergence of Irreversible Drug Resistance via a Potentially Reversible Tolerant State

Because they exhibit a high degree of phenotypic plasticity [31], cancer cells can switch on cell-autonomous traits such as persistence and quorum sensing when stressed [32]. Cancer cells exhibiting a persister trait are slow growing and can give rise to tolerant cells that, as discussed above, play an important role in the emergence of true drug resistance [33,34]. To comprehend how drug resistance may evolve from an intermediate tolerant state, it is helpful to view cancer from Waddington's [35] epigenetic landscape perspective (Figure 2). The concept of a "landscape" represents a high-dimensional state space where each phenotype acts as an "attractor" determined by the underlying PIN and is buffered against environmental fluctuations. Cellular PINs are organized following scale-free (rather than random) configurations. Therefore, PINs follow a power law distribution, wherein a few nodes (referred to as hubs) have numerous edges (connections), while most nodes have few or very few edges. Scale-free networks are resilient to random node failures; however, they are susceptible to targeted attacks on hubs. PINs serve as the main conduit of information flow with crucial roles in cellular decision-making [17,36]. PINs can determine the fate that a cell can realize and can robustly establish its phenotype because they are minimally frustrated [37]. Frustration is defined as the inability of the system to simultaneously minimize the competing interaction energies between its components [38]. In cancer cells, PIN frustration can be a viable mechanism of achieving phenotypic plasticity besides epigenetic changes. As discussed in the next paragraph, PIN frustration is driven by a special class of proteins with high structural flexibility and an ability to interact with multiple partners.

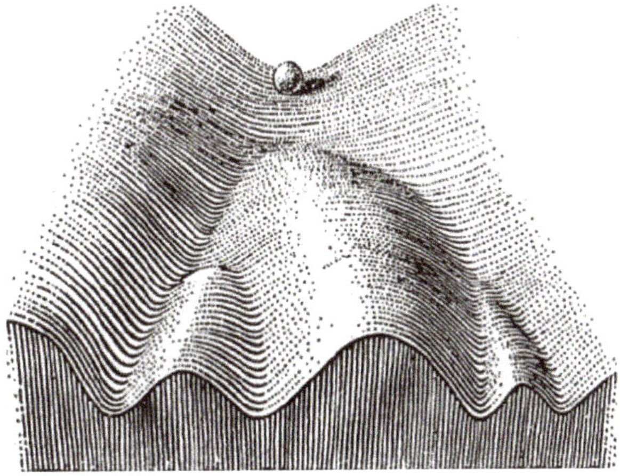

Figure 2. Schematic illustration of Waddington's epigenetic landscape [35]. The ball rolling down the hill (the x axis) represents a pluripotent cell that differentiates as it rolls down the valleys. The fate of the cell is decided by the attractors that reside at the bottom of the hill (the y axis). The valleys are separated by ridges that preclude transdifferentiation [39].

Intrinsically disordered proteins (IDPs) are proteins that lack a rigid 3D structure and exist as ensembles of interconverting conformations [40]. Because they are highly malleable, IDPs interact with multiple partners and thus occupy hub positions in PINs. Noise contributed in part by the conformational dynamics of IDPs ('conformational noise') plays an important role in cell-fate specification [17,18,36]. Nonetheless, in response to stress, especially in conjunction with post-translational modifications such as phosphorylation, IDPs engage in promiscuous interactions and drive phenotypic transitions by PIN rewiring [17,18,36,41]. This heuristic can uncover new attractors in the system—including "cancer attractors," defined as hidden stable states of PINs [42,43]—and cause phenotypic switching. Upon stress withdrawal, IDPs reconfigure the PIN to return to the original phenotype, highlighting the reversible nature of phenotypic switching. However, if stressful conditions persist, chronic stress can result in persistent network frustration, which is relieved by acquiring specific DNA mutations and/or other genetic alterations, making the phenotypic change permanent [32,44]. Thus, it follows that, non-genetic mechanisms can eventually lead to acquired resistance via irreversible genetic changes at the individual cell level. Further, as discussed below, this stepwise trajectory to drug resistance also highlights an unprecedented opportunity to preclude or delay it by controlling stress (by manipulating drug dose/time) experienced by the tolerant tumor cells.

6. Intermittent or 'Adaptive' Therapy—An Eco-Evolutionary Principles-Based Therapeutic Strategy to Preclude or Delay Onset of Drug Resistance

The rationale for this treatment strategy is based on the principles of ecology and evolution. Within the tumor microenvironment (TME), cancer cells reside with several other cell types that cohabit in this space. By producing growth factors and proinflammatory cytokines to promote angiogenesis, these cells create an ecosystem that enables the malignant cell population to grow and flourish. Therefore, group behavior, an emergent property defined as the collective actions performed by the individuals in the group as a whole, imposes costs and benefits to the participating individuals that can be recast as a game pay-off matrix. Thus, evolutionary game theory, which provides an elegant conceptual framework to capture the frequency-dependent nature of ecosystem dynamics, can be used to model tumor progression and dynamics. In fact, game theory can also be leveraged to discern the games cancer cells play by either cooperating or competing in the absence or presence of stress (selective pressure). Therefore, treatment options that consider the strategies cancer cells adopt to deal with drug effects have been developed and are referred to as intermittent or 'adaptive' therapy [45].

Typically, such treatment protocols call for initial therapies to induce adaptive changes in the tumor environment such that the proliferation of resistant clones is markedly suppressed for extended periods. In this paradigm, it is recommended that therapy is administered in small doses to attenuate tumor growth but has just enough to improve the symptoms. In other words, it is recommended that a minimal dose of the treatment (that is necessary and not at the maximum tolerated) must be used to achieve the desired result. Furthermore, treatment is administered intermittently (in alternate cycles) rather than continuously (given at every scheduled time) so that a drug-sensitive tumor population will be sustained at the expense of the resistant ones. In addition, drug combinations/epigenetic modifiers may be used in the intermittent/adaptive therapy regimen if necessary. Although, in this treatment strategy, the tumor is not completely eradicated, and it is likely that the tumor progresses between treatments; it is also likely that the tumor cells will continue to be sensitive to therapy and therefore delay or may even preclude the onset of drug-resistant disease (Figure 3) and thus, prolong overall survival.

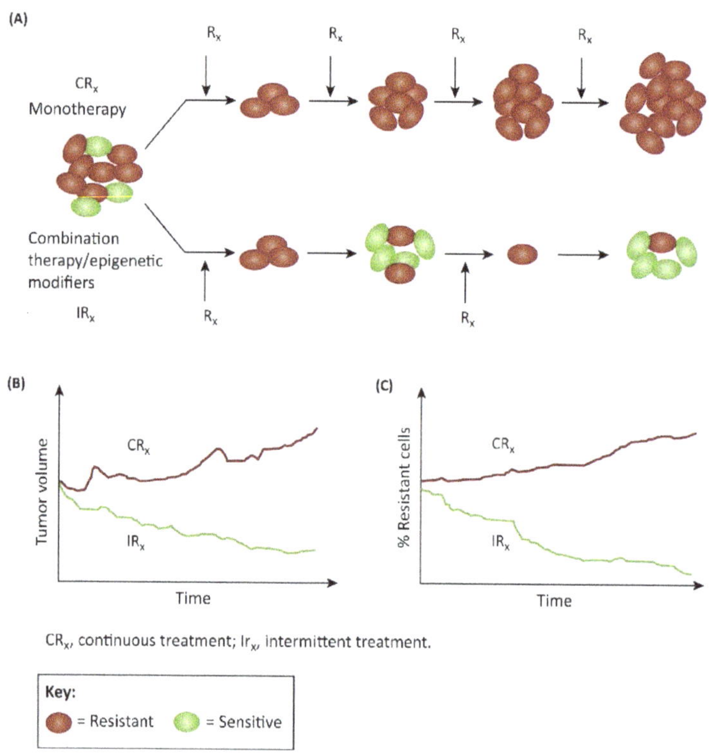

Figure 3. Continuous Monotherapy versus Intermittent Combination Therapy. (**A**) In continuous monotherapy, the idea is to eradicate the tumor as quickly as possible. However, this strategy can give rise to resistance, and resistant cells are expected to propagate over time (top). By contrast, combination therapy applied intermittently (bottom) could induce 'adaptive strategies' to change the tumor environment in such a way that the proliferation of the resistant clones can be suppressed for prolonged periods of time. Therapy is applied in small doses to reduce the tumor population only sufficient enough to improve the symptoms. Furthermore, treatment is intermittent so that drug-sensitive cells will proliferate at the expense of the resistant ones. (**B**,**C**) Although the tumor will increase in size between treatments, the extant tumor cells will continue to be sensitive to therapy [21].

Indeed, intermittent therapy for prostate and breast cancer [46], melanoma [47], rectal cancer [48], and pediatric sarcomas [49] are currently being evaluated in the clinic with promising results. In some other cancers, e.g., non-small cell lung cancer, our preclinical data obtained using a Team Medicine approach also indicate that such strategies may prove successful as well [50]. They could also have a significant impact on mutant KRAS inhibitors, such as sotorasib, which was recently approved for lung cancer treatment and is promising but is already being reported to develop resistance [51–56].

Despite the promise and initial success, more research is warranted to gain a deeper understanding of intermittent/adaptive therapy and of the side effects, if any, especially when drug combinations are to be used. For example, in one study, it was reported that when a tumor is sensitive to two or more drugs, the simultaneous application of these drugs could result in the emergence of cells resistant to both therapies. However, when these drugs were applied one at a time, a subpopulation of cells was sensitive to one or the other drug, delaying the emergence of double-resistant cell clones [57]. Conversely, in another study on lung cancer, it was observed that concurrent targeting of multiple kinases, rather than a single kinase, resulted in the complete (100%) inhibition of tumor growth. The latter strategy was only effective when intermittent and not continuous therapy was administered. One possibility for this dramatic inhibition is likely due to the lack of adaptability of the tumor cells to the changing fitness threshold imposed by selection [58].

Two randomized trials have investigated intermittent dosing regimens with BRAF and MEK inhibitors for the treatment of BRAF-mutated advanced malignant melanoma. Gonzalez-Cao et al. [59] reported lower median progression-free survival (PFS; 6.9 months versus 16.2 months; $p = 0.079$) with the intermittent use of vemurafenib and cobimetinib when compared to continuous dosing in 70 patients with treatment-naïve advanced melanoma. No statistical difference was observed for overall survival (OS) or in objective response rates (OSS). In another randomized, open-label, phase two trial, comparing continuous versus intermittent BRAF and MEK inhibition in patients with BRAF-mutated melanoma, Algazi et al. [60] reported that continuous dosing was associated with a statistically significant improvement in median PFS compared with intermittent dosing (9.0 months versus 5.5 months, $p = 0.064$, pre-specified two-sided $\alpha = 0.2$). Even though there was a PFS difference between the two groups, no differences were observed in the OS and ORR. This could possibly be due to the finding that intermittent dosing was associated with longer survival after progression (HR 0.76; 80% CI 0.78 to 1.00). Maio et al. [61] improved efficacy with intermittent MEK inhibition when combined with anti-PD-1 immunotherapy (pembrolizumab) in patients with advanced or metastatic BRAF-mutated solid tumors (36% colorectal cancer and 10% melanoma). ORR was reported to be 8% effective with concurrent and 28% effective with the intermittent dosing groups, respectively. Several trials have investigated the role of intermittent androgen deprivation therapy (ADT) in the treatment of advanced prostate cancer. These trials have reported that intermittent ADT has similar clinical outcomes when compared to continuous ADT with no statistically significant differences in OS, cancer-specific survival, or PFS [62–64]. However, intermittent ADT is associated with an improved quality of life and a lower risk of adverse events [63–65]. Ongoing trials are now investigating intermittent ADT in combination with additional therapies, such as radiation or immunotherapy, that can potentially further increase the time of systemic treatment [66]. Thus, the clinical trials so far, which have investigated intermittent dosing regimens, have yielded mixed results, highlighting the complexity of translation preclinical studies into human trials and the challenges of selecting the optimal dosing regimen. Future trials should focus on exploring this approach in biomarker-selected populations, as well as elucidating which subgroups of patients may benefit most from this approach.

Since intermittent therapy relies on drug-sensitive cells to suppress the proliferation of the tolerant cells, the success of such therapies is dependent on the initial population of the two cell types within the tumor. The hypothetical scenario shown in Figure 4, where the three models represent the three patients, may help better appreciate the underlying nuances. Here, Patient 1, with the highest number of sensitive cells compared to tolerant cells, will have more prolonged progression-free survival in response to drug treatment. The second patient with an equal number of sensitive and tolerant populations will have shorter progression-free survival, while Patient 3, with the highest number of tolerant cells, will have the shortest progression-free survival. The purpose of intermittent therapy is to prevent the development of drug refractory-resistant clones and is based on the presumption that the sensitive cells, in the absence of the drug, will compete with resistant cells and grow faster to suppress the growth of resistant cells. Thus, the more percentage of sensitive cells, the better the response to the intermittent therapy, and, thus, in scenarios like model 1 and model 2, the success rate will be higher compared to model 3. Moreover, the inherent assumption in intermittent therapy, that the sensitive and tolerant cell types are competitive, may not hold universally. Depending on the tissue/cancer type, more complex ecological relationships may exist among the different cell types (such as cooperation or competition, depending on the stress level/drug dosage) that may require more fine-tuned, dynamic adjustments to drug schedules/dosages as the therapy progresses and the tumor phenotypic landscape evolves.

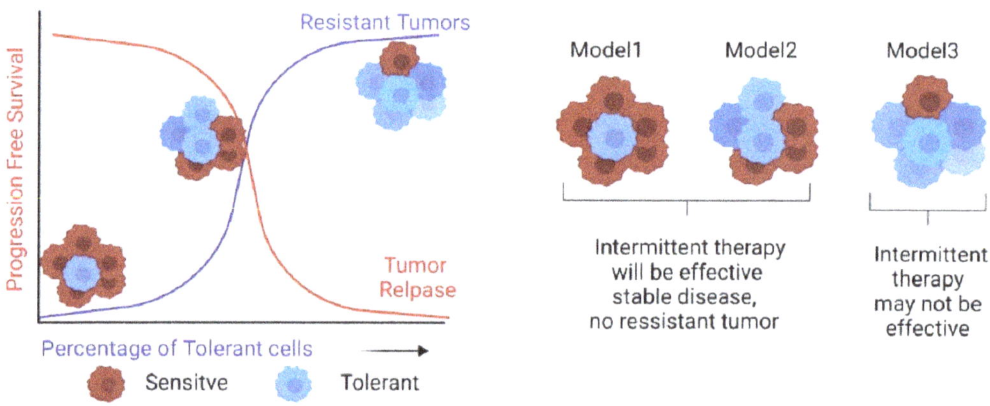

Figure 4. The cartoon representing the importance of tumor heterogeneity on therapeutic approach, continuous verses intermittent.

Thus, it follows that, before initiating a therapeutic strategy, a detailed genetic (and possibly transcriptomic and epigenetic) analysis of the patient's tumor is imperative. For example (Figure 5), patients with an RAF mutation are likely to respond better to BRAF inhibitors, while patients with BRAF and MEK mutations may respond less to the same inhibitors, and patients with MEK and PI3K mutations will not respond to BRAF inhibitors at all. Likewise, patients with activated RTK signaling are unlikely to respond to the BRAF inhibitor drug treatment effectively, as the tumor will likely take advantage of the bypass signaling through the AKT -mTOR pathway to overcome the drug effect. Therefore, pre-trial validation of the mutations in the tumor by NGS and the expression of the activated signaling need to be determined so that patients with a similar mutational profile or expression status may be grouped into a specific cohort. As illustrated in Figure 5, patients in each of the four cohorts can be treated with continuous or intermittent therapy, and the statistical significance can be derived to determine the best therapeutic approach. However, comparing intermittent versus continuous treatment between the two cohorts will give insignificant information. Figure 5 also suggests the role of Team Medicine, where basic help from scientists can identify those signaling pathways that need to be targeted for effective therapy through experiments, clinicians and bioinformaticians can help to validate the basic study by analyzing thousands of public data and help to look for specific mutations that can contribute to the pathways, molecular pathologists can help in determining the expression of these proteins in the tumor biopsies, and finally, clinicians, aided by cumulative information and precise mathematical models, can design the drug treatment strategy.

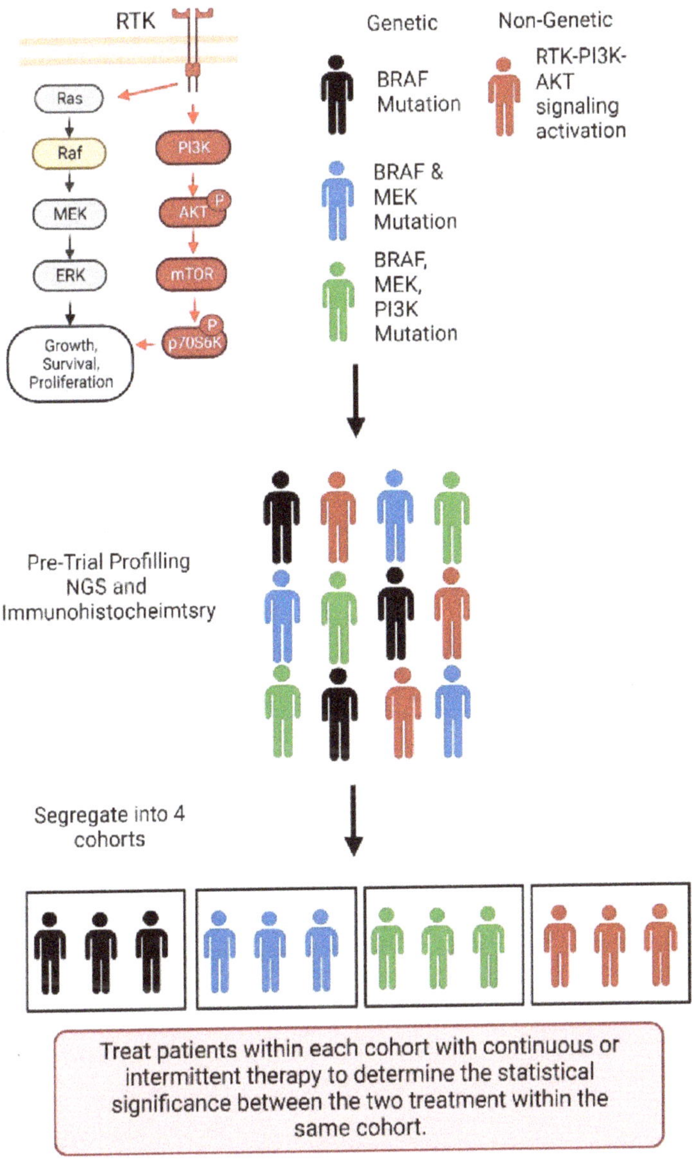

Figure 5. A schematic representing the pretreatment preparation for choosing the best treatment strategy. RTK, receptor tyrosine kinase; Ras, Ras protooncogene; Raf. Raf protooncogene, serine/threonine kinase; MEK, MAP kinase-ERK kinase; ERK, extracellular regulated MAP kinase; PI3K, phosphatidylinositol 3-kinase; AKT, AKT serine/threonine kinase 1; mTOR, Mechanistic Target Of Rapamycin Kinase; BRAF, B-Raf Proto-Oncogene, Serine/Threonine Kinase; NGS, next generation sequencing.

7. Concluding Remarks

There exist multiple mechanisms that regulate phenotypic switching and drug resistance, even within a given cancer type. Furthermore, although promising in some cases, challenges still remain with regard to intermittent therapy, as discussed above. Thus, a better understanding of the mechanisms can help us to design the most effective therapeutic approach. Nonetheless, from the foregoing, it is obvious that these exciting developments in medical oncology, which expound the virtues of modern translational research, can only be made possible by a true Team Medicine approach, including basic scientists and clinicians. By incorporating treatment strategies based on the principles of ecology and evolution in clinical protocols, and by reaching out to patients who frequent those hospitals that are part of the network formed by academic centers rather than the academic centers themselves, we can enhance the precision in which we deliver personalized medicine to all our patients, regardless of their economic status or their ability to access advanced medical centers. We trust that our integrated efforts at the City of Hope, in conjunction with the cancer treatment centers of America, shall serve as a good example to those who wish to adopt this paradigm. Last but not least, lowering the dose of the drug and its frequency (because of intermittent rather than continuous therapy) can also have a significant impact on lowering the toxicity and undesirable side effects of the drugs while lowering the financial burden carried by the patient and insurance providers [67].

Author Contributions: Conceptualization, P.K. (Prakash Kulkarni) and R.S.; resources, R.S.; writing—original draft preparation, P.K. (Prakash Kulkarni) and R.S.; writing—review and editing, A.M. (Atish Mohanty), S.B., S.S. (Sharad Singhal), L.G., S.R. (Sravani Ramisetty), T.M., B.M., S.M., J.M., N.G., P.K. (Pauline Kim), R.B., S.R. (Swapnil Rajurkar), S.S. (Shanmuga Subbiah), T.T., D.N., A.M. (Amartej Merla), S.V.K., T.P., P.B., B.T., P.V., S.S. (Sagun Shrestha) and B.L., R.G., P.L.R., F.M.S., E.P.; visualization, P.K. (Prakash Kulkarni) and R.S.; supervision, P.K. (Prakash Kulkarni) and R.S.; project administration, P.K. (Prakash Kulkarni) and R.S. All authors have read and agreed to the published version of the manuscript.

Funding: This work was supported in part by Robert and Kathleen Henderson via the Robert and Kathleen Henderson Lung Cancer Research Accelerator Fund at the City of Hope and by William and Anna Tenenblatt via The William & Anna Tenenblatt Foundation.

Institutional Review Board Statement: Not applicable for studies not involving humans or animals.

Informed Consent Statement: Not applicable for studies not involving humans.

Data Availability Statement: Not applicable.

Conflicts of Interest: The authors declare no conflict of interest.

References

1. Sung, H.; Ferlay, J.; Siegel, R.L.; Laversanne, M.; Soerjomataram, I.; Jemal, A.; Bray, F. Global Cancer Statistics 2020: GLOBOCAN Estimates of Incidence and Mortality Worldwide for 36 Cancers in 185 Countries. *CA Cancer J. Clin.* **2021**, *71*, 209–249. [CrossRef] [PubMed]
2. Available online: https://www.cancer.org/research/cancer-facts-statistics/global.html (accessed on 1 September 2022).
3. Salgia, R.; Kulkarni, P. Integrating Clinical and Translational Research Networks—Building Team Medicine. *J. Clin. Med.* **2020**, *9*, 2975. [CrossRef] [PubMed]
4. Hansford, S.; Huntsman, D.G. Boveri at 100: Theodor Boveri and genetic predisposition to cancer. *J. Pathol.* **2014**, *234*, 142–145. [CrossRef] [PubMed]
5. McKusick, V.A. Marcella O'Grady Boveri (1865–1950) and the chromosome theory of cancer. *J. Med. Genet.* **1985**, *6*, 431–440. [CrossRef]
6. Boveri, T. *Zur Frage der Entstehung Maligner Tumoren*; Verlag von Gustav Fischer: Jena, Germany, 1914.
7. Boveri, T. *The Origin of Mnalignzatnt Tunors. (Translated by M Boveri.)*; Williams and Wilkins: Baltimore, MD, USA, 1929.
8. Vogelstein, B.; Kinzler, K.W. Cancer genes and the pathways they control. *Nat. Med.* **2004**, *10*, 789–799. [CrossRef]
9. Vogelstein, B.; Kinzler, K.W. The Path to Cancer—Three Strikes and You're Out. *N. Engl. J. Med.* **2015**, *373*, 1895–1898. [CrossRef]
10. Laland, K.; Uller, T.; Feldman, M.; Sterelny, K.; Müller, G.B.; Moczek, A.; Jablonka, E.; Odling-Smee, J.; Wray, G.A.; Hoekstra, H.E.; et al. Does evolutionary theory need a rethink? *Nature* **2014**, *514*, 161–164. [CrossRef]
11. Waarts, M.R.; Stonestrom, A.J.; Park, Y.C.; Levine, R.L. Targeting mutations in cancer. *J. Clin. Investig.* **2022**, *132*. [CrossRef]

12. Salgia, R.; Pharaon, R.; Mambetsariev, I.; Nam, A.; Sattler, M. The improbable targeted therapy: KRAS as an emerging target in non-small cell lung cancer (NSCLC). *Cell Rep. Med.* **2021**, *2*, 100186. [CrossRef]
13. Pagliarini, R.; Shao, W.; Sellers, W.R. Oncogene addiction: Pathways of therapeutic response, resistance, and road maps toward a cure. *EMBO Rep.* **2015**, *16*, 280–296. [CrossRef]
14. Bell, C.C.; Gilan, O. Principles and mechanisms of non-genetic resistance in cancer. *Br. J. Cancer* **2020**, *122*, 465–472. [CrossRef] [PubMed]
15. Shlyakhtina, Y.; Moran, K.; Portal, M. Genetic and Non-Genetic Mechanisms Underlying Cancer Evolution. *Cancers* **2021**, *13*, 1380. [CrossRef] [PubMed]
16. Sharma, S.V.; Lee, D.Y.; Li, B.; Quinlan, M.P.; Takahashi, F.; Maheswaran, S.; McDermott, U.; Azizian, N.; Zou, L.; Fischbach, M.A.; et al. A chromatin-mediated reversible drug-tolerant state in cancer cell subpopulations. *Cell* **2010**, *141*, 69–80. [CrossRef] [PubMed]
17. Kulkarni, P.; Achuthan, S.; Bhattacharya, S.; Jolly, M.K.; Kotnala, S.; Leite, V.B.P.; Mohanty, A.; Orban, J.; Roy, S.; Rangarajan, G.; et al. Protein conformational dynamics and phenotypic switching. *Biophys. Rev.* **2021**, *13*, 1127–1138. [CrossRef]
18. Kulkarni, V.; Kulkarni, P. Intrinsically disordered proteins and phenotypic switching: Implications in cancer. *Prog. Mol. Biol. Transl. Sci.* **2019**, *166*, 63–84. [CrossRef]
19. Bowler, E.H.; Wang, Z.; Ewing, R.M. How do oncoprotein mutations rewire protein-protein interaction networks? *Expert Rev. Proteom.* **2015**, *12*, 449–455. [CrossRef]
20. Paliouras, M.; Zaman, N.; Lumbroso, R.; Kapogeorgakis, L.; Beitel, L.K.; Wang, E.; Trifiro, M. Dynamic rewiring of the androgen receptor protein interaction network correlates with prostate cancer clinical outcomes. *Integr. Biol.* **2011**, *3*, 1020–1032. [CrossRef]
21. Salgia, R.; Kulkarni, P. The Genetic/Non-genetic Duality of Drug 'Resistance' in Cancer. *Trends Cancer* **2018**, *4*, 110–118. [CrossRef]
22. Brauner, A.; Fridman, O.; Gefen, O.; Balaban, N.Q. Distinguishing between resistance, tolerance and persistence to antibiotic treatment. *Nat. Rev. Microbiol.* **2016**, *14*, 320–330. [CrossRef]
23. Balaban, N.Q.; Helaine, S.; Lewis, K.; Ackermann, M.; Aldridge, B.; Andersson, D.I.; Brynildsen, M.P.; Bumann, D.; Camilli, A.; Collins, J.J.; et al. Definitions and guidelines for research on antibiotic persistence. *Nat. Rev. Microbiol.* **2019**, *17*, 441–448. [CrossRef]
24. Kumar, R.; Chaudhary, K.; Gupta, S.; Singh, H.P.; Kumar, S.; Gautam, A.; Kapoor, P.; Raghava, G.P.S. Cancer DR: Cancer drug resistance database. *Sci. Rep.* **2013**, *3*, srep01445. [CrossRef] [PubMed]
25. Jolly, M.K.; Kulkarni, P.; Weninger, K.; Orban, J.; Levine, H. Phenotypic Plasticity, Bet-Hedging, and Androgen Independence in Prostate Cancer: Role of Non-Genetic Heterogeneity. *Front. Oncol.* **2018**, *8*, 50. [CrossRef] [PubMed]
26. Greaves, M.; Maley, C.C. Clonal evolution in cancer. *Nature* **2012**, *481*, 306–313. [CrossRef]
27. Zheng, Q.; Zhang, M.; Zhou, F.; Zhang, L.; Meng, X. The Breast Cancer Stem Cells Traits and Drug Resistance. *Front. Pharmacol.* **2021**, *11*, 599965. [CrossRef] [PubMed]
28. Ibragimova, M.; Tsyganov, M.; Litviakov, N. Tumour Stem Cells in Breast Cancer. *Int. J. Mol. Sci.* **2022**, *23*, 5058. [CrossRef]
29. Walcher, L.; Kistenmacher, A.-K.; Suo, H.; Kitte, R.; Dluczek, S.; Strauß, A.; Blaudszun, A.-R.; Yevsa, T.; Fricke, S.; Kossatz-Boehlert, U. Cancer Stem Cells—Origins and Biomarkers: Perspectives for Targeted Personalized Therapies. *Front. Immunol.* **2020**, *11*, 1280. [CrossRef]
30. Huang, S. Reconciling Non-Genetic Plasticity with Somatic Evolution in Cancer. *Trends Cancer* **2021**, *7*, 309–322. [CrossRef]
31. Gomez, K.; Rabadan, R. A persistent look at how tumours evade therapy. *Nature* **2021**, *596*, 491–493. [CrossRef]
32. Bhattacharya, S.; Mohanty, A.; Achuthan, S.; Kotnala, S.; Jolly, M.K.; Kulkarni, P.; Salgia, R. Group Behavior and Emergence of Cancer Drug Resistance. *Trends Cancer* **2021**, *7*, 323–334. [CrossRef]
33. Ramirez, M.; Rajaram, S.; Steininger, R.J.; Osipchuk, D.; Roth, M.A.; Morinishi, L.S.; Evans, L.; Ji, W.; Hsu, C.-H.; Thurley, K.; et al. Diverse drug-resistance mechanisms can emerge from drug-tolerant cancer persister cells. *Nat. Commun.* **2016**, *7*, 10690. [CrossRef]
34. Oren, Y.; Tsabar, M.; Cuoco, M.S.; Amir-Zilberstein, L.; Cabanos, H.F.; Hütter, J.-C.; Hu, B.; Thakore, P.I.; Tabaka, M.; Fulco, C.P.; et al. Cycling cancer persister cells arise from lineages with distinct programs. *Nature* **2021**, *596*, 576–582. [CrossRef] [PubMed]
35. Waddington, C.H. *The Strategy of the Genes*; Geo Allen & Unwin: London, UK, 1957.
36. Mahmoudabadi, G.; Rajagopalan, K.; Getzenberg, R.H.; Hannenhalli, S.; Rangarajan, G.; Kulkarni, P. Intrinsically disordered proteins and conformational noise: Implications in cancer. *Cell Cycle* **2013**, *12*, 26–31. [CrossRef] [PubMed]
37. Tripathi, S.; Kessler, D.A.; Levine, H. Biological Networks Regulating Cell Fate Choice are Minimally Frustrated. *Phys. Rev. Lett.* **2020**, *125*, 088101. [CrossRef] [PubMed]
38. Parra, R.G.; Schafer, N.P.; Radusky, L.G.; Tsai, M.-Y.; Guzovsky, A.B.; Wolynes, P.G.; Ferreiro, D.U. Protein Frustratometer 2: A tool to localize energetic frustration in protein molecules, now with electrostatics. *Nucleic Acids Res.* **2016**, *44*, W356–W360. [CrossRef] [PubMed]
39. Kulkarni, P. Intrinsically Disordered Proteins: Insights from Poincaré, Waddington, and Lamarck. *Biomolecules* **2020**, *10*, 1490. [CrossRef]

40. Kulkarni, P.; Leite, V.B.P.; Roy, S.; Bhattacharyya, S.; Mohanty, A.; Achuthan, S.; Singh, D.; Appadurai, R.; Rangarajan, G.; Weninger, K.; et al. Intrinsically disordered proteins: Ensembles at the limits of Anfinsen's dogma. *Biophys. Rev.* **2022**, *3*, 011306. [CrossRef]
41. Liu, Z.; Miller, D.; Li, F.; Liu, X.; Levy, S.F. A large accessory protein interactome is rewired across environments. *Elife* **2020**, *9*, e62365. [CrossRef]
42. Huang, S.; Ernberg, I.; Kauffman, S. Cancer attractors: A systems view of tumors from a gene network dynamics and developmental perspective. *Semin. Cell Dev. Biol.* **2009**, *20*, 869–876. [CrossRef]
43. Li, Q.; Wennborg, A.; Aurell, E.; Dekel, E.; Zou, J.-Z.; Xu, Y.; Huang, S.; Ernberg, I. Dynamics inside the cancer cell attractor reveal cell heterogeneity, limits of stability, and escape. *Proc. Natl. Acad. Sci. USA* **2016**, *113*, 2672–2677. [CrossRef]
44. Fitzgerald, D.M.; Hastings, P.; Rosenberg, S.M. Stress-Induced Mutagenesis: Implications in Cancer and Drug Resistance. *Annu. Rev. Cancer Biol.* **2017**, *1*, 119–140. [CrossRef]
45. Gatenby, R.A.; Brown, J.S. The Evolution and Ecology of Resistance in Cancer Therapy. *Cold Spring Harb Perspect. Med.* **2020**, *10*, a040972. [CrossRef] [PubMed]
46. Park, D.S.; Luddy, K.A.; Robertson-Tessi, M.; O'Farrelly, C.; Gatenby, R.A.; Anderson, A.R. Searching for Goldilocks: How Evolution and Ecology Can Help Uncover More Effective Patient-Specific Chemotherapies. *Cancer Res.* **2020**, *80*, 5147–5154. [CrossRef] [PubMed]
47. Kavran, A.J.; Stuart, S.A.; Hayashi, K.R.; Basken, J.M.; Brandhuber, B.J.; Ahn, N.G. Intermittent treatment of BRAFV600E melanoma cells delays resistance by adaptive resensitization to drug rechallenge. *Proc. Natl. Acad. Sci. USA* **2022**, *119*, e2113535119. [CrossRef]
48. Felder, S.I.; Fleming, J.B.; Gatenby, R.A. Treatment-induced evolutionary dynamics in nonmetastatic locally advanced rectal adenocarcinoma. *Adv. Cancer Res.* **2021**, *151*, 39–67. [CrossRef] [PubMed]
49. Reed, D.R.; Metts, J.; Pressley, M.; Fridley, B.L.; Hayashi, M.; Isakoff, M.S.; Loeb, D.M.; Makanji, R.; Roberts, R.D.; Trucco, M.; et al. An evolutionary framework for treating pediatric sarcomas. *Cancer* **2020**, *126*, 2577–2587. [CrossRef]
50. Nam, A.; Mohanty, A.; Bhattacharya, S.; Kotnala, S.; Achuthan, S.; Hari, K.; Srivastava, S.; Guo, L.; Nathan, A.; Chatterjee, R.; et al. Dynamic Phenotypic Switching and Group Behavior Help Non-Small Cell Lung Cancer Cells Evade Chemotherapy. *Biomolecules* **2021**, *12*, 8. [CrossRef] [PubMed]
51. Dunnett-Kane, V.; Nicola, P.; Blackhall, F.; Lindsay, C. Mechanisms of Resistance to KRASG12C Inhibitors. *Cancers* **2021**, *13*, 151. [CrossRef]
52. Tanaka, N.; Lin, J.J.; Li, C.; Ryan, M.B.; Zhang, J.; Kiedrowski, L.A.; Michel, A.G.; Syed, M.U.; Fella, K.A.; Sakhi, M.; et al. Clinical Acquired Resistance to KRASG12C Inhibition through a Novel KRAS Switch-II Pocket Mutation and Polyclonal Alterations Converging on RAS-MAPK Reactivation. *Cancer Discov.* **2021**, *11*, 1913–1922. [CrossRef]
53. Koga, T.; Suda, K.; Fujino, T.; Ohara, S.; Hamada, A.; Nishino, M.; Chiba, M.; Shimoji, M.; Takemoto, T.; Arita, T.; et al. KRAS Secondary Mutations That Confer Acquired Resistance to KRAS G12C Inhibitors, Sotorasib and Adagrasib, and Overcoming Strategies: Insights From the In Vitro Experiments. *J. Thorac. Oncol.* **2021**, *16*, 1321–1332. [CrossRef]
54. Addeo, A.; Banna, G.; Friedlaender, A. KRAS G12C Mutations in NSCLC: From Target to Resistance. *Cancers* **2021**, *13*, 2541. [CrossRef]
55. Reck, M.; Carbone, D.; Garassino, M.; Barlesi, F. Targeting KRAS in non-small-cell lung cancer: Recent progress and new approaches. *Ann. Oncol.* **2021**, *32*, 1101–1110. [CrossRef] [PubMed]
56. Xue, J.Y.; Zhao, Y.; Aronowitz, J.; Mai, T.T.; Vides, A.; Qeriqi, B.; Kim, D.; Li, C.; de Stanchina, E.; Mazutis, L.; et al. Rapid non-uniform adaptation to conformation-specific KRAS(G12C) inhibition. *Nature* **2020**, *577*, 421–425. [CrossRef] [PubMed]
57. Enriquez-Navas, P.M.; Wojtkowiak, J.W.; Gatenby, R.A. Application of Evolutionary Principles to Cancer Therapy. *Cancer Res.* **2015**, *75*, 4675–4680. [CrossRef] [PubMed]
58. Xue, Y.; Martelotto, L.; Baslan, T.; Vides, A.; Solomon, M.; Mai, T.T.; Chaudhary, N.; Riely, G.J.; Li, B.T.; Scott, K.; et al. An approach to suppress the evolution of resistance in BRAFV600E-mutant cancer. *Nat. Med.* **2017**, *23*, 929–937. [CrossRef]
59. Gonzalez-Cao, M.; Casas, C.M.D.L.; Oramas, J.; Berciano-Guerrero, M.A.; de la Cruz, L.; Cerezuela, P.; Arance, A.; Muñoz-Couselo, E.; Espinosa, E.; Puertolas, T.; et al. Intermittent BRAF inhibition in advanced BRAF mutated melanoma results of a phase II randomized trial. *Nat. Commun.* **2021**, *12*, 7008. [CrossRef]
60. Algazi, A.P.; Othus, M.; Daud, A.I.; Lo, R.S.; Mehnert, J.M.; Truong, T.-G.; Conry, R.; Kendra, K.; Doolittle, G.C.; Clark, J.I.; et al. Continuous versus intermittent BRAF and MEK inhibition in patients with BRAF-mutated melanoma: A randomized phase 2 trial. *Nat. Med.* **2020**, *26*, 1564–1568. [CrossRef]
61. Maio, M.; Carlino, M.S.; Joshua, A.M.; McWhirter, E.; Ribas, A.; Ascierto, P.A.; Miller, W.H., Jr.; Butler, M.O.; Ferrucci, P.F.; Zielinski, R.R.; et al. KEYNOTE-022: Pembrolizumab with trametinib in patients with BRAF wild-type melanoma or advanced solid tumours irrespective of BRAF mutation. *Eur. J. Cancer* **2022**, *160*, 1–11. [CrossRef]
62. Hussain, M.; Tangen, C.M.; Berry, D.L.; Higano, C.S.; Crawford, E.D.; Liu, G.; Wilding, G.; Prescott, S.; Sundaram, S.K.; Small, E.J.; et al. Intermittent versus Continuous Androgen Deprivation in Prostate Cancer. *N. Engl. J. Med.* **2013**, *368*, 1314–1325. [CrossRef]
63. Crook, J.M.; O'Callaghan, C.J.; Duncan, G.; Dearnaley, D.P.; Higano, C.S.; Horwitz, E.M.; Frymire, E.; Malone, S.; Chin, J.; Nabid, A.; et al. Intermittent androgen suppression for rising PSA level after radiotherapy. *N. Engl. J. Med.* **2012**, *367*, 895–903. [CrossRef]

64. Schulman, C.; Cornel, E.; Matveev, V.; Tammela, T.L.; Schraml, J.; Bensadoun, H.; Warnack, W.; Persad, R.; Salagierski, M.; Veiga, F.G.; et al. Intermittent Versus Continuous Androgen Deprivation Therapy in Patients with Relapsing or Locally Advanced Prostate Cancer: A Phase 3b Randomised Study (ICELAND). *Eur. Urol.* **2016**, *69*, 720–727. [CrossRef]
65. Tsai, H.-T.; Pfeiffer, R.M.; Philips, G.K.; Barac, A.; Fu, A.Z.; Penson, D.; Zhou, Y.; Potosky, A.L. Risks of Serious Toxicities from Intermittent versus Continuous Androgen Deprivation Therapy for Advanced Prostate Cancer: A Population Based Study. *J. Urol.* **2017**, *197*, 1251–1257. [CrossRef] [PubMed]
66. Perera, M.; Roberts, M.J.; Klotz, L.; Higano, C.S.; Papa, N.; Sengupta, S.; Bolton, D.; Lawrentschuk, N. Intermittent versus continuous androgen deprivation therapy for advanced prostate cancer. *Nat. Rev. Urol.* **2020**, *17*, 469–481. [CrossRef] [PubMed]
67. Mason, N.T.; Burkett, J.M.; Nelson, R.S.; Pow-Sang, J.M.; Gatenby, R.A.; Kubal, T.; Peabody, J.W.; Letson, G.D.; McLeod, H.L.; Zhang, J. Budget Impact of Adaptive Abiraterone Therapy for Castration-Resistant Prostate Cancer. *Am. Health Drug Benefits* **2021**, *14*, 15–20. [PubMed]

MDPI
St. Alban-Anlage 66
4052 Basel
Switzerland
www.mdpi.com

Journal of Clinical Medicine Editorial Office
E-mail: jcm@mdpi.com
www.mdpi.com/journal/jcm